ATH
STAT.
IBRARY

Infinite Groups 1994

Infinite Groups 1994

Proceedings of the International Conference,
held in Ravello, Italy, May 23–27, 1994

Editors

Francesco de Giovanni
Martin L. Newell

Walter de Gruyter · Berlin · New York 1996

Editors

Francesco de Giovanni
Dipartimento di Matematica
e Applicazioni
Università di Napoli
I-80126 Napoli
Italy

Martin L. Newell
Department of Mathematics
University College
Galway
Ireland

1991 Mathematics Subject Classification: 20-06

⊗ Printed on acid-free paper which falls within the guidelines of the ANSI to ensure permanence and durability.

Library of Congress Cataloging-in-Publication Data

Infinite groups 1994 : proceedings of the international conference held in Ravello, Italy, May 23–27, 1994 / editors: Francesco de Giovanni, Martin L. Newell.
p. cm.
ISBN 3-11-014332-1 (alk. paper)
1. Ininite groups--Congresses. I. De Giovanni, Francesco. II. Newell, Martin L., 1939- .
QA178.I53 1996 95-38030
512'.2--dc20 CIP

Die Deutsche Bibliothek – Cataloging-in-Publication Data

Infinite Groups ... : proceedings of the international conference ...
- Berlin ; New York : de Gruyter.
1994. Held in Ravello, Italy, May 23–27, 1994. - 1996
ISBN 3-11-014332-1

Printed in Germany.
Typeset using the authors' TeX files: I. Zimmermann, Freiburg. – Printing: Gerike GmbH, Berlin.
Binding: Lüderitz & Bauer GmbH, Berlin. – Cover design: Thomas Bonnie, Hamburg.

Preface

We present with pleasure this volume containing most of the talks delivered at the Ravello Conference held in May 1994. The vast scope and diversity of the contributions reflect the wide range of current investigations and developments in the theory of Infinite Groups. It also confirms our conviction that an International Conference on this topic was timely and that our joint decision to organise one was well worth while. Although the decision was ours, we acknowledge that the labour and burden of dealing with all the necessary arrangements was borne in most part by others in the Department of Mathematics in Napoli. For their heroic efforts we wish to thank most sincerely Rara Celentani, Giovanni Cutolo, Ulderico Dardano, Silvana Rinauro and Gianni Vincenzi.

It was extremely gratifying that the conference attracted such an impressive world-wide list of participants. The international renown and proven ability of our main speakers combined with the particular charm of the Ravello venue, were important factors. But also the range of interesting topics covered by contributors in the parallel afternoon sessions sustained the stimulating momentum and resulted in many late-night heated discussions in the Piazza. We thank them all for making the venture such a memorable success.

With particular appreciation we wish to gratefully acknowledge the financial support generously provided by the "Consiglio Nazionale delle Ricerche", the "Dipartimento di Matematica e Applicazioni dell'Università di Napoli Federico II", the "Ente Provinciale per il Turismo di Salerno", the "International Science Foundation", the "Istituto Italiano per gli Studi Filosofici", the "Ministero dell'Università e della Ricerca Scientifica e Tecnologica" and the "Università di Napoli Federico II".

To the civic authorities in Ravello we extend our appreciation for their hospitable welcome and courteous attention to the needs of so many foreign visitors.

We are confident that the publications contained in these collected proceedings will prove an important and significant contribution to the development of the study of infinite groups. Many of the deep questions raised will chart the way forward and provide the impetus for further years of mathematical endeavour. To Manfred Karbe and the staff of Walter de Gruyter & Co. we record our deepest appreciation, not only for their support, encouragement and cooperation while preparing this volume, but also for the provision of the book exhibition which provided a special focal point of interest during the conference.

In October we were deeply saddened by the sudden demise of one of our main speakers Professor Brian Hartley. His loss to the mathematical community is a great one. Few could have imagined that the relaxed, unassuming, good humoured personality who gave so much of his time and energy promoting the general conviviality of the occasion, would be so shortly taken from us. His ingenious techniques and mathematical insight provide a lasting source of inspiration for others to follow.

Finally we record our enormous debt of gratitude to Dr. Giovanni Cutolo who spent many hours checking and editing written contributions. We include of course the gallant band of referees who diligently assessed the papers and thank them for their professional evaluations and the promptness of their replies.

Every effort has been made to ensure that these proceedings contain an accurate reproduction of the articles submitted, but we beg your indulgence should any omissions nevertheless occur.

Francesco de Giovanni
Martin L. Newell

Table of Contents

F. de Giovanni G. Zappa

K. W. Gruenberg

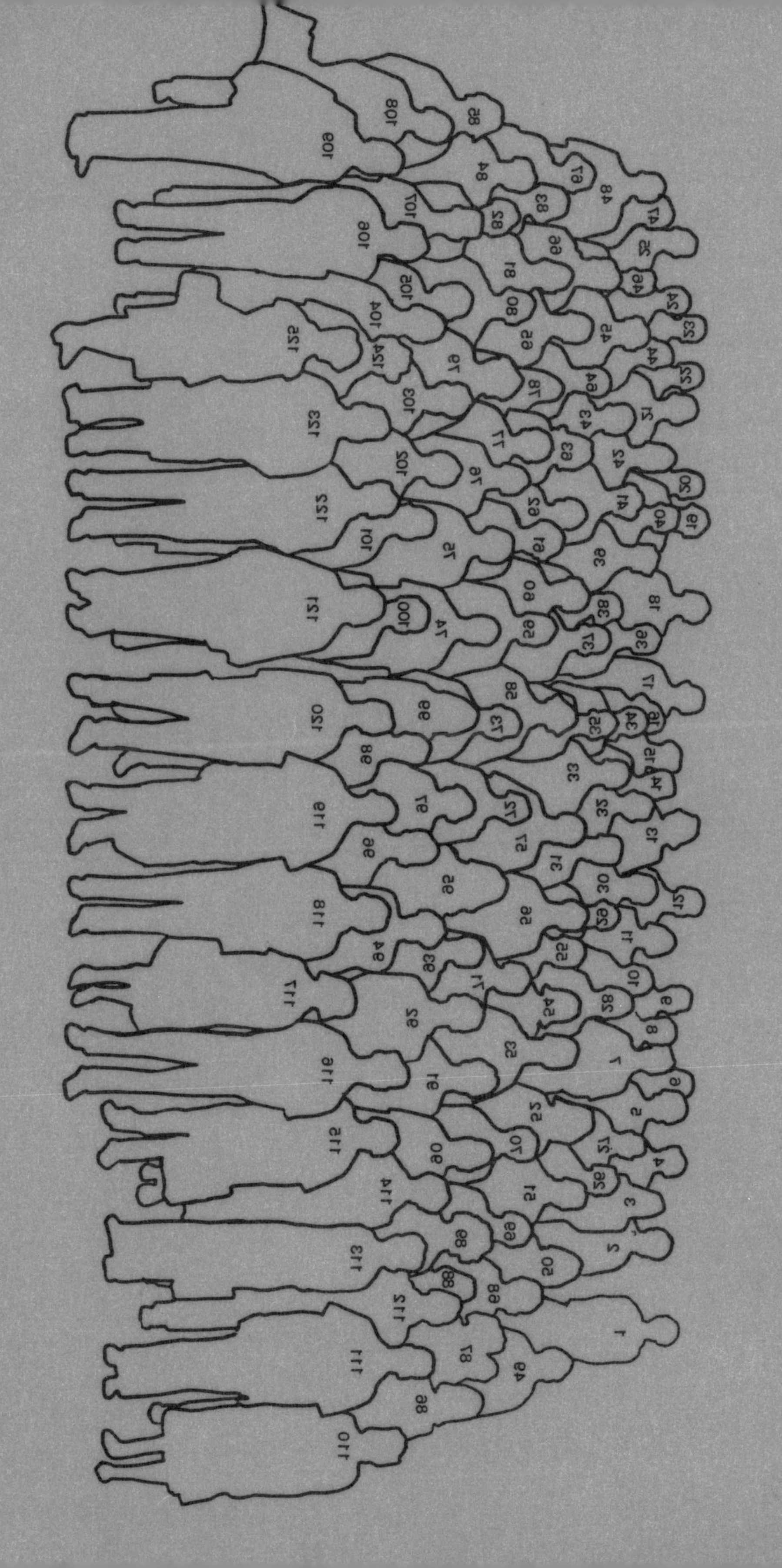

1
2
3
4
5
6
7
8
9
10
11
12
13
14
15
16
17
18
19
20
21
22
23
24
25
26
27
28
29
30
31
32
33
34
35
36
37
38
39
40
41
42
43
44
45
46
47
48
49
50
51
52
53
54
55
56
57
58
59
60
61
62
63
64
65
66
67
68
69
70
71
72
73
74
75
76
77
78
79
80
81
82
83
84
85
86
87
88
89
90
91
92
93
94
95
96
97
98
99
100
101
102
103
104
105
106
107
108
109
110
111
112
113
114
115
116
117
118
119
120
121
122
123
124
125

1. Y. Chen
2. E. Zelmanov
3. C. Martínez
4. M. J. Tomkinson
5. M. Barlotti
6. A. Lichtman
7. R. A. Bryce
8. A. Lucchini
9. N. Gavioli
10. G. Parmeggiani
11. F. Dalla Volta
12. C. M. Scoppola
13. B. Aupetit
14. C. Cassidy
15. R. Ayoub
16. C. Ayoub
17. L. A. Kurdachenko
18. H. Heineken
19. O. H. Kegel
20. A. G. R. Stewart
21. S. Tobin
22. R. Dark
23. J. McDermott
24. H. Smith
25. G. Ellis
26. F. Catino
27. F. Napolitani
28. P. Soules
29. Ya. D. Polovickii
30. M. Hopfinger
31. L. S. Kazarin
32. E. Jabara
33. P. Shumyatsky
34. R. D. Blyth
35. V. Shpilrain
36. J. Plotkin
37. V. Obraztsov
38. M. S. Lucido
39. J. Ward
40. F. Haug
41. R. E. Phillips
42. F. Leinen
43. O. Puglisi
44. A. Strojnowski
45. V. J. Sushchanskii
46. B. Höfling
47. J. S. Wilson
48. J. Buckley
49. J. C. Lennox
50. M. R. Celentani
51. M. Karbe
52. I. Zimmermann
53. W. Nickel
54. M. C. Tamburini
55. L. Serena
56. C. M. Campbell
57. E. F. Robertson
58. G. L. Walls
59. T. Fay
60. E. Plotkin
61. L. G. Kovács
62. S. E. Schuur
63. L. S. Spiezia
64. S. M. Vovsi
65. N. Vavilov
66. I. Ya. Subbotin
67. V. Fedri
68. B. Plotkin
69. F. Grunewald
70. R. Göbel
71. V. Pannone
72. V. I. Senashov
73. D. Dikranjan
74. D. Burde
75. J. Tůma
76. D. J. S. Robinson
77. L. Ribes
78. M. Morigi
79. W. Kappe
80. Ya. P. Sysak
81. O. Dickenschied
82. B. Bruno
83. M. Maj
84. G. Corsi
85. J. Wiegold
86. Mrs. Segal
87. D. Segal and his daughter
88. M. C. Cirino Groccia
89. T. Landolfi
90. F. Menegazzo
91. C. Casolo
92. M. Mainardis
93. R. Piemontino
94. A. L. Gilotti
95. C. Nicotera
96. G. Vincenzi
97. A. Mann
98. A. Russo
99. C. K. Gupta
100. J. B. Riles
101. K. W. Gruenberg
102. O. Macedońska
103. P. A. Zalesskii
104. E. Raptis
105. U. Tiberio
106. D. Varsos
107. B. Hartley
108. L. C. Kappe
109. W. M. Holubowski
110. S. Rinauro
111. U. Dardano
112. G. Cutolo
113. G. Zappa
114. G. Casadio Zappa
115. S. Rao
116. M. L. Newell
117. C. Tibiletti
118. G. Zacher
119. F. de Giovanni
120. B. Amberg
121. P. Plaumann
122. W. Herfort
123. S. Andreadakis
124. T. Hurley
125. Mrs. Andreadakis

B. Hartley M. Newell

J. Wiegold D. J. S. Robinson

L. A. Kurdachenko H. Heineken

F. Menegazzo D. Dinkranjan

J. Wilson E. Zelmanov

Some Results and Problems about Factorized Groups

Bernhard Amberg

1991 Mathematics Subject Classification: 20D40, 20F19

1. Introduction

In the last years the structure of groups $G = AB$ which are the product of two of its subgroups A and B have received considerable attention. The most important results about mainly infinite such factorized groups are presented in the monograph [7]; see also [17] and the survey articles [2], [3], [35] and [42]. Among the central problems that were considered are the following:

(1) Let the group $G = AB$ be the product of two subgroups A and B and suppose that A and B satisfy some finiteness condition $\mathfrak{X}$. When does the factorized group G have the same finiteness condition $\mathfrak{X}$?

(2) Let the group $G = AB = AM = BM$ be the product of two subgroups A and B and a normal subgroup M of G and suppose that A, B and M satisfy some nilpotency condition $\mathfrak{N}$. When does the factorized group G have the same nilpotency condition $\mathfrak{N}$?

(3) Let the group $G = AB$ be the product of two subgroups A and B which satisfy some nilpotency condition $\mathfrak{N}$. What can be said about the structure of the factorized group G, when does it satisfy some solubility requirement? Which subgroups such as the Hirsch–Plotkin radical and the Fitting subgroup of $G = AB$ can be written as a product of a subgroup of A and a subgroup of B?

The following article can be considered as an appendix to [7], to which we refer as a general reference. In particular we will discuss some new results with respect to the above problems which have been found since the appearence of this book and would have had an influence to our writing if we had known them at that time.

Offprint from
Infinite Groups 94, Eds.: de Giovanni/Newell

2. Finiteness Conditions

2.1. General Products

Consider a group $G = AB$ which is the product of two subgroups A and B and suppose that $\mathfrak{X}$ is a group theoretical property which is inherited by subgroups, epimorphic images and extensions. If A and B are $\mathfrak{X}$-groups, when does it follow that G likewise has the group theoretical property $\mathfrak{X}$? This problem is particularly interesting when $\mathfrak{X}$ is some finiteness condition.

The main finiteness conditions that were considered are the following:

A group G has ***min*** if it satisfies the minimum condition on subgroups. G has ***max*** if it satisfies the maximum condition on subgroups. G is a ***minimax group*** if it has a finite series whose factors satisfy the minimum or maximum condition for subgroups. G has ***finite Prüfer rank*** $r = r(G)$ if every finitely generated subgroup of G can be generated by r elements and r is the least such number. G has ***finite abelian section rank*** if every elementary abelian p-section of G is finite for every prime p. G has ***finite torsion-free rank*** if it has a series of finite length whose factors are periodic or infinite cyclic. The number of infinite cyclic factors in any such series is an invariant of G called its ***torsion-free rank*** $r_0(G)$. G is ***periodic*** if and only if it has torsion-free rank $r_0(G) = 0$.

The following theorem announced by Ivanov in [30] shows that without further assumptions none of the usual finiteness conditions is inherited by products of groups (see also [31], Theorem 34).

Theorem 2.1. *Let p and q be primes not less that 10^{78}, and let A and B be infinite groups all of whose non-trivial proper subgroups have order p and q, respectively. Then every countable group X can be embedded in a product $G = AB$ of A and B such that $A \cap B = 1$.*

Simple examples show that Theorem 2.1 has the following consequence.

Corollary 2.2. *None of the following finiteness conditions is inherited by products of groups: "min", "max", "minimax", "finite Prüfer rank", "finite abelian section rank", "periodic", "finite torsion-free rank".*

Using wreath products with respect to infinite sequences of permutation groups, Suchansky has also constructed residually finite groups $G = AB$ which are the product of two locally finite subgroups A and B, but which are not locally finite. Here G may be a p-group for some prime p, but also may have elements of infinite order; furthermore A may be a p-group and B a q-group for primes p and q with $p \neq q$ (see [47] and [48]).

There are many papers about the factorizations of finite simple groups. Factorizations of semisimple algebraic groups and of Lie groups were considered by Onishchik in [38] and [39], Sections 5 and 14. In Belyaev [15] it is for instance shown that every countable locally finite simple group which is not a finitary linear group has many factorizations as a product of two residually finite groups.

2.2. Hyperabelian-by-Finite and Radical-by-Finite Groups

The situation is completely different if the group $G = AB$ satisfies some solubility requirement. For hyperabelian-by-finite groups the results obtained so far can be summarized in the following general theorem to which several authors have contributed; for details see [7], Chapters 3 and 4.

Theorem 2.3. *Let $\mathfrak{X}$ be one of the following finiteness conditions "min", "max", "minimax", "finite Prüfer rank", "finite abelian section rank", "finite torsion-free rank" or "being a π-group for some set of primes π". If the hyperabelian-by-finite group $G = AB$ is the product of two $\mathfrak{X}$-subgroups A and B, then G is an $\mathfrak{X}$-group.*

The special case of Theorem 2.3 that every hyperabelian-by-finite product $G = AB$ of two subgroups A and B with finite torsion-free rank has likewise finite torsion-free rank was recently proved by Sysak in [53]. By [7], Theorem 4.1.8, the torsion-free rank satisfies the inequality $r_0(G) \leq r_0(A) + r_0(B) - r_0(A \cap B)$ in this case, but it is unknown whether equality holds here (see [7], section 4.1, in particular Question 8).

It is an interesting question for which other classes of generalized soluble groups a similar theorem holds. For instance, it would seem plausible that Theorem 2.3 extends to radical-by-finite groups. In fact, it follows from a result of Chernikov that a radical-by-finite product of two subgroups with minimum condition on subgroups likewise satisfies the minimum condition on subgroups (see [7], Theorem 3.2.12). Furthermore, Theorem 4 of Sysak [54] implies that a radical-by-finite product of two periodic groups is periodic. But it seems to be unknown at present whether similar results hold for the other finiteness conditions. In particular it is open whether a radical group $G = AB$ which is the product of two subgroups A and B with maximum condition likewise satisfies the maximum condition ([7], Section 4.4, Question 9). It is not difficult to see that we may reduce the consideration to the case that there exists a locally nilpotent normal subgroup M of G such that $G = AM = BM$.

2.3. Locally Soluble Groups

Since soluble products of polycyclic groups are polycyclic by Theorem 2.3, it is also clear that every locally soluble product of two groups with maximum condition on subgroups likewise satisfies the maximum condition on subgroups. A similar result holds for the minimum condition, since every product of two subgroups with minimum condition on subgroups satisfies the minimum condition for normal subgroups (see

[7], Lemma 1.2.6), and a locally soluble group with minimum condition for normal subgroups is known to be hyperabelian. On the other hand, Sysak has given examples to show that even a locally finite-soluble product of two subgroups with finite abelian section rank need not have finite abelian section rank. A general construction of such groups can be found in [54], Theorem 3.

In the following $\pi(X)$ denotes the set of primes p for which there is an element of order p in the group X.

Theorem 2.4 (Sysak [49]). *For every prime q there exists a locally finite-soluble group G of the form $G = A \ltimes V = AA^{\alpha}$, where V is a minimal normal subgroup of G which is an elementary abelian q-subgroup of G, A is a residually finite q'-subgroup of G and α is a locally inner automorphism of order q of G, which stabilizes the series $1 \triangleleft V \triangleleft G$; in addition, it may be assumed that A satisfies one of the following two conditions:*

(a) $\pi(A) = \pi(q^n - 1)$ *for any given natural number $n \geq 2$,*

(b) $q = 2$ *and all Sylow subgroups of A are finite and abelian.*

If, in particular, the group A in Theorem 2.4 satisfies condition (b), then the group G is the product of two subgroups with finite abelian section rank, but has infinite q-rank for the prime q. If the group A satisfies condition (a) and $q = 2$ and $2^n - 1$ is a Mersenne prime or $q = 3$ and $n = 2$, then G is the product of two p-subgroups, but itself has elements of two different prime orders.

A group satisfies the ***weak minimum condition for subgroups*** if every descending chain of subgroups S_i has only finitely many infinite indices $|S_{i+1} : S_i|$. The weak minimum condition for normal subgroups is defined similarly. It is well-known that a locally soluble minimax group is soluble and satisfies the weak minimum condition for subgroups. The following lemma is elementary.

Lemma 2.5 (Amberg [1]). *If the group $G = AB$ is the product of two subgroups A and B with weak minimum condition for subgroups, then G satisfies the weak minimum condition for normal subgroups.*

It follows from Theorem 2.1 that the group G in Lemma 2.5 the group G need not satisfy the weak minimum condition for subgroups. But for locally soluble groups the following can be said.

Theorem 2.6 (Amberg and Sysak [12]). *If the locally soluble group $G = AB$ is the product of two minimax subgroups A and B, then G is a soluble minimax group.*

Comments on the proof of Theorem 2.6. Recall that the standard polynomial of degree n is the polynomial

$$S_n(x_1, \ldots, x_n) = \sum_{\pi \in \mathrm{Sym}(n)} (\operatorname{sgn} \pi) x_{\pi 1} \ldots x_{\pi n}.$$

The ring R satisfies the standard polynomial of degree n if $S_n(r_1, \ldots, r_n) = 0$ for all elements $r_1, r_2, \ldots, r_n$ of R. It is easy to see that the property that a ring satisfies the standard polynomial for some degree n is inherited by subrings, factor rings and cartesian products. The Theorem of Amitsur and Levitzki says that the ring $M_n(R)$ of $n \times n$-matrices with coefficients in the commutative ring R satisfies the standard polynomial of degree $2n$. The proof of Theorem 2.8 depends on the following proposition about the endomorphism ring of an abelian group of finite Prüfer rank which is of independent interest.

Proposition 2.7. *The endomorphism ring of an abelian group of finite Prüfer rank r satisfies the standard polynomial of degree $2r$.*

The proof of Theorem 2.6 reduces to the case of a hyperabelian group by the following theorem. Note that a group G is ***residually of bounded finite Prüfer rank*** if there exist normal subgroups N_i of G with $\bigcap N_i = 1$ and a positive integer k such that the Prüfer ranks r_i of the factor groups G/N_i satisfy $r_i \leq k$ for every i in the index set I. It is also used that a locally soluble subgroup of the multiplicative group of a ring with a polynomial identity is hyperabelian.

Theorem 2.8 (Amberg and Sysak[12]). *If the locally soluble group G is residually of bounded Prüfer rank, then G is hyperabelian.*

Using similar methods it can also be proved that every locally soluble product of two groups with finite Prüfer rank has likewise finite Prüfer rank. But it seems to be unknown at present whether a locally soluble product of two groups with finite torsion-free rank has likewise finite torsion-free rank.

Brookes and Smith have shown in [16] that if a group G is a product of two normal subgroups, each of which is locally soluble-of-finite-Prüfer-rank, then G is locally soluble if and only if it is locally of finite Prüfer rank. Since by a theorem of Robinson every finitely generated soluble group with finite Prüfer rank is a minimax group (see [41], Theorem 10.38), this implies that a locally soluble product of two normal locally minimax groups is a locally minimax group. It would be interesting to know whether this extends to arbitrary locally soluble products of locally minimax groups.

3. Triply Factorized Groups

3.1. Connections with Group Algebras

In the investigation of factorized groups one often encounters triply factorized groups of the form $G = AB = AM = BM$ where A, B and M are subgroups of the group G and M is normal in G. In many cases the group G in addition has the "intersection property" that $A \cap M = B \cap M = 1$, so that A and B are complements of M in G.

The following result establishes an interesting connection of triply factorized groups with group algebras; for a proof see [54], Theorem 1.

Theorem 3.1 (Sysak [51]). *Let G be a group and M a right kG-module for some commutative ring k. Then the following conditions are equivalent:*

(1) *In the augmentation ideal $\Delta_k G$ there exists a right ideal $\mathfrak{p}$ of kG such that $\Delta_k G = G - 1 + \mathfrak{p}$, and the kG-module M is isomorphic with the kG-module $\Delta_k G/\mathfrak{p}$.*

(2) *In the semi-direct product $G \ltimes M$ there exists a subgroup H such that $G \ltimes M = H \ltimes M = GH$.*

Moreover, for every subgroup H of G the right ideal $\mathfrak{p}$ can be chosen such that $G \cap H = G \cap (1 + \mathfrak{p})$.

In this situation also the following result of [13] can be proved.

Proposition 3.2. *Let G be a hyperabelian locally minimax group and k be a field of characteristic $p > 0$. Let $\Delta_k G = G - 1 + \mathfrak{p}$ for some right ideal $\mathfrak{p}$ of the group algebra kG, and let $G_0 = (1 + \mathfrak{p}) \cap G$. Then the maximal normal p'-subgroup of G is contained in G_0.*

It can be shown that Proposition 3.2 extends to hyperabelian groups with finite torsion-free rank, if and only if, for every hyperabelian product $G = AB$ with finite torsion-free rank the rank equality $r_0(G) = r_0(A) + r_0(B) - r_0(A \cap B)$ holds.

Proposition 3.2 has the following consequence, which may be useful when products of two locally minimax groups are studied and which perhaps also extends to hyperabelian products of groups with finite torsion-free rank. A group G is called ***locally minimax*** if every finitely generated subgroup of G is a minimax group.

Corollary 3.3. *Let the group $G = AB = AM = BM$ be the product of two locally minimax subgroups A and B and a normal elementary abelian p-subgroup M with $A \cap M = B \cap M = 1$. Then the maximal normal p'-subgroup of A is normal in G.*

3.2. Hyperabelian Groups with Finite Torsion-Free Rank

In Chapter 6 of [7] it is shown that under certain finiteness conditions a triply factorized group $G = AB = AK = BK$ of two subgroups A and B and a normal subgroup K satisfies some nilpotency condition if the three subgroups A, B and K satisfy this same nilpotency condition. For instance, Amberg, Franciosi and de Giovanni have shown that a group $G = AB = AK = BK$ with finite abelian section rank is locally nilpotent (and hence hypercentral) if A, B and K are locally nilpotent ([7], Theorem 6.3.8). In the proof of this cohomological arguments play a role. The proof of the following more general result uses Theorem 3.1.

Theorem 3.4 (Amberg and Sysak [13]). *Let the hyperabelian group $G = AB = AM = BM$ be the product of three locally nilpotent subgroups A, B and M, where M is normal in G. If G has finite torsion-free rank, then G is locally nilpotent.*

If the hyperabelian group G in Theorem 3.4 does not have finite torsion-free rank, then G need not be locally nilpotent, even if A, B and M are abelian; see [7], Theorem 6.1.1. By Theorem 2.4 there exists a locally finite-soluble group $G = AB = AM = BM$ where M is a minimal normal subgroup of G and so an elementary abelian q-subgroup for some prime q and A and B are locally finite p-subgroups for some prime p, but G is not locally nilpotent. Since G is a periodic radical group, this example shows that Theorem 3.4 also becomes false when the factorized group G is no longer hyperabelian. It also follows that the Hirsch–Plotkin radical of a radical and locally finite-soluble product of two locally nilpotent subgroups need not be factorized, so that Question 11 (a) of [7] has a negative answer.

Recall that the ***Hirsch–Plotkin series*** of the group G is defined by

$$R_0 = 1, \quad R_{\alpha+1}/R_\alpha \text{ is the Hirsch–Plotkin radical of } G/R_\alpha$$

for every ordinal α, and for limit ordinals λ

$$R_\lambda = \bigcup_{\beta<\lambda} R_\beta.$$

Theorem 3.4 has the following consequence which is proved as in the case of a hyperabelian group with finite abelian section rank (see [7], Corollary 6.3.9).

Corollary 3.5. *Let the hyperabelian group $G = AB$ with finite torsion-free rank be the product of two locally nilpotent subgroups A and B. Then each term of the Hirsch–Plotkin series of G is factorized. In particular the Hirsch–Plotkin radical $R = R(G)$ of G is factorized, i.e. satisfies $R = (A \cap R)(B \cap R)$ and $A \cap B \subseteq R$.*

Comments on the Proof of Theorem 3.4. The proof is by induction on the torsion-free rank of M. Suppose first that the torsion-free rank of M is trivial, i.e. M is periodic.

The hypotheses of the theorem are inherited by epimorphic images. But local nilpotency is a property defined by the finitely generated subgroups of a group. To characterize this group class "from above", the following definition turns out to be useful.

A group G is a ***p*-group over its subgroup *S*** if for every g in G there is a p-power p^n such that g^{p^n} is in S.

The relevance of this concept in our situation can be seen from the following lemma.

Lemma 3.6. *Let the hyperabelian group $G = AB = AM = BM$ be the product of two locally nilpotent subgroups A and B and a normal p-subgroup M of G. If G is a p-group over $A \cap B$, then G is locally nilpotent.*

Using this lemma it is not difficult to see that it is sufficient to prove Theorem 3.4 for a periodic normal subgroup M of G in the case when M is an elementary abelian p-group for some prime p. For this the following lemma is crucial.

Lemma 3.7. *Let G be a hyperabelian group such that $G = AM = BM = AB$, where M is an elementary abelian normal p-subgroup of G and A and B are locally nilpotent groups with finite torsion-free rank. Then G is a p-group over $A \cap B$.*

In the proof of this Theorem 3.1 and Proposition 3.2 and the following lemmas play a role.

Lemma 3.8. *Let G be a locally nilpotent group and k a field of characteristic $p > 0$ and let $\mathfrak{p}$ be a right ideal of the group algebra kG. Let $G_0 = (1 + \mathfrak{p}) \cap G$ and let G be a p-group over G_0. Then for every finitely generated subring V of $\Delta_k(G)$ there exists a positive integer m such that $V^m \subseteq \mathfrak{p}$.*

Lemma 3.9. *Let G be a hyperabelian and locally nilpotent group with finite torsion-free rank and let k be a field of characteristic $p > 0$. Suppose that $\Delta_k(G) = G - 1 + \mathfrak{p}$ for some right ideal $\mathfrak{p}$ of the group algebra kG. Then G is a p-group over $G_0 = (1 + \mathfrak{p}) \cap G$.*

If Theorem 3.4 is false in the general case choose a counterexample where the torsion-free rank of M is minimal. Then we have seen that M is not periodic, so that the maximal periodic subgroup $T = T(M)$ is properly contained in M. It can be shown that also G/T is a counterexample, so that we may assume that M is torsion-free and hence nilpotent. We may even suppose that M is torsion-free abelian. If $T(G)$ is the maximum periodic normal subgroup of G, it follows that $M \cap T(G) = 1$. Hence $G/T(G)$ is not locally nilpotent, and we may suppose that G has no non-trivial periodic normal subgroups. In particular $C_A(M)$ is torsion-free. Since the periodic subgroups of $A/C_A(M)$ are finite, it follows that G has finite Prüfer rank, and so G is locally nilpotent by the above-mentioned theorem of Amberg, Franciosi and de Giovanni (see [7], Theorem 6.3.8). This final contradiction proves Theorem 3.4.

Theorem 3.4 raises the following questions about a hyperabelian group $G = AB = AM = BM$ with finite torsion-free rank which is the product of two subgroups A and B and a normal subgroup M of G. Note that these questions have positive answers if G has finite abelian section rank (see [5], [6] and [20]):

(a) If A, B and M are (locally nilpotent)-by-$\mathfrak{X}$, is it true that also G (locally nilpotent)-by-$\mathfrak{X}$, where $\mathfrak{X}$ is the class of finite, Chernikov, polycyclic, minimax, periodic groups or the class of groups with finite Prüfer rank or finite abelian section rank?

(b) If A, B and M are locally polycyclic, is it true that also G locally polycyclic?

(c) If A, B and M are FC-hypercentral (FC-nilpotent), is it true that G likewise FC-hypercentral (FC-nilpotent)?

3.3. *FC*-Nilpotent Groups

Another result about triply factorized groups is the following theorem concerning FC-nilpotent groups. Recall that a group G is ***FC-nilpotent*** if the upper FC-central series of G reaches G after a finite number of steps.

Theorem 3.10 (Amberg, Franciosi and de Giovanni [8]). *Let the FC-nilpotent group $G = AB = AM = BM$ be the product of two hypercentral subgroups A and B and a nilpotent normal subgroup M of G. Then G is hypercentral.*

It is well-known that for FC-nilpotent groups the classes of hypercentral and of locally nilpotent groups coincide. It seems likely that every FC-hypercentral group $G = AB = AM = BM$ which is the product of three hypercentral subgroups A, B and M, where M is normal in G, is likewise hypercentral.

Theorem 3.10 has the following consequence on the "outside problem" (see section 7 of [7]).

Corollary 3.11. *Let the FC-nilpotent group $G = AB$ be the product of a nilpotent proper subgroup A and a proper hypercentral subgroup B. Then either A or B is contained in a proper normal subgroup of G. If A and B are distinct hypercentral subgroups, then either A or B is contained in a proper ascendant subgroup of G.*

3.4. Connections with Radical Rings

An associative ring R is called ***radical*** if it coincides with its Jacobson radical, which means that R forms a group under the operation $a \circ b = a + b + ab$ for all a and b in R. This group is called the ***adjoint group*** R° of R. If the radical ring R is embedded in the usual way into a ring R_1 with identity, then R° is isomorphic with the subgroup $1 + R$ of the group of units of R_1. Examples of such radical rings are the nilpotent rings and even the nil rings. Recall that an associative ring R is a ***nil ring*** if every element $a \in R$ is ***nilpotent***, i.e. there exists a positive integer n such that $a^n = 0$. The ring R is ***nilpotent*** if $R^n = 0$ for some positive integer n.

The adjoint group $A = R^\circ$ operates on the additive group $M = R^+$ of R via the rule $m^a = m + ma$ for all a in A and m in M. Let $G(R) = A \ltimes M$ be the semidirect product of A with M and identify A and M with its subgroups $\{(a, 0) \mid a \in A\}$ and $\{(0, m) \mid m \in M\}$, respectively. If $B = \{(r, r) \mid r \in R\}$ is the diagonal subgroup of $G(R)$, then the ***associated group*** of R is

$$G(R) = A \ltimes M = B \ltimes M = AB.$$

Then $G = AB$ and $A \cap B = 1$, where the subgroups A and B are isomorphic with R° and the normal subgroup M of $G(R)$ is isomorphic with R^+. This construction gives examples of triply factorized groups with the intersection property.

What can be said about the structure of the adjoint group of a radical ring? The following theorem, which generalizes a result of Watters in [56] for the maximum condition, gives a relation between the three groups R°, R^+ and $G(R)$ of a radical ring.

Theorem 3.12 (Amberg and Dickenschied [4]). *Let $\mathfrak{X}$ be a class of minimax groups which is closed under the forming of subgroups, epimorphic images and extensions. The following conditions of the radical ring R are equivalent:*

(i) *The additive group R^+ is an $\mathfrak{X}$-group.*

(ii) *The associated group $G(R)$ is an $\mathfrak{X}$-group.*

(iii) *The adjoint group R° is an $\mathfrak{X}$-group.*

In this case R is a nilpotent ring.

The additive group of the (commutative) radical subring

$$R = \left\{\tfrac{u}{v} \mid u \text{ is an even and } v \text{ is an odd integer}\right\}$$

of the ring of rational numbers has Prüfer rank 1, but R° has infinite torsion-free rank and R is not even a nil ring. Hence there is no complete analogue of Theorem 3.12 for the other finiteness conditions. However, the following theorem shows that these finiteness conditions are inherited from the adjoint group R° of a radical ring R to the additive group R^+ and that they imply some nilpotency conditions of the ring R.

Theorem 3.13 (Amberg and Dickenschied [4]). *Let R be a radical ring. Then the following holds:*

(a) *If R° has finite torsion-free rank n, then also $r_0(R^+) = n$ and R is a nil ring.*

(b) *If R° has finite abelian subgroup rank, then so does R^+, and R is a two-sided T-nilpotent ring of class $cl(R) \leq \omega + r_0(R^+)$.*

(c) *If R° has finite Prüfer rank, then so does R^+, and $r(R^+)$ is bounded by a function which only depends on $r(R^\circ)$.*

Here a ring R is called ***two-sided T-nilpotent of class*** $cl(R) = \alpha$ if $B_\alpha(R) = R$ and α is the least ordinal with this property, where the ***transfinite two-sided annihilator series*** of a ring R is defined by

$$B_0(R) = 0, \qquad B_{\alpha+1}(R) = \{a \in R \mid aR + Ra \subseteq B_\alpha(R)\}$$

for each ordinal α and for each limit ordinal λ

$$B_\lambda(R) = \bigcup_{\beta<\lambda} B_\alpha(R).$$

For nil-rings and for radical rings whose additive group is periodic more can be said, in particular the converse of Theorem 3.13 holds in these cases (see [19]). The question arises which is the exact bound for the Prüfer rank in Theorem 3.13 (c). Is perhaps $r(R^+) \leq 2r(R^\circ)$?

3.5. On Products of Abelian Groups

If the group $G = AB = AM = BM$ is the product of three abelian subgroups A, B and K where K is normal in G such $A \cap M = B \cap M = 1$, then there exists a radical ring R such that the group G is the associated group $G(R)$ which can be constructed as the semidirect product of the additive group $R^+ \simeq K$ of R by the adjoint group $R^\circ \simeq A$ of R (see [7], Proposition 6.1.4). Using this the following theorem has been proved which answers [2], Question IX (a) in the affirmative.

Theorem 3.14 (Sysak [52]). *Let the group $G = AB$ be the product of two abelian subgroups A and B.*

(a) *The Fitting subgroup of G is factorized if and only if the Hirsch–Plotkin radical of G is factorized.*

(b) *If A or B has finite torsion-free rank, then the Fitting subgroup and the Hirsch–Plotkin radical of G are factorized.*

4. Products of Nilpotent Groups

4.1. Finite Products of Nilpotent Groups

A well-known theorem of Kegel and Wielandt states that a finite group $G = AB$ which is the product of two nilpotent subgroups A and B is soluble (see [7], Theorem 2.4.3). In order to determine the structure of such groups it is interesting to know which subgroups of G are conjugate to a subgroup S that inherits the factorization. This question is discussed in detail in [10]. A subgroup S of a factorized group $G = AB$ is called ***prefactorized*** if $S = (A \cap S)(B \cap S)$; it is called ***factorized*** if in addition $A \cap B$ is contained in S. For instance it follows from Corollary 3.5 that the Fitting subgroup of a finite product of two nilpotent groups is always factorized. On the other hand, it can be shown that unlike the Fitting subgroup the commutator subgroup, the Frattini subgroup, the centre and the hypercentre of a finite product of nilpotent groups are in general not prefactorized.

The following elementary lemma is easily proved by using the theory of P.Hall on finite soluble groups.

Lemma 4.1. *Let the finite and soluble group $G = AB$ be the product of two subgroups A and B whose orders are relatively prime. Then every subgroup of G has a factorized conjugate. In particular every subnormal subgroup of G is factorized.*

Recall that a set Σ of Hall subgroups of a finite soluble group G is a ***Hall system*** of G if Σ contains a Hall π-subgroup of G for every set of primes π and $HK = KH$ for every pair H and K of subgroups in Σ.

Lemma 4.2. *If the finite soluble group $G = AB$ is the product of two nilpotent subgroups A and B, then the system Σ of all $A_\pi B_\pi$ where π is a set of primes, is a Hall system of G which reduces into every prefactorized subgroup S of G, i.e. the set of subgroups $S \cap A_\pi B_\pi$ is a Hall system of S.*

Using this the following uniqueness statement about pronormal subgroups can be proved. Recall that a subgroup P of a group G is ***pronormal*** if P and P^g are conjugate in $\langle P, P^g \rangle$ for every element g in G.

Proposition 4.3. *A pronormal subgroup of a finite group $G = AB$ which is the product of two nilpotent subgroups A and B has at most one prefactorized conjugate.*

For abnormal subgroups we even obtain the following existence theorem. A subgroup S of a group G is ***abnormal*** if $g \in \langle S, S^g \rangle$ for every element g in G.

Theorem 4.4 (Amberg and Höfling [10]). *Every abnormal subgroup of a finite group $G = AB$ which is the product of two nilpotent subgroups A and B has a unique factorized conjugate.*

The proof of Theorem 4.4 is by induction on the order of the group G. It is shown that a minimal counterexample is a nonabelian primitive group and so the orders of A and B are relatively prime by [7], Lemma 2.5.2.

A class $\mathfrak{H}$ of finite soluble groups is a ***Schunck class*** if a group G is an $\mathfrak{H}$-group whenever its primitive epimorphic images are $\mathfrak{H}$-groups. For instance every saturated formation is a Schunck class. The characteristic $\mathrm{Char}(\mathfrak{H})$ is the set of all primes p such that the cyclic group of order p is an $\mathfrak{H}$-group. Furthermore, a subgroup Y of a group G is ***$\mathfrak{H}$-maximal*** if for every $\mathfrak{H}$-subgroup Y^* of G with $Y \subseteq Y^*$ it follows that $Y = Y^*$.

Theorem 4.5 (Amberg and Höfling [10]). *Let $\mathfrak{H}$ be a Schunck class and let the finite group $G = AB$ be the product of two nilpotent subgroups A and B. If H is an $\mathfrak{H}$-maximal subgroup of G, then*

(a) *if $\pi(A) \cap \pi(B) \subseteq \mathrm{Char}(\mathfrak{H})$, then H has a factorized conjugate in G,*

(b) *if $\mathfrak{H}$ is a saturated formation, then H has a prefactorized conjugate.*

As a consequence of Proposition 4.3 and Theorem 4.5 we obtain the following generalization of a result of Heineken; see [7], Theorem 2.5.15. Recall that a subgroup P of a finite group G is an ***$\mathfrak{H}$-projector*** of G if the factor group PN/N is $\mathfrak{H}$-maximal in G/N for every normal subgroup N of G.

Corollary 4.6. *Let $\mathfrak{H}$ be a Schunck class and let the finite group $G = AB$ be the product of two nilpotent subgroups A and B.*

(a) *If $\pi(A) \cap \pi(B) \subseteq \mathrm{Char}(\mathfrak{H})$, then G has a unique factorized $\mathfrak{H}$-projector.*

(b) (Heineken [26]) *If $\mathfrak{H}$ is a saturated formation, then G has a unique prefactorized $\mathfrak{H}$-projector.*

The following corollary of Theorem 4.5 extends earlier results of Kegel, Pennington and Peterson; see [7], Theorem 2.5.10.

Corollary 4.7 (Amberg and Fransman [9]). *Let $\mathfrak{H}$ be a Schunck class of finite groups. Let the finite group $G = AB = AC = BC$ be the product of two nilpotent subgroups A and B and an $\mathfrak{H}$-subgroup C. If $\pi(A) \cap \pi(B) \subseteq \mathrm{Char}(\mathfrak{H})$, then G is a soluble $\mathfrak{H}$-group.*

In [10] an example is given that for Fitting classes $\mathfrak{F}$ the $\mathfrak{F}$-injectors of a finite product of nilpotent groups need not be prefactorized in general. However it is also shown there that for Fitting classes $\mathfrak{F}$ every product of two finite nilpotent groups has a unique prefactorized (factorized) $\mathfrak{F}$-injector if and only if every product of two finite nilpotent groups has a prefactorized (factorized) $\mathfrak{F}$-radical.

Some results of this section have been extended in the forthcoming dissertation of B. Höfling to certain locally finite-soluble products of two locally nilpotent subgroups.

4.2. On the Derived Length

It is a well-known open question whether the derived length of a finite group $G = AB$ which is the product of two nilpotent subgroups A and B is bounded by a function of the nilpotency classes c of A and d of B (see [7], Question 5). Gross proved that the Frattini factor group $G/\Phi(G)$ has derived length at most $c+d$, and this of course means that the subgroup $G^{(c+d-1)}$ is nilpotent; see [7], Theorem 2.5.5. The following theorem extends this result to radical groups with finite abelian section rank. By $\mathfrak{S}_1$ we denote the class of radical groups with finite abelian section rank containing elements of only finitely many distinct prime orders.

Theorem 4.8 (Franciosi and de Giovanni [23]). *Let the radical group $G = AB$ with finite abelian section rank be the product of two nilpotent subgroups A and B with nilpotency classes c and d, respectively. Then the subgroup $G^{(c+d-1)}$ is locally nilpotent. If G is an $\mathfrak{S}_1$-group, then $G^{(c+d-1)}$ is nilpotent.*

Pennington has shown that if the finite group $G = AB$ is the product of two nilpotent subgroups A and B with nilpotency classes c and d, then the $(c+d)$-th term of the derived series of G is a nilpotent π-group where $\pi = \pi(A) \cap \pi(B)$; see [7], Theorem 2.5.3. This result can be extended to periodic radical products of two nilpotent groups.

Theorem 4.9 (Franciosi and de Giovanni [23]). *Let the periodic radical group $G = AB$ be the product of two nilpotent subgroups A and B with nilpotency classes c and d, respectively. Then the subgroup $G^{(c+d)}$ is a π-group, where $\pi = \pi(A) \cap \pi(B)$. In particular, if $\pi(A) \cap \pi(B) = \emptyset$, the group G is soluble with derived length at most $c+d$.*

More on the derived length of finite soluble factorized groups can be found in Kazarin [34].

4.3. Hirsch–Plotkin Radical and Fitting Subgroup

It follows for instance from Corollary 3.5 that the Fitting subgroup of a finite product of two nilpotent groups is factorized. In this section we mention some related results on more general classes of soluble products of nilpotent groups.

The first theorem shows that if $G = AB$ is a radical group with finite abelian section rank which is factorized by two nilpotent subgroups A and B, then the centres $Z(A)$ and $Z(B)$ act trivially on the second factor of the Hirsch–Plotkin series of G.

Theorem 4.10 (Franciosi and de Giovanni [23]). *Let the radical group $G = AB$ with finite abelian section rank be the product of two nilpotent subgroups A and B. If R is the Hirsch–Plotkin radical of G and H/R is the Hirsch–Plotkin radical of G/R, then $[Z(A), H] \subseteq R$.*

Theorem 4.10 has the following consequence.

Corollary 4.11. *Let the radical group $G = AB$ with finite abelian section rank be the product of two nilpotent subgroups A and B, and let R be the Hirsch–Plotkin radical of G. Then the subgroups $Z(A)R$ and $Z(B)R$ are subnormal in G with defect at most* 2.

However, it can be shown that there exists a finite group $G = AB$ which is the product of two nilpotent subgroups A and B with Fitting subgroup $F = F(G)$ such that $Z(A)F$ and $Z(B)F$ are both not normal in G, and the subgroups AF and BF are both not subnormal in G. Corresponding to Theorem 4.10 we have the following result.

Theorem 4.12 (Franciosi and de Giovanni [23]). *Let the $\mathfrak{S}_1$-group $G = AB$ be the product of two nilpotent subgroups A and B. If F is the Fitting subgroup of G and L/F is the Fitting subgroup of G/F, then $[Z(A), L] \subseteq F$.*

It can be deduced from Theorem 4.12 that if the $\mathfrak{S}_1$-group $G = AB$ is the product of two nilpotent subgroups A and B, then the subgroups $Z(A)F$ and $Z(B)F$ are subnormal of defect at most 2. Moreover, $[Z(A), Z(B)]$ is a nilpotent subnormal subgroup of G.

5. Locally Finite Products of Locally Nilpotent Groups

5.1. Minimal Normal Subgroups

Robinson and Stonehewer have shown in [43] that every chief factor of a product $G = AB$ of two abelian subgroups A and B is centralized by A or B; see [7], Theorem 6.6.9. Moreover, Stonehewer proved in [45] that if $G = AB$ is finite and A and B are nilpotent, then for every minimal normal subgroup N of G one of the subgroups AN and BN is nilpotent. The following theorem extends these results to periodic radical groups factorized by two locally nilpotent subgroups.

Theorem 5.1 (Franciosi, de Giovanni and Sysak [24]). *Let the periodic radical group $G = AB$ be the product of two locally nilpotent subgroups A and B, at least one of which is hyperabelian. If N is a minimal normal subgroup of G, then one of the subgroups AN and BN is locally nilpotent. Moreover, if AN is locally nilpotent, then $Z(A)$ centralizes N, and if both AN and BN are locally nilpotent the subgroup N is contained in the centre of G.*

Theorem 5.1 has the following consequence for trifactorized groups.

Corollary 5.2. *Let the periodic radical group $G = AB = AC = BC$ be the product of three locally nilpotent subgroups A, B and C, at least two of which are hyperabelian. Then G is locally nilpotent.*

It follows from Theorem 2.4 that in these results the hypothesis that the factors are hyperabelian cannot be omitted.

The next result shows that Theorem 4.10 also holds for periodic hyperabelian groups.

Theorem 5.3 (Franciosi, de Giovanni and Sysak [24]). *Let the periodic radical group $G = AB$ be the product of two locally nilpotent subgroups A and B, at least one of which is hyperabelian. If R is the Hirsch–Plotkin radical of G and H/R is the Hirsch–Plotkin radical of G/R, then $[Z(A), H] \subseteq R$.*

It follows from Theorem 5.3 that in this situation the subgroups $Z(A)R$ and $Z(B)R$ are subnormal with defect at most 2. Moreover, the subgroup $[Z(A), Z(B)]$ is contained in the Hirsch–Plotkin radical of G and so is locally nilpotent.

It is also possible to extend the theorem of Gross to periodic radical groups.

Theorem 5.4 (Franciosi, de Giovanni and Sysak [24]). *Let the periodic radical group $G = AB$ be the product of two nilpotent subgroups A and B with nilpotency classes c and d, respectively. Then the subgroup $G^{(c+d-1)}$ is locally nilpotent. In particular, the Hirsch–Plotkin series of G has length at most $c + d$.*

5.2. An Extension of the Theorem of Kegel and Wielandt

In which ways does the Theorem of Kegel and Wielandt extend to infinite groups, in particular to locally finite groups? Is for example every locally finite product of two locally nilpotent subgroups always locally soluble? Kegel has shown in [36] that every linear product of two nilpotent groups is soluble and that every locally finite product of two locally nilpotent FC-groups is locally soluble. Moreover, Kazarin proved in [33] that if the locally finite group $G = AB$ is the product of an abelian group A and a locally nilpotent group B, then G is locally soluble.

Theorem 5.5 (Franciosi, de Giovanni and Sysak [25]). *Let the locally finite group $G = AB$ be the product of two hypercentral subgroups A and B, at least one of which is an FC-group. If R is the Hirsch–Plotkin radical of G, then G/R is hyperabelian. In particular, G is a radical group.*

Theorem 5.5 implies another generalization of Pennington's theorem on finite products of nilpotent groups of coprime order (see [7], Theorem 2.5.3, and Theorem 4.9 above).

Corollary 5.6. *Let the locally finite groups $G = AB$ be the product of two nilpotent subgroups A and B, with classes c and d, respectively. If one of the two subgroups A or B is an FC-group, then the subgroup $G^{(c+d)}$ is a π-group, where $\pi = \pi(A) \cap \pi(B)$. In particular, if $\pi(A) \cap \pi(B) = \emptyset$, the group G is soluble with derived length at most $c + d$.*

6. Some Further Results

6.1. On the Rank of a Product of Two Finite p-Groups

It is known that the Prüfer rank of a finite p-group $G = AB$ which is the product of two subgroups A and B whose Prüfer ranks are bounded by r, is bounded by a polynomial function of r; see [7], Theorem 4.3.5. Although no bound is given there explicitly, the analysis of the proof of this theorem leads to polynomial bounds of relatively high degree. It seems to be unknown whether there is a linear bound. An attempt to give better bounds is made in [11].

The following theorem gives a bound for the normal rank of $G = AB$ which will immediately give a bound for the Prüfer rank of G essentially by a theorem of Thompson, [29], Satz 12.3, p. 343. Recall that the ***normal rank*** $r_n(X)$ of a group X is the maximum of the minimal number of generators of each normal subgroup of X. If α is a real number, then $\lceil \alpha \rceil = m$ is the smallest integer such that $\alpha \leq m$. δ_{ij} denotes the Kronecker symbol.

Theorem 6.1 (Amberg and Kazarin [11]). *Let the finite p-group $G = AB$ be the product of two subgroups A and B. Let $r^* = \min\{r(A), r(B)\}$ and $r_1 = r(A)+r(B)$. Then the normal rank $r_n(G)$ satisfies the following inequality:*

$$r_n(G) \le r^*(\lceil \log_p r_n(G)\rceil + 1 + \lceil \log_p r^*\rceil \lceil \log_2 2r^*\rceil + \delta_{2p}) + r_1.$$

The inequality in Theorem 6.1 implies that for any $\varepsilon > 0$ and for sufficiently large r^* we have the following almost linear bound

$$r_n(G)^{1-\varepsilon} \le r^*(3 + \lceil \log_p r^*\rceil \lceil \log_2 2r^*\rceil) + r_1.$$

Therefore the bound in Theorem 6.1 for the normal rank of G and so also the bound for its Prüfer rank is close to being linear. Even if this result may not be best possible, it will be useful in the study of the structure of finite products of groups with low rank.

If the two subgroups A and B are abelian, Theorem 6.1 can be improved as follows.

Theorem 6.2 (Amberg and Kazarin [11]). *Let the finite p-group $G = AB$ be the product of two abelian subgroups A and B. Let $r^* = \min\{r(A), r(B)\}$ and $r_1 = r(A) + r(B)$. Then the normal rank $r_n(G)$ of G satisfies the inequality*

$$r_n(G) \le r^*\lceil \log_p r_n(G)\rceil + r_1.$$

If the finite p-group $G = AB$ is the product of two cyclic subgroups A and B, then it follows from Theorem 6.2 that $r_n = r_n(G) \le \lceil \log_p r_n\rceil + 2$. This implies $r_n \le 3$ and even $r_n \le 2$ for $p > 3$. Note however that there exists a finite 2-group of normal rank 3 which is a product of two of its cyclic subgroups. (see [29], Aufgabe 28, p. 341).

6.2. More on the Hirsch–Plotkin Radical

The symmetric group of degree 4 is the product of a subgroup isomorphic with the symmetric group of degree 3 and a cyclic group of order 4, but its Fitting group $F = F(G)$ is not factorized, i.e. $F \neq AF \cap BF$. But even in this general situation something more can be said. Recall that by [7], Lemma 2.1.2, for the FC-centre of an arbitrary product $G = AB$ of two subgroups A and B the following holds: If A_1, B_1 and F_1 are the FC-centres of A, B and G, then $F_1 = A_1F_1 \cap B_1F_1$. Similar statements hold for the centre and the hypercentre.

The following theorem is first proved for the special case of a finite soluble product of two subgroups.

Theorem 6.3 (Franciosi and de Giovanni [21]). *Let the soluble-by-finite group $G = AB$ with finite abelian section rank be the product of two subgroups A and B, and let R be the Hirsch–Plotkin radical of G. Then $R = A_0R \cap B_0R$, where A_0 and B_0 are the Hirsch–Plotkin radicals of A and B, respectively.*

If the two subgroups A and B in Theorem 6.3 are locally nilpotent, then it follows in particular that the Hirsch–Plotkin radical of G is factorized. Similarly, the following result implies that the Fitting subgroup of a soluble $\mathfrak{S}_1$-group which is the product of two nilpotent subgroups, is factorized.

Theorem 6.4 (Franciosi and de Giovanni [21]). *Let the soluble-by-finite $\mathfrak{S}_1$-group $G = AB$ be the product of two subgroups A and B, and let F be the Fitting subgroup of G. Then $F = A_0F \cap B_0F$, where A_0 and B_0 are the Fitting subgroups of A and B, respectively.*

By [7], Theorem 6.1.2, there exists a triply factorized group $G = AB = AK = BK$, where A, B and K are torsion-free abelian subgroups and K is normal in G, but G is not locally nilpotent. Hence Theorem 6.3 does not hold in general. But the question arises whether the above theorems can be extended to soluble-by-finite groups with finite torsion-free rank, in particular to periodic groups.

A result of Johnson [32] says that for every arbitrary finite soluble group $G = AB$ which is the product of two subgroups A and B and for every set of primes π the maximal π-subgroups satisfy $O_\pi(A) \cap O_\pi(B) \subseteq O_\pi(G)$. While this is false for finite groups in general it even holds for arbitrary periodic hyperabelian groups with no perfect subgroups and no infinite elementary abelian groups involved in $O_\pi(A)$.

In [46] Sozutov has shown that there exist groups $G = AB$ which are the product of two normal subgroups A and B in which every subgroup generated by $n \geq 2$ elements is nilpotent, but G does not have this property. Thus, in particular, the product of two normal 2-nilpotent groups need not be 2-nilpotent. Observe however that by the theorem of Hirsch and Plotkin the product of two normal locally nilpotent subgroups is always locally nilpotent.

6.3. Groups Factorized by Finitely Many Subgroups

By the theorem of Kegel and Wielandt every finite group $G = A_1A_2 \dots A_n$ which is the product of pairwise permutable nilpotent subgroups A_i is soluble (see [7], Corollary 2.4.4). It was proved in [36] that a linear group $G = A_1A_2 \dots A_n$ which is the product of pairwise permutable locally nilpotent subgroups A_i is soluble in each of the following cases: (1) G is finitely generated, (2) $n = 2$, (3) $n \geq 3$ and the product of any three of the subgroups A_i is soluble. Kegel asked in [36], Remark to Theorem 2.8, whether there are insoluble linear groups over a field which are the product of finitely many pairwise permutable locally nilpotent subgroups. Such groups can in fact be constructed in the following way.

Let R be an associative ring and R_1 the ring obtained from R by adjoining an identity 1 if R is without one and $R_1 = R$ otherwise. Denote by $G_2(R)$ the group of all invertible 2×2-matrices over R_1 that are congruent to the identity matrix modulo R, and by $t_{ij}(s)$ the transvection with the element s in the position (i, j). Then the following can be verified directly.

Proposition 6.5 (Sysak [50]). *Let R be a radical ring and let A be the subgroup of $G_2(R)$ consisting of all its diagonal matrices, $B = t_{12}(-1)At_{12}(1)$, and $C = t_{21}(-1)At_{21}(1)$. Then $G_2(R) = ABC$ and the subgroups A, B, C pairwise commute, i.e. $AB = BA$, $AC = CA$ and $BC = CB$.*

As a consequence of Proposition 6.5 we obtain an affirmative answer to Kegel's question. Note that periodic insoluble groups $G = ABC$ which are the product of three pairwise permutable abelian subgroups A, B and C with Prüfer rank 1 have already been constructed by Marconi (see [7], Proposition 7.6.3).

Corollary 6.6. *Let F be a field containing a radical subring $R \neq 0$ (or, equivalently, having a non-periodic multiplicative group). Then $G_2(R)$ is an insoluble linear group over F which is a product of three pairwise permutable abelian subgroups.*

Proposition 6.5 also implies a negative answer to a question of Huppert in [29], p. 725, who asked whether for each $n \geq 1$ the derived length of a product of n pairwise permutable cyclic subgroups does not exceed n. For $p > 2$ the existence of such groups has already been established by Vasil'ev in [55], but no concrete examples are given there.

Corollary 6.7. *For any prime p and any natural number d there exists a finite p-group that is decomposable into a product of three pairwise permutable cyclic subgroups, and whose derived length is greater than d.*

Note that, on the other hand, a product of pairwise permutable cyclic subgroups A_i is supersoluble and abelian-by-finite if the A_i are infinite cyclic (see [28]).

For linear groups over a commutative ring the following generalization of the theorems of Kegel and Wielandt is announced in [50].

Theorem 6.8 (see Sysak [50]). *Let the linear group $G = A_1A_2 \ldots A_n$ over a commutative ring R with an identity be the product of pairwise permutable locally nilpotent subgroups A_i. Then in each of the following cases the group G is locally soluble:*

(a) $n = 2$.

(b) *$n \geq 3$ and the product of any three of the subgroups A_i is locally soluble.*

(c) *Each subring of R is locally nilpotent.*

(d) *Each of the subgroups A_i, $1 \leq i \leq n-1$, has finite torsion-free rank.*

By results of Zaitsev this implies

Corollary 6.9. *Let the linear group $G = A_1A_2 \ldots A_n$ over a commutative ring R with an identity be the product of pairwise permutable locally nilpotent subgroups A_i. Then the torsion-free rank of G does not exceed the sum of the torsion-free ranks of the subgroups A_i, $1 \leq i \leq n$.*

A special case of this is

Corollary 6.10. *A linear group over a commutative ring with an identity which is the product of finitely many pairwise permutable periodic locally nilpotent subgroups is periodic and locally soluble.*

References

[1] Amberg, B., Factorizations of infinite groups. Habilitationsschrift, Universität Mainz 1973.

[2] —, Infinite factorized groups. In: Groups - Korea 1988, Lecture Notes in Math. 1398, Springer-Verlag, Berlin 1989.

[3] —, Triply factorized groups. In: Groups - St. Andrews 1989, London Math. Soc. Lecture Note Ser. 159, 1–13, Cambridge University Press 1991.

[4] —, Dickenschied, O., On the adjoint group of a radical ring. Bull. Canad. Math. Soc. To appear.

[5] —, Franciosi, S., de Giovanni, F., Groups with an FC-nilpotent triple factorization. Ricerche Mat. 36 (1987), 103–114.

[6] —, —, —, Triply factorized groups. Comm. Algebra 18 (1990), 789–809.

[7] —, —, —, Products of Groups. Clarendon Press, Oxford 1992.

[8] —, —, —, FC-nilpotent products of hypercentral groups. Forum Math. 7 (1995), 307–316.

[9] —, Fransman, A., Products of groups and group classes. Israel J. Math. 87 (1994), 9–18.

[10] —, Höfling, B., On finite products of nilpotent groups. Arch. Math. (Basel) 63 (1994), 1–8.

[11] —, Kazarin, L. S., On the rank of a finite product of two p-groups. In: Groups - Korea 1994, 9–14, Walter de Gruyter, Berlin 1995.

[12] —, Sysak, Y. P., Locally soluble products of minimax groups. In: Groups - Korea 1994, 1–8, Walter de Gruyter, Berlin 1995.

[13] —, Sysak, Y. P., Groups with finite torsion-free rank which have a locally nilpotent triple factorization. J. Algebra. To appear.

[14] Andrukhov, A. M., Uniform products of infinite cyclic groups, In: Groups with systems of complemented subgroups. Akad. Nauk Ukrain. Inst. Mat. Kiev 1972.

[15] Belyaev, V. V., Locally finite simple groups as a product of two inert subgroups. Algebra i Logika 31 (1992), 360–368 = Algebra and Logic 31 (1992), 216–221.

[16] Brookes, C. J. B., Smith, H., A remark on products of locally soluble groups. Bull. Austral. Math. Soc. 30 (1984), 175–177.

[17] Chernikov, N. S., Groups which are Products of Permutable Subgroups. Nauk Dumka, Kiev 1987.

[18] —, Factorizations of groups by almost BFC-groups. In: Infinite groups and connections with algebraic structures. Akad. Nauk Ukrain. Inst. Mat. Kiev (1993), 367–387.

[19] Dickenschied O., On the adjoint group of some special radical rings. To appear.

[20] Franciosi, S., de Giovanni, F., On products of locally polycyclic groups. Arch. Math. (Basel) 55 (1990), 417–421.

[21] —, —, On the Hirsch–Plotkin radical of a factorized group. Glasgow Math. J. 4 (1992), 193–199.

[22] —, —, On periodic subgroups of factorized groups. Arch. Math. (Basel) 61 (1993), 313–318.

[23] —, —, On products of nilpotent groups. Ricerche Mat. To appear.

[24] —, —, Sysak, Y. P., On locally finite groups factorized by locally nilpotent subgroups. J. Pure Appl. Algebra. To appear.

[25] —, —, —, An extension of the Kegel–Wielandt Theorem to locally finite groups. Glasgow Math. J. To appear.

[26] Heineken, H., Products of finite nilpotent groups. Math. Ann. 287 (1990), 643–652.

[27] —, The product of two finite nilpotent subgroups and its Fitting series. Arch. Math. (Basel) 59 (1992), 209–214.

[28] —, Lennox, J. C., A note on products of abelian groups. Arch. Math. (Basel) 41 (1983), 498–501.

[29] Huppert, B., Endliche Gruppen I. Springer, Berlin 1967.

[30] Ivanov, S. V., General products of certain groups. Proc. XIth All-Union Symposium on Group Theory, Sverdlovsk 1989, 147–148.

[31] —, Ol'shanskii, A. Y., Some applications of graded diagrams in combinatorial group theory. In: Groups-St. Andrews 1989, London Math. Soc. Lecture Note Ser. 159, 258–308, Cambridge University Press 1991.

[32] Johnson, P. M., A property of factorizable groups. Arch. Math. (Basel) 60 (1993), 414–419.

[33] Kazarin, L. S., On a problem of Szép. Math. USSR Izv. 28 (1987), 467–495.

[34] —, Soluble products of groups. In: Infinite Groups 1994, Proc. Internat. Conference, 111–123, Walter de Gruyter, Berlin–New York 1995.

[35] —, Kurdachenko, L. A., Finiteness conditions and factorizations in infinite groups. Uspehi Mat. Nauk 47(1992), 75–114 = Russian Math. Surveys 47 (1992), 81–126.

[36] Kegel, O. H., On the solubility of some factorized linear groups. Illinois J. Math. 9 (1965), 535–547.

[37] Kramer, O.-U., On the theory of soluble factorizable groups. Bull. Austral. Math. Soc. 15 (1976), 97–110.

[38] Onishchik, A. L., Parabolic factorizations of semisimple algebraic groups. Math. Nachr. 104 (1981), 315–329.

[39] —, Topology of transitive transformation groups. Barth, Leipzig 1994.

[40] Parmeggiani, G., Sulla serie di Fitting del prodotto di due gruppi nilpotent finiti. Rend. Sem. Mat. Univ. Padova 91 (1994), 273–278.

[41] Robinson, D. J. S., Finiteness Conditions and Generalized Soluble Groups. Springer, Berlin 1972.

[42] —, Infinite factorized groups. Rend. Sem. Mat. Fis. Milano 53 (1983), 347–355.

[43] —, Stonehewer, S. E., Triple factorizations by abelian groups. Arch. Math. (Basel) 60 (1993), 223–232.

[44] Shunkov, V. P., Groups that are decomposable into a uniform product of p-groups. Dokl. Akad. Nauk SSSR 154 (1964), 542–544 = Soviet Math. Dokl. (1964), 147–149.

[45] Stonehewer, S. E., On finite dinilpotent groups. J. Pure Appl. Algebra 88 (1993), 239–244.

[46] Sozutov, A. I., On the nil radical in groups. Algebra i Logika 30 (1991), 102–105 = Algebra and Logic 30 (1991), 70–72.

[47] Suchansky, V. I., Wreath products and general products of groups. Dokl. Akad. Nauk SSSR 316 (1991), 1054–1058 = Soviet Math. Dokl. 43 (1991), 239–242.

[48] —, Wreath products and factorizations of groups. Contemp. Math. 131 (1992), 401–407.

[49] Sysak, Y. P., Products of infinite groups. Akad. Nauk Ukrain. Inst. Mat. Kiev, Preprint 82.53 (1982).

[50] —, Linear groups that are decomposable into products of locally nilpotent subgroups. Uspehi Mat. Nauk 45 (1990), 165–166.

[51] —, Radical modules and their application in the theory of factorized groups. Akad. Nauk Ukrain. Inst. Mat. Kiev, Preprint 89.18 (1993).

[52] —, On a question of B. Amberg. Ukrain. Mat. Z. 45 (1994), 457–461.

[53] —, Radical modules over hyperabelian groups of finite torsion-free rank. To appear.

[54] —, Some examples of factorized groups and their relation to ring theory. In: Infinite Groups 1994, Proc. Internat. Conference, 257–269, Walter de Gruyter, Berlin–New York 1995.

[55] Vasil'ev, V. G., Solvability length of uniform products of cyclic groups. Algebra i Logika 17 (1978), 260–266 = Algebra and Logic 17 (1978), 180–184.

[56] Watters, J. F., On the adjoint group of a radical ring. J. London Math. Soc. 43 (1968), 725–729.

Normalizers of Subnormal Subgroups

James C. Beidleman, Martyn R. Dixon and Derek J. S. Robinson

1991 Mathematics Subject Classification: 20E15

The Wielandt Subgroup

In 1958 Wielandt [15] introduced the subgroup which bears his name, the intersection of all the normalizers of subnormal subgroups of a group G. This is denoted by

$$W(G);$$

it is, of course, a characteristic subgroup of G. Clearly $W(G)$ is a T-group, i.e. every subnormal subgroup is normal. Also, G is a T-group precisely when $G = W(G)$.

The ***upper Wielandt series*** of G is defined by iteration in the usual way:

$$W_0(G) = 1, \ W_{\alpha+1}(G)/W_\alpha(G) = W(G/W_\alpha(G)),$$

and, if λ is a limit ordinal,

$$W_\lambda(G) = \bigcup_{\alpha<\lambda} W_\alpha(G).$$

The smallest ordinal α for which $W_\alpha(G) = G$, (if it exists), is called the ***Wielandt length*** of G.

Since its introduction, the Wielandt subgroup has attracted a good deal of attention from group theorists. We begin with a survey of what has been discovered over the last thirty five years.

Known Results

1. (Wielandt [15]). *$W(G)$ contains all non-cyclic simple subnormal subgroups of G, and also all minimal normal subgroups of G that satisfy* min-n*, the minimal condition on normal subgroups.*

Offprint from
Infinite Groups 94, Eds.: de Giovanni/Newell

From this we obtain at once

2. *A finite group has finite Wielandt length.*

This result can be extended to certain infinite groups. It was shown by Robinson [12] and Roseblade [14] that the Wielandt subgroup of a group with min-*sn*, the minimal condition on subnormal subgroups, is unexpectedly large.

3. *If G is a group with* min-*sn, then $G/W(G)$ is finite.*

Combining this with Wielandt's result, we obtain at once

4. *If G is a group with* min-*sn, then G has finite Wielandt length.*

On the other hand, the Wielandt subgroup of an infinite group can easily be trivial: for example $W(D_\infty) = 1$. Here it is worth noting that Cossey [6] has proved that a polycyclic group has finite Wielandt length if and only if it is finite-by-nilpotent.

5. *The Wielandt length of soluble groups.*

If G is a soluble group with finite Wielandt length ℓ, then the derived length of G is at most 2ℓ. This is because soluble T-groups are metabelian ([11]). Sharper bounds for the derived length have been obtained by Camina [4], Bryce and Cossey [3], and Casolo [5]. Indeed the last named author showed that *the derived length is at most $\frac{1}{3}(5\ell + 2)$ and the nilpotent length is at most $\ell + 1$.*

Obviously the Wielandt subgroup of a group always contains the centre. For certain types of soluble group these subgroups are quite close in the following sense.

6. *If G is a finitely generated soluble-by-finite group with finite (Prüfer) rank, then $W(G)$ is contained in the FC-centre $FC(G)$. Also $W(G)/Z(G)$ is finite if G is either polycyclic and metanilpotent or abelian-by-finite* (Cossey [6], Brandl, Franciosi and de Giovanni [2]).

IT-Groups

A group G is called an ***IT-group*** if every infinite subnormal subgroup is normal. Thus T-groups and all finite groups are *IT*-groups. The class of *IT*-groups was introduced in 1985 by Franciosi and de Giovanni [7]. They studied infinite soluble *IT*-groups and obtained detailed information on their structure; for example, these groups have derived length at most 3.

In a subsequent paper Heineken [9] considered infinite *IT*-groups in general and was able to obtain some surprising structural information about them; for example, an infinite *IT*-group G has a normal T-subgroup M such that G/M is a residually finite T-group and is either soluble or a finite extension of an abelian group of finite exponent. Heineken's paper also contains an interesting construction for IT-groups to which we shall return later.

The articles by Franciosi and de Giovanni and by Heineken have been our motivation for the introduction and study of an infinite version of the Wielandt subgroup which we call the generalized Wielandt subgroup.

The Generalized Wielandt Subgroup

For an arbitrary group G we define the ***generalized Wielandt subgroup***

$$IW(G)$$

to be the intersection of all the normalizers of the infinite subnormal subgroups of G, or G if there are no such subgroups. Obviously $IW(G)$ is an IT-group, and G is an IT-group if and only if $G = IW(G)$.

Since the group G might have a large number of finite subnormal subgroups, one might suspect that $IW(G)$ could be much larger than $W(G)$. On the other hand, the results of Franciosi and de Giovanni and of Heineken suggest that IT-groups may not be so different from T-groups. So there is conflicting evidence.

In fact it is the main conclusion of our work that for "most" groups G the subgroups $W(G)$ and $IW(G)$ are equal. Indeed the condition $W(G) \neq IW(G)$ imposes severe restrictions on the abelian normal subgroups of G. Also for arbitrary G the structure of the factor group $IW(G)/W(G)$ is quite restricted. Thus the conclusion would seem to be that in the partially ordered set of subnormal subgroups of an infinite group the finite subnormals are relatively unimportant.

It is our intention here to give a description of our main results with only a few details of the proofs. Full proofs will appear in [1]. We also take the opportunity to point out some consequences of the results and further directions for research.

An Illustration

In order to illustrate the power of the assumption $W \neq IW$, let us consider a group G with a non-torsion abelian normal subgroup A. Suppose that $W(G) \neq IW(G)$, so that there is a finite subnormal subgroup H and an element g of $IW(G)$ such that $H \neq H^g$. Now A contains an element a of infinite order. Then $B_0 = \langle a^H \rangle$ is a finitely generated infinite abelian subgroup, so for some $m > 0$ the subgroup $B = B_0^m$ is non-trivial and free abelian.

Now $B \operatorname{sn} G$, i.e. B is subnormal in G, and B is normalized by H. Therefore $HB \operatorname{sn} G$, and of course HB is infinite. Bearing in mind that $g \in IW(G)$, we conclude that $HB = (HB)^g = H^g B$. But $H \operatorname{sn} HB$, H is finite and B is torsion-free, so

$[H, B] = 1$ and H is the torsion-subgroup of HB. Thus we obtain the contradiction $H = H^g$.

This argument shows that if $W(G) \neq IW(G)$, then an abelian normal subgroup A of G must be torsion. The next step — which we shall not carry out here — is to argue that A has only finitely many non-trivial primary components, and that each of these has finite rank. By using similar arguments one can prove the following basic result.

Proposition. *If G is a group such that $W(G) \neq IW(G)$, then every abelian normal subgroup of G is Prüfer-by-finite.*

Here" Prüfer" means a group of some type p^∞, and the terminology "Prüfer-by-finite" is intended to include "finite" as a possibility. After this result one proceeds to pin down the structure of the soluble subnormal subgroups of G. The principal result is the following.

Theorem 1 ([1]). *Let G be a group such that $W(G) \neq IW(G)$. Then*

(i) *the Baer radical of G is Prüfer-by finite, and hence is nilpotent;*

(ii) *the subsoluble radical of G is an extension of a Prüfer-by-finite group by a torsion-free abelian group.*

Recall here that a ***subsoluble group*** is a group with an ascending series of subnormal subgroups whose factors are abelian. Every group has a unique largest subnormal subsoluble subgroup, the ***subsoluble radical*** ([10]).

The Controlling Subgroup *S(G)*

It turns out that there is a subgroup of the subsoluble radical which controls the relative behavior of $W(G)$ and $IW(G)$. For an arbitrary group G we define

$$S(G)$$

to be the subgroup generated by all the finite soluble subnormal subgroups of G. This is easily seen to be a characteristic, locally finite and soluble subgroup of G.

The first indication of the importance of $S(G)$ is given by

Theorem 2 ([1], Theorem 3). *Let G be any group and let d be a positive integer. Then $IW(G)$ normalizes almost all subnormal subgroups H of G such that $|H \cap S(G)| \leq d$.*

Corollary 1. *Let H be a subnormal subgroup of a group G which is not normalized by $IW(G)$. Then $|G : N_G(H)|$ is finite and $\langle H^G \rangle$ is finite.*

To see this, note that $IW(G)$ cannot normalize any H^g, $g \epsilon G$, and $|H \cap S(G)| = |H^g \cap S(G)|$. Hence H has only a finite number of conjugates in G. Of course H itself is finite, so $\langle H^G \rangle$, being generated by finitely many finite subnormal subgroups, is finite.

Corollary 2. *If $S(G)$ is finite, then $IW(G)$ normalizes almost all the subnormal subgroups of G.*

Corollary 3. *If G is any group, then $IW(G)/W(G)$ is residually finite.*

To prove this, let R be the intersection of the normalizers of all the subnormal subgroups of G which are *not* normalized by $IW(G)$. Then $R \lhd G$ and G/R is residually finite. Finally $R \cap IW(G) = W(G)$.

The Structure of $IW(G)/W(G)$

We come now to the main theorem which shows how the subgroup $S(G)$ controls the structure of the factor $IW(G)/W(G)$.

Theorem 3 ([1]). *Let G be an arbitrary group, and write $S = S(G)$, $W = W(G)$ and $I = IW(G)$.*

(i) *If S is not Prüfer-by-finite, then $I = W$.*

(ii) *If S is Prüfer-by-finite and infinite, then I/W is a soluble T-group.*

(iii) *If S is finite, then I/W is finite.*

This generalizes the result of Heineken on IT-groups mentioned above. It should be kept in mind that very precise information is available on the structure of soluble T-groups (see [11]). On the other hand, it is interesting that the structure of I/W is arbitrary when S is finite. This is seen from an easy example. Let H be a finite non-cyclic simple group and let K be any finite group; then $G = H \ wr \ K$ is finite, so $G = IW(G)$. On the other hand, $W(G)$ coincides with the base group, so $IW(G)/W(G) \simeq K$.

Groups with $IW(G)/W(G)$ Infinite

In [9] Heineken constructed an example of an IT-group G such that $G/W(G)$ is an infinite elementary abelian p-group. Here we indicate how to modify his construction to obtain an IT-group G such that $G/W(G)$ contains an element of infinite order.

Construction. Let p be a prime and let X_i, Y_i, $i = 1, 2, \ldots$, be two sequences of finite groups satisfying the following requirements:

(a) $Z(X_i) \leq X_i' = X_i''$ and $Z(Y_i) \leq Y_i' = Y_i''$;

(b) $Z(X_i) = \langle u_i \rangle$ and $Z(Y_i) = \langle v_i \rangle$ are cyclic of order p^i;

(c) $X_i'/Z(X_i)$ and $Y_i'/Z(Y_i)$ are non-cyclic simple groups;

(d) $(X_i)_{ab} = \langle x_i X_i' \rangle$ and $(Y_i)_{ab} = \langle y_i Y_i' \rangle$ are cyclic of order p^{2i};

(e) $C_{X_i}(X_i'/Z(X_i)) = Z(X_i)$ and $C_{Y_i}(Y_i'/Z(Y_i)) = Z(Y_i)$.

For example, we could take X_i and Y_i to be suitable factors of $GL_{p^{2i}}(q^{m_i})$ where $m_i > 0$ and q is a prime such that $q^{m_i} \equiv 1 \pmod{p^{2i}}$.

Next let

$$V_i = \langle s_i, t_i \mid s_i^{p^{2i}} = 1 = t_i^{p^{2i}},\ [s_i, t_i] \in Z(V_i) \rangle,$$

a group of order p^{6i}.

Now form $X_i \times Y_i \times V_i$ and its factor

$$D_i = \langle X_i', Y_i', x_i s_i, y_i t_i \rangle / \langle u_i v_i^{-1}, u_i d_i^{-p^i} \rangle$$

where $d_i = [s_i, t_i]$. Put

$$H = D_1 \times D_2 \times \cdots / \langle d_i^{-1} d_{i+1}^{p^2} \mid i = 1, 2, \ldots \rangle$$

and observe that conjugation in D_i by $x_i y_i t_i$ leads to an automorphism α of H with infinite order.

Finally put

$$G = \langle \alpha \rangle \ltimes H.$$

A short computation reveals that $IW(G) = G$, so G is an IT-group, and

$$G/W(G) \simeq \mathbb{Z} \oplus \operatorname*{Dr}_{i=1,2,\ldots} (\mathbb{Z}_{p^i} \oplus \mathbb{Z}_{p^i}).$$

Here of course $S(G)$ is of type p^∞.

Some Further Results

We note some consequences of the main results. The first of these involves the ***upper generalized Wielandt series***; this is the ascending series defined by

$$IW_0(G) = 1,\ IW_{\alpha+1}(G)/IW_\alpha(G) = IW(G/IW_\alpha(G))$$

and

$$IW_\lambda(G) = \bigcup_{\alpha<\lambda} IW_\alpha(G)$$

if λ is a limit ordinal.

Theorem 4. *Let G be any group. Then for each ordinal α there is an integer $n(\alpha) \geq 0$ such that*

$$W_\alpha(G) \leq IW_\alpha(G) \leq W_{\alpha+n(\alpha)}(G).$$

Also $n(\alpha) = 0$ if α is a limit ordinal.

Proof. It suffices to show that $IW(G) \leq W_{m+1}(G)$ for some $m \geq 0$ since transfinite induction can be used. Assume that $IW(G) \not\leq W_2(G)$; then $IW(G/W(G)) \neq W(G/W(G))$, so the Baer radical of $G/W(G)$ is Prüfer-by-finite, by Theorem 1. Suppose that $IW(G)/W(G)$ is infinite; then this factor is residually finite and soluble, by Theorem 3 and Corollary 3. Now the Fitting subgroup of $IW(G)/W(G)$ must be Prüfer-by-finite, whence it is finite. Hence we may conclude that

$$IW(G)/W(G) \leq W_m(G/W(G))$$

for some $m \geq 0$. Thus $IW(G) \leq W_{m+1}(G)$ as claimed. □

Corollary 1. *Let G be any group and denote by $\ell(G)$ and $i\ell(G)$ the lengths of the upper Wielandt and upper generalized Wielandt series. Then*

$$i\ell(G) \leq \ell(G) \leq i\ell(G) + n(G)$$

for some integer $n(G) \geq 0$.

Corollary 2. *The upper Wielandt series and the upper generalized Wielandt series have the same limit.*

Profinite Groups

One can define the Wielandt subgroup and the generalized Wielandt subgroup of a profinite group by restricting to closed subnormal subgroups. Thus if G is a profinite group, the Wielandt subgroup $W(G)$ is the intersection of the normalizers of the closed subnormal subgroups of G. Similarly the generalized Wielandt subgroup $IW(G)$ is the intersection of the normalizers of the closed infinite subnormal subgroups of G.

Let H be a closed subgroup of G. Then $H = \bigcap_N HN$ where N ranges over all the open normal subgroups of G. This implies that $N_G(H) = \bigcap_N N_G(HN)$. But $N_G(HN)$ contains an open subgroup, so it is open and hence closed; therefore $N_G(H)$ is closed. It follows that $W(G)$ and $IW(G)$ are closed topologically characteristic subgroups of G.

Theorem 5. *If G is an infinite profinite group, then $W(G) = IW(G)$.*

Proof. Let H be any closed subnormal subgroup of G and write $I = IW(G)$. If N is an open normal subgroup of G, then $|G : N|$ is finite, so HN is infinite and $HN = (HN)^I$. But $H = \bigcap_N HN$, so $H^I = H$, and $I = W(G)$. □

Automorphisms Fixing Subnormal Subgroups

One soon observes that what is critical in the proofs of our results is the existence of automorphisms that fix all infinite subnormal subgroups but not all finite subnormal subgroups. This suggests consideration of the following subgroups:

$$\mathrm{Aut}_{sn}(G) = \{\alpha \in \mathrm{Aut}\, G \mid H^\alpha = H,\ \forall H \; sn\, G\}$$

and

$$\mathrm{Aut}_{isn}(G) = \{\alpha \in \mathrm{Aut}\, G \mid H^\alpha = H,\ \forall H \; sn\, G,\ H \text{ infinite}\}.$$

The conjugation map $G \to \mathrm{Aut}\, G$ induces an embedding

$$IW(G)/W(G) \to \mathrm{Aut}_{isn}(G)/\mathrm{Aut}_{sn}(G).$$

Thus the condition $\mathrm{Aut}_{sn}(G) \neq \mathrm{Aut}_{isn}(G)$ is weaker than $W(G) \neq IW(G)$. Most of the results mentioned above hold when $W(G)$ is replaced by $\mathrm{Aut}_{sn}(G)$ and $IW(G)$ by $\mathrm{Aut}_{isn}(G)$

Thus the analogue of Theorem 1 is

Theorem 1*. *Let G be a group such that* $\mathrm{Aut}_{sn}(G) \neq \mathrm{Aut}_{isn}(G)$. *Then*

(i) *the Baer radical of G is Prüfer-by-finite*;

(ii) *the subsoluble radical of G is an extension of a Prüfer-by-finite group by a torsion-free abelian group.*

From here on we take a somewhat different point of view and ask what can be said about the structure of the groups $\mathrm{Aut}_{sn}(G)$ and $\mathrm{Aut}_{isn}(G)$. When G is subsoluble, it is not hard to see that these groups are soluble.

Theorem 6. *Let G be a subsoluble group. Then $\mathrm{Aut}_{sn}(G)$ is metabelian, and, if G is infinite, $\mathrm{Aut}_{isn}(G)$ is soluble with derived length at most* 3.

Proof. Let B denote the Baer radical of G. Then $C_G(B) \leq B$ since G is subsoluble. Put $X = \mathrm{Aut}_{sn}(G)$ and $Y = \mathrm{Aut}_{isn}(G)$; then X induces power automorphisms in B, so $[B, X'] = 1$. Since $C_G(B) \leq B$, it follows that $[G, X'] \leq B$ and thus $X'' = 1$. Now assume that G is infinite, so B is infinite. We can assume that $X \neq Y$ and so it follows from Theorem 1* that B is Prüfer-by-finite; let D denote the maximum divisible subgroup of B. Since D is a Prüfer group, Y induces power automorphisms in D, and the same is true of B/D. Therefore $[B, Y''] = 1$. Also $[G, Y''] \leq B$, so $Y''' = 1$. □

This generalizes part of Theorem A of Franciosi and de Giovanni [8]. A more surprising result is that $\mathrm{Aut}_{sn}(G)$ is always T-by-soluble when G is a finite group. More precisely there is

Theorem 7 ([13]). *Let G be a subsoluble-by-finite group and put $X = \mathrm{Aut}_{sn}(G)$. Then the group $X/X \cap \mathrm{Inn}(G)$ is soluble of derived length at most* 4. *Moreover the derived length is at most* 3 *if $H^1(G/S, Z(S)) = 0$ where S is the subsoluble radical of G.*

Here the bound 3 cannot be improved since $\mathrm{Out}(D_4(3)) \simeq S_4$. Of course the proof of Theorem 7 uses the Schreier Conjecture and hence the classification of finite simple groups.

On the other hand, there is no such result for

$$\mathrm{Aut}_n(G),$$

the subgroup of automorphisms that fix every normal subgroup of G. In fact

Theorem 8 ([13]). *Let F be an arbitrary finite group. Then there is a finite semisimple group G such that $\mathrm{Aut}_n\, G/\mathrm{Inn}\, G$ has a subgroup isomorphic with F.*

So the conclusion would seem to be that, as far as automorphisms of a finite group are concerned, the subnormal subgroups are much more important than the normal subgroups.

References

[1] Beidleman, J. C., Dixon, M. R., Robinson, D. J. S., The generalized Wielandt subgroup of a group Canad. J. Math. To appear.

[2] Brandl, R., Franciosi, S., de Giovanni, F., On the Wielandt subgroup of infinite soluble groups. Glasgow Math. J. 32 (1990), 121–125.

[3] Bryce, R. A., Cossey, J., The Wielandt subgroup of a finite soluble group. J. London Math. Soc. (2) 40 (1989), 244–256.

[4] Camina, A. R., The Wielandt length of a finite group. J. Algebra 15 (1970), 142–148.

[5] Casolo, C., Soluble groups with finite Wielandt length. Glasgow Math. J. 31 (1989), 329–334.

[6] Cossey, J., The Wielandt subgroup of a polycyclic group. Glasgow Math. J. 33 (1991), 231–234.

[7] Franciosi, S., de Giovanni, F., Groups in which every infinite subnormal subgroup is normal. J. Algebra 96 (1985), 566–580.

[8] —, —, On automorphisms fixing subnormal subgroups of soluble groups. Atti Acc. Naz. Lincei Rend. Cl. Sci. Fis. Mat. Natur. (8) 82 (1988), 217–222.

[9] Heineken, H., Groups with restrictions on their infinite subnormal subgroups. Proc. Edinburgh Math. Soc. 31 (1988), 231–241.

[10] Phillips, R. E., Combrink, C. R., A note on subsolvable groups. Math. Z. 92 (1966), 349–352.

[11] Robinson, D. J. S., Groups in which normality is a transitive relation. Proc. Cambridge Philos. Soc. 60 (1964), 21–38.

[12] —, On the theory of subnormal subgroups. Math. Z. 89 (1965), 30–51.

[13] —, Automorphisms fixing every subnormal subgroup of a finite group. Arch. Math. (Basel). To appear.

[14] Roseblade, J. E., On certain subnormal coalition classes. J. Algebra 1 (1964), 132–138.

[15] Wielandt, H., Über den Normalisator der subnormalen Untergruppen. Math. Z. 69 (1958), 463–465.

Isomorphisms between Homeomorphic Connected Compact-Like Abelian Groups*

Dikran N. Dikranjan and Vladimir V. Uspenskij

Abstract. According to Scheinberg's theorem two connected compact abelian groups are isomorphic as topological groups whenever they are homeomorphic. This was proved to hold true also for some non-compact groups in a recent paper of Rodinò and the first named author [DR] (namely, the precompact abelian groups G which coincide with the subgroup of *all* metrizable elements of their completion). Groups with this property are pseudocompact, so that the question arose whether Scheinberg's theorem extends to *all* pseudocompact groups, i. e. whether pseudocompact, connected abelian group which are homeomorphic as topological spaces are isomorphic also as topological groups [DR]. Here we answer negatively this question in the smaller class of countably compact groups by showing that homeomorphic countably compact connected abelian groups need not be isomorphic even as abstract groups. We also discuss the extension of Scheinberg's theorem to pseudocompact groups G which are topologically invariant in their completion $\hat{G}$.

1991 Mathematics Subject Classification: Primary 54D25; Secondary 22C05, 54A35

1. Introduction

It is usual to invoke finiteness-like conditions in the theory of infinite groups. A natural way to do it is to impose compact-like topologies on the group (the pro-finite topology being a leading example). The interplay between the topology and the algebraic structure of a topological group has two aspects. The first one concerns the impact of the ***algebraic structure*** on the topological one. It is illustrated by the following classical theorem of van der Waerden [vW] (see also [CRb]): a compact connected semisimple Lie group admits a unique precompact topology, namely the given compact metrizable group topology. (For further discussion in this direction, for example on the uniqueness of the compact group topology of the p-adic integers etc., see §5.)

In the opposite direction one can ask to what extent the ***topology*** of a (compact-like) topological group determines the algebraic structure of the group. For non-

* Research supported by a Humboldt Research Fellowship and by funds MURST60%.

Offprint from
Infinite Groups 94, Eds.: de Giovanni/Newell

connected groups such a relation may totally fail, for example every compact metrizable totally disconnected group is homeomorphic to $\{0, 1\}^{\omega}$, hence this homogeneous topological space admits many (totally different from algebraic point of view) group structures which make it a topological group.

For compact connected groups there are positive results in this direction. For example, two compact connected simple Lie groups are isomorphic if they are homeomorphic (or even if they have the same homotopy type) [BB, Theorem 9.3]. On the other hand, "simple" cannot be omitted here: there exist two semisimple compact connected Lie groups which are homeomorphic but not isomorphic [BB, Theorem 9.4]. There is no such an example in the abelian case: according to Scheinberg's well-known result [S] two connected compact abelian groups are isomorphic as *topological groups* whenever they are homeomorphic. We sketch an easy proof of this fact [S]: if G is a compact connected abelian group, then the Pontryagin dual $\hat{G}$ is canonically isomorphic to the cohomology group $H^1(G, \mathbb{Z})$ (this is obvious if G is a torus, and the general case follows since every compact connected abelian group is a projective limit of tori), and thus G is the dual of a discrete group which depends only on the topology of the space G.

Scheinberg's theorem can be extended to some non-compact groups. For example, a similar assertion was proved to hold true for the special case of the subgroup of metrizable elements (see §3 for the definition) of compact abelian groups in [DR] (i. e. for precompact groups G in which every element is metrizable and G contains all metrizable elements of its completion). Groups with this property are pseudocompact, so that this stimulated the following question forwarded in [DR].

Question 1.1 ([DR]). Is it true that two connected pseudocompact abelian groups are isomorphic as topological groups whenever they are homeomorphic?

As we noted, Scheinberg's theorem cannot be extended for non-abelian compact connected groups [BB, Theorem 9.4]. Since any non-abelian connected compact group generates the variety of all groups, one cannot hope to "approximate" abelian by a weaker property. This shows that the only essential resource to extend Scheinberg's theorem remains the compactness condition as in Question 1.1. We give here a strongly negative answer to 1.1 within much smaller classes of compact-like groups (in particular, countably compact groups) by showing that even isomorphism as abstract groups is not available in this case (see §4).

In §3 we prove a positive result. Namely, a slight modification of the proof of Theorem 3.1 in [DR] shows that the answer to 1.1 is "Yes" within the class of pseudocompact groups G which are functorial in their completion $\hat{G}$ (see Definition 3.1).

2. Preliminaries

In this work we consider only Tikhonov spaces, in particular, all topological groups are Hausdorff.

We recall first several generalizations of compactness for topological groups. A topological group G is ***precompact*** if a finite number of left translates of each non-empty open subset of G cover G. In every precompact group G the left and right uniformities coincide and the completion $\hat{G}$ is compact. A topological space X is ***pseudocompact*** if every real-valued continuous function defined on X is bounded. A topological space X is ***initially*** γ-compact, where γ is an infinite cardinal, if every open cover of X with less than γ elements has a finite subcover. Initially ω_1-compact spaces are usually called ***countably compact***. Clearly, a space is compact if and only if it is initially γ-compact for every infinite cardinal γ, and countably compact spaces are pseudocompact. Pseudocompact topological groups are precompact by a well known result of Comfort and Ross [CR].

We denote by $\mathbb{N}$ and $\mathbb{Z}$ the sets of naturals and integers, respectively, by $\mathbb{Q}$ the rationals, by $\mathbb{R}$ the reals, by $\mathbb{T}$ the circle group $\mathbb{R}/\mathbb{Z}$, by $\mathfrak{c}$ the cardinality of continuum. For a precompact topological group G we denote by $\hat{G}$ the Weil completion of G.

Let G be an abelian group and A be a subset of G. We denote by $\langle A \rangle$ the subgroup of G generated by A.

3. Extension of Scheinberg's Theorem to Functorial Subgroups of Compact Abelian Groups

Definition 3.1. Let $\mathfrak{C}$ be a full subcategory of the category of topological groups. A ***functorial subgroup*** in $\mathfrak{C}$ is defined by assigning to each $G \in \mathfrak{C}$ a subgroup $\mathfrak{r}(G)$ such that whenever $f : G \to H$ is a continuous homomorphism in $\mathfrak{C}$ one has $f(\mathfrak{r}(G)) \subseteq \mathfrak{r}(H)$.

The guiding example of a functorial subgroup should be the connected component of a topological group. Another example can be obtained in the following way. Let α be a cardinal. An element x of a topological group G is said to be of ***character*** α if the character of the cyclic subgroup $\langle x \rangle$ of G is α ([DR]). The elements of countable character are the metrizable elements introduced by Wilcox [W] for locally compact groups. Denote by $M_\alpha(G)$ the subset of elements of G of character less than α. Clearly $M_\alpha(G) \subseteq M_\beta(G)$ if $\alpha \leq \beta$. If G is precompact, $M_\omega(G)$ coincides with the subset of torsion elements of G. In case G is abelian $M_\alpha(G)$ is a subgroup of G, and this defines a functorial subgroup in the category of precompact abelian groups ([DR]). Wilcox [W] proved that for a compact abelian group G the subgroup $M_{\omega_1}(G)$ (and so $M_\alpha(G)$ for each $\alpha > \omega$) is dense and pseudocompact.

Further information concerning functorial subgroups of topological groups can be found in [DPS] (see also [D]).

A slight modification of the proof of Theorem 3.1 in [DR] gives the following result for pseudocompact groups G which are functorial subgroups of their completion $\hat{G}$.

Theorem 3.2. *Let τ be a functorial subgroup in the category of precompact abelian groups and G, H be connected pseudocompact abelian groups such that $G = \tau(\hat{G})$ and $H = \tau(\hat{H})$. Then G and H are homeomorphic if and only if G and H are isomorphic as topological groups.*

Proof. Assume that G and H are homeomorphic. Thus also their Stone–Čech compactifications βG and βH are homeomorphic. According to [CR] $\beta G = \hat{G}$ and $\beta H = \hat{H}$. Furthermore, the completions $\hat{G}$ and $\hat{H}$ are connected. Since homeomorphic compact connected abelian groups are isomorphic as topological groups (see [S]), there exists an isomorphism of topological groups $j : \hat{G} \to \hat{H}$. By the functoriality of G and H the isomorphism j sends G to H. □

Call a precompact group G ***rigid*** if every topological isomorphism j of $\hat{G}$ satisfies $j(G) \subseteq G$.

Question 3.3. Is it possible to extend Scheinberg's result to rigid pseudocompact abelian groups?

4. The Example

Theorem 4.1. *There exist two connected countably compact abelian groups which are homeomorphic as topological spaces but not isomorphic (even) as abstract groups.*

Proof. Set $K = \mathbb{T}^2$ and consider the subgroups $A = \langle (i, i) \rangle \cong \mathbb{Z}(4) = \mathbb{Z}/4\mathbb{Z}$ and $B = \langle (1, -1), (-1, 1) \rangle \cong \mathbb{Z}(2)^2$ of K (here we think of $\mathbb{T}$ as the multiplicative group of complex numbers z such that $|z| = 1$, so that $i^2 = -1$). There exists a homeomorphism $h : K \to K$ such that $h(A) = B$ and $h(1) = 1$. Set $G = K^{\omega_1}$ and for a subgroup H of K denote by G_H the subgroup of G consisting of those functions $f : \omega_1 \to K$ such that $|\{\beta < \omega_1 : f(\beta) \notin H\}| \leq \omega$. Then G_A and G_B are homeomorphic. Since both G_A and G_B contain the Σ-product $S = \Sigma K^{\omega_1}$, they are pseudocompact, thus connected (since G is connected). Assume that there exists an *algebraic* isomorphism $j : G_A \to G_B$. Now note that $G_A = A^{\omega_1} + S$ and $G_B = B^{\omega_1} + S$ and in both cases the subgroup S is the maximal divisible subgroup. Then $j(S) = S$ so that also $G_A/S \cong G_B/S$, but G_B/S is a group of exponent 2, while $G_A/S \cong \mathbb{Z}(4)^{\omega_1}/(\Sigma\mathbb{Z}(4)^{\omega_1})$ is not - a contradiction. □

Remark 4.2. (a) The group $\mathbb{T}^2$ cannot be replaced by $\mathbb{T}$ in the above argument since any two finite subgroups of $\mathbb{T}$ of the same cardinality (i. e. homeomorphic) are isomorphic.

(b) One can prove by the above argument that for each $\gamma > \omega$ there exists two connected initially γ-compact abelian groups which are homeomorphic as topological spaces but not isomorphic (even) as abstract groups. It suffices to take now the power K^γ and define $G_H = \{f \in K^\gamma : |\{\beta < \gamma : f(\beta) \notin H\}| < \gamma\}$ and then take G_A and G_B with A and B as before. Now every subset of less than γ elements of G_A (and G_B) is covered by a compact subset of G_A, which gives initial γ-compactness. One can show that for every *algebraic* homomorphism $f : G_B \to G_A$, $|\operatorname{coker} f| = |G_A| = 2^\gamma$. If $2^{<\gamma} < 2^\gamma$ then also for every *algebraic* homomorphism $j : G_A \to G_B$, $|\ker j| = |G_A| = 2^\gamma$, but in case $2^{<\gamma} < 2^\gamma$ one can find an *algebraic* monomorphism $j : G_A \to G_B$.

(c) The groups in item (b) are γ-compact, have cardinality 2^γ and weight γ. Clearly, a non-compact group G with $\gamma = w(G) < |G|$ can be at most γ-compact, so that this is the best possible result (from the point of view of compactness) for groups G with $w(G) < |G|$. Changing slightly the above construction one can get also two homeomorphic non-isomorphic γ-compact connected abelian groups G, H with $w(G) = |G| = 2^\gamma$ ($= w(H) = |H|$ obviously). To this end take now cardinals $\gamma \leq \sigma$ and consider the subgroup $\tilde{G} = \{f \in K^\sigma : |\{\beta < \gamma : f(\beta) \neq 0\}| \leq \gamma\}$ of K^σ. Now, for a subgroup H of K consider the following subgroup of $\tilde{G}$: $\tilde{G}_H = \{f \in \tilde{G} : |\{\beta < \gamma : f(\beta) \notin H\}| < \gamma\}$. Taking G_A and G_B with A and B as before we have $|G_A| = |G_B| = \sigma 2^\gamma$, so it suffices to choose $\sigma \geq 2^\gamma$.

(d) None of the groups constructed above is rigid, so that one can extend Question 3.3 to countably compact groups as well.

The γ-compact groups G produced by our method have cardinality (at least) 2^γ and satisfy $|G| \geq w(G)$. This leaves open the following

Question 4.3. Let γ be a cardinal with $2^{<\gamma} < 2^\gamma$ and G, H be homeomorphic γ-compact non-compact connected abelian groups. Are these groups isomorphic as (abstract) groups if one of the following conditions holds:

a) $\gamma = |G| = |H| < w(G) = w(H)$,

b) $|G| = |H| \leq 2^{<\gamma}$.

5. Epilogue

Here we discuss the "inverse problem": when two topological groups isomorphic as abstract groups are homeomorphic. Obviously, one can just take one group with two group topologies and look at how much they differ. It is well known that the group $\mathbb{Z}_p$ of p-adic integers admits a unique compact group topology, it can be shown similarly that the groups $G = \prod_p \mathbb{Z}_p^{k_p}$, where each k_p is finite, admit a unique compact group topology, namely that with basic nbds of 0 the subgroups nG, $n \in \mathbb{N}$ (see Orsatti [O1] and [O2], or [DPS]). It is shown in [DSh] that $\mathbb{Z}_p$ admits many non-homeomorphic pseudocompact topologies (namely, for each cardinal σ satisfying $\omega < \sigma \leq 2^{2^\omega}$ there exists a connected pseudocompact group topology of weight σ on $\mathbb{Z}_p$ which may be chosen also to have or not the property "locally connected"). This suggests the following

Question 5.1. Does $\mathbb{Z}_p$ admit a countably compact group topology beyond the p-adic one? How many non-homeomorphic ones?

Obviously, if a topology as in Question 5.1 exists, then it should be non-metrizable. Furthermore, such a topology cannot be zero-dimensional. In fact, the only pseudocompact zero-dimensional group topology on $\mathbb{Z}_p$ is the p-adic one since such a topology must be precompact and have a base of open subgroups (p-adic topology is the only one with this property).

Acknowledgments. It is a pleasure to thank G. Itzkowitz who suggested the reference to [BB].

References

[BB] Baum P. F, Browder W., The cohomology of quotients of classical groups. Topology 3 (1965), 305–336.

[CRb] Comfort W. W., Robertson, L. C., Images and quotients of $SO(3, \mathbb{R})$: Remarks on a theorem of van der Waerden. Rocky Mountain J. Math. 17 (1987), 1–13.

[CR] Comfort W. W, Ross, K. A., Pseudocompactness and uniform continuity in topological groups. Pacific J. Math. 16 (1966), 483–496.

[D] Dikranjan, D., Radicals and duality, General Algebra 1988 (R. Mlitz, ed.). Elsevier Science Publishers B. V. (North Holland), 1990, 45–64.

[DPS] —, Prodanov, Iv., Stoyanov, L., Topological groups: characters, dualities and minimal group topologies. Pure and Applied Mathematics 130, Marcel Dekker Inc., New York–Basel 1989.

[DR] —, Rodinò, N., Isomorphisms between finite powers of connected, totally minimal, pseudocompact, abelian groups. Talk given at Topology Conference, Lecce (Italy), 1990, Suppl. Rend. Circ. Mat. Palermo Ser.II 29 (1992), 385–398.

[DSh] —, Shakhmatov, D., Pseudocompact topologies on groups. Topology Proc. 17 (1992), 335–342.

[E] Engelking, R., General Topology. Warszawa, PWN, 1977.

[HR] Hewitt, E., Ross, K. A., Abstract harmonic analysis I. Springer-Verlag, Berlin–Heidelberg–New York 1963.

[O1] Orsatti, A., Una caratterizzazione dei gruppi abeliani compatti o localmente compatti nella topologia naturale. Rend. Sem. Mat. Univ. Padova 39 (1967), 219–225.

[O2] —, Sui gruppi abeliani ridotti che ammettono una unica topologia compatta. Rend. Sem. Mat. Univ. Padova 43 (1970), 341–347.

[OR] —, Rodinò , N., Isomorphisms between powers of compact abelian groups. Topology Appl. 23 (1985), 271–277.

[S] Scheinberg, S., Homeomorphism and isomorphism of abelian groups. Canad. J. Math. 26 (1974), 1515–1519.

[vW] van der Waerden, B. L., Stetigkeitssätze für halbeinfache Liesche Gruppen. Math. Z. 36 (1933), 780–786.

[W] Wilcox, H., Pseudocompact groups. Pacific J. Math. 19 (1966), 365–379.

Applications of Abelian Groups and Model Theory to Algebraic Structures

Manfred Dugas and Rüdiger Göbel

1991 Mathematics Subject Classification: 20E22, 20E36

1. Introduction

This paper grew out of a survey talk given at "Infinite Groups 1994" in Ravello, Italy. Surprisingly the stormy developments in abelian groups which began in 1974 with the appearence of Shelah's fundamental paper [71] seems to be almost unobserved in infinite non-commutative group theory. The paper in question did not only solve the Whitehead problem — which turned out to be undecidable in ZFC — it also provided a combinatorial tool for constructions of (abelian) groups, which is called the "Shelah-elevator", [39]. Clearly such tools should have some effect on other areas of algebra and "real group theory" in particular. The impact should either derive directly from abelian group theory, which more often supports module theory in general than group theory, or it should derive from a transfer of combinatorial, model theoretic tools originally developed for abelian groups. It was the intention of the authors to establish these links in the last decade and we would like to report on this program which is still going on. We want to follow both ways of passing from abelian groups to algebra having particular problems in group theory in mind. The general algebraic concept is discussed in Section 2 and it will be applied to fields in Section 3, to representation theory in Section 4 and to group theory in Section 5 and the last section is devoted to modules over Nagata-type rings. The recent results in abelian groups and the combinatorial methods are specially designed for investigating automorphism groups (or endomorphism monoids) of algebraic structures.

Following the aim of the Ravello-Conference we will concentrate on applications to group theory for the rest of this introduction. We will provide three different types of nasty abelian groups which will serve as building blocks for groups: First we indicate the construction of a 2-divisible, 2-torsion-free abelian group G with an alternating

Offprint from
Infinite Groups 94, Eds.: de Giovanni/Newell

bilinear form f having prescribed endomorphisms

$$\operatorname{End} G = R \oplus f_G.$$

Here f_G is a two-sided ideal of endomorphisms coming from the bilinear form and $R = \mathbb{Z}H$ is a given group ring. Coding G into a nilpotent group N of class 2, the famous Baer–Lazard-Theorem allows us to prescribe the automorphism group of N up to its stabilizer of the central series, see (5.2). A short discussion of earlier results is given in Section 5. We will provide a nasty abelian p-group having only very particular endomorphisms and homomorphisms. This group is plugged infinitely often into a locally finite p-group, which after this treatment is unable to allow essential automorphisms: The group has a trivial center and its automorphisms are all inner. For this treatment we define a p-adic wreath product sitting between the ordinary wreath product and a complete (= cartesian) wreath product. This notion seems to be more useful in group theory than the complete wreath product because homomorphisms often become continuous in a natural sense. The constructed groups provide an answer to a question raised by Ph. Hall in his Cambridge lectures in 1966 (see Thomas [76]). Moreover we are able to prescribe countable outer automorphism groups of such locally finite p-groups. A slight modification of these methods gives universal, locally finite groups (so-called ULF-groups) with prescribed outer automorphism groups. This result uses tools from Thomas [77] and allows us to drop the GCH-hypothesis from his theorem [77]. It illustrates that there is no hope to characterize automorphism groups of ULF-groups, a question raised in [47]. Finally we construct generalized E-rings, a question promoted by R. S. Pierce for his structure theory on particular torsion-free abelian groups of finite rank. Our aim is different: In order to extend results of Zalesskii's [83] in finding certain torsion-free nilpotent groups of class 2 with prescribed outer automorphism groups, it turns out that generalized E-rings are the right candidates. Recall that a ring R is an E-ring if $R = \operatorname{End}_{\mathbb{Z}} R^+$, i.e. the multiplication is given with the addition. Subrings of $\mathbb{Q}$ are the classical examples for E-rings. If H is any group, we will construct a ring R such that $\operatorname{End}_{\mathbb{Z}} R^+ = RH$ and $\operatorname{Aut} R = H$, where RH is the group ring of H over R. Observe that R becomes an E-ring if we choose $H = 1$. We find a proper class of E-rings. An obvious extension of Zalesskii's matrix-construction provides nilpotent groups of class 2 with various properties concerning their automorphisms, see Section 5.

We must add a few remarks about the tools for constructing these nasty abelian groups used above and in Section 3, 4 and 5. One of the most disturbing facts about group constructions is the following wrong attempt to produce examples: Inductively choose a chain $G_i \subset G_{i+1}$ of groups taking care of all the properties you want at each individual step i. Then take the union G of this infinite (countable or larger) chain and observe that all your efforts were useless and have no effect on G. A method taking care of unions is needed. The first useful tool in this direction is Jensen's $\diamond$ which unfortunately needs additional set theory as GCH and more. This prediction principle was discovered by R. Jensen [44] in Gödel's constructible universe. Fortunately the diamond principle can be weakened to hold in ordinary set theory. This is what we call Shelah's Black Box, another combinatorial tool next to Shelah's elevator. An elementary but tickly proof of this result of Shelah's [70], can be found in [14], and variants of the Black Box are given in [31], see also Shelah [70]. It can be used to predict automorphisms on the union while working on certain layers of the chains, and this is what we need to construct nasty abelian and non-abelian groups or fields and rings.

2. The General Strategy

Let $\mathcal{C}$ be a fixed category. If M, M' are objects in $\mathcal{C}$, then we are interested in the structure of $\mathrm{Mor}_{\mathcal{C}}(M, M') = \mathrm{Mor}(M, M')$, the set of all morphisms from M to M'. In particular we will consider $\mathrm{Mor}(M, M)$, which is denoted by $\mathrm{End}\, M$. Clearly $\mathrm{End}\, M$ is a monoid by composition of morphisms and $\mathrm{End}\, M$ is called the endomorphism monoid of M, cf. [45], [54]. If morphisms of M have inverses in $\mathcal{C}$, then $\mathrm{End}\, M$ is a group and it is customary to denote $\mathrm{End}\, M$ by $\mathrm{Aut}\, M$, and $\mathrm{Aut}\, M$ is called the automorphism group of M. Depending on $\mathcal{C}$, $\mathrm{End}\, M$ may have additional structure induced by M; this may be an addition and scalar multiplication, making $\mathrm{End}\, M$ into a ring or an algebra or some topology making $\mathrm{End}\, M$ into a topological monoid or a topological ring. Most famous example are matrix algebras (of vector spaces) or the Hasse-topology on automorphism groups of fields (see Hasse [43]) and the finite topology on the endomorphism ring of an abelian group (see Fuchs [36]). Hence $\mathcal{C}$ leads to some derived category $\mathcal{C}'$ containing all $\mathrm{End}\, M$ with $M \in \mathcal{C}$. In order to investigate the degree of complexity of some category $\mathcal{C}$, many papers deal with the inverse "Noether problem" of Galois theory. Here we will investigate similar, but different inverse problems, which lead to "realization theorems":

We want to find the range of the functor $\mathrm{End}_{-} : \mathcal{C} \to \mathcal{C}'$

This approach needs some fine tuning which can be explained in the "trivial case" e.g. $\mathrm{End}\, M = 1$, which is non-trivial work in most cases. Like in Jonsson [45] (p. 48, Definition 2.6.1) many authors call a structure M a rigid object of $\mathcal{C}$ if $|\mathrm{End}\, M| = 1$, i.e. the identity is the only morphism in M. This definition has the disadvantage that many categories have no rigid objects: groups $\neq 1$ are never rigid, and fields of characteristic $p > 0$ are excluded as well. Hence we will modify the notion of rigid objects which leads to an extension of several notions considered differently in special categories and provides important hints how to attack inverse problems mentioned above.

An endomorphism $\sigma \in \mathrm{End}\, M$ is called ***inessential*** (in $\mathcal{C}$) if any monomorphism $\alpha : M \to A$ in $\mathcal{C}$ gives rise to an endomorphism $\widehat{\sigma} \in \mathrm{End}\, A$ such that $\alpha\widehat{\sigma} = \sigma\alpha$. We have a commutative diagram

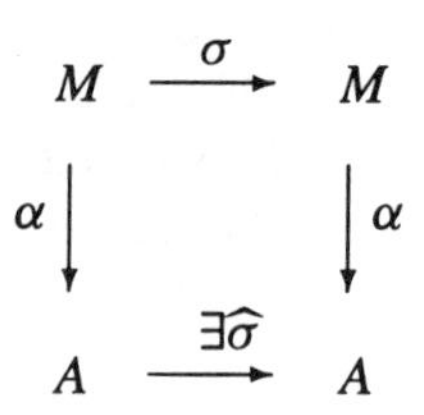

and say that $\sigma \in \operatorname{Iness} M$, see [23]. We borrowed this notion from Corner [13] where he considers a special class of abelian groups and introduces some ideal $\operatorname{Iness} M$ having the above property. Closely related to $\operatorname{Iness} M$ is a subset $\operatorname{Ines} M$ of $\operatorname{Iness} M$ consisting of all those $\sigma \in \operatorname{End} M$ where the above $\widehat{\sigma}$ takes values in M. We have $\sigma = \alpha\widehat{\sigma}$ and the following commutative diagram holds.

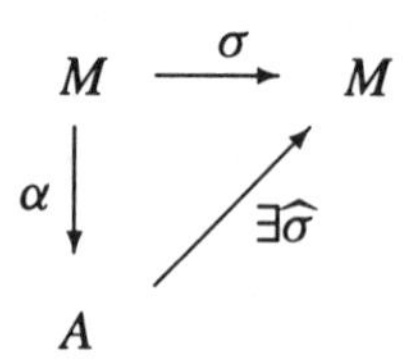

Obviously $\operatorname{Ines} M \subseteq \operatorname{Iness} M \subseteq \operatorname{End} M$ and in many cases $\operatorname{Ines} M$ will be an ideal of $\operatorname{End} M$, which is useful to determine the structure of $\operatorname{Iness} M$. However, if morphisms are bijections then $\operatorname{Ines} M$ is empty and we only deal with $\operatorname{Iness} M$. Finally we call M rigid ($\mathcal{C}$-rigid) with respect to the category $\mathcal{C}$ if $\operatorname{Iness} M = \operatorname{End} M$ where $\operatorname{Iness} M$ and $\operatorname{End} M$ are calculated in $\mathcal{C}$. If we want to emphasize that morphisms considered in $\mathcal{C}$ are automorphisms (endomorphisms) we call those rigid objects auto-rigid or endo-rigid respectively. We will see that auto-rigid groups have no outer automorphism and fields are endo-rigid if the only endomorphisms are Frobenius homomorphisms. Moreover we want to mention that Shelah [72] constructed endo-rigid boolean algebras with c.c.c. for all cardinals $\lambda^{\aleph_0}$ with $\lambda > \aleph_0$. The inessential endomorphisms are the "endomorphisms of the scheme" given on p. 87 in [72].

Guided by the general frame above, by classical result in algebra and advances of recent years in abelian group theory we are now ready to consider the following problems for concrete categories $\mathcal{C}$ with objects M, M'.

(A) (i) *Find an internal characterization of* $\operatorname{Ines(s)} M$*!*

(ii) *Determine a large class of rigid objects! Can we find* 2^κ *rigid objects of cardinality* κ*?*

(iii) *Can every object be embedded into a rigid object?*

(B) *Give a description of the derived categories* $\mathcal{C}'$ *with objects (if the quotient makes sense):* $\operatorname{End} M / \operatorname{Ines} M$ *or* $\operatorname{End} M / \operatorname{Iness} M$ *or* $\operatorname{Iness} M / \operatorname{Ines} M$ *respectively.*

(C) *Prove "realization theorems".*

(i) *If $E \in \mathcal{C}'$, then determine those $M \in \mathcal{C}$ for which* $\operatorname{End} M/\operatorname{Iness} M \simeq E$ *(or* $\operatorname{End} M/\operatorname{Ines} M \simeq E$ *respectively).*

(ii) *How many non-isomorphic E's with* (i) *and of fixed cardinality do exist?*

(D) *Determine rigid classes $\{G_i \in \mathcal{C},\ i \in I\}$ with $|G_i| = |G_j|$ and* $\operatorname{Mor}(G_i, G_j) = \operatorname{Iness}(G_i, G_j)$ *for all $i \neq j$.*

We begin with field theory.

3. Field Theory

If F is a field, then $\operatorname{End} F$ denotes the monoid of all non-zero endomorphisms $\varphi : F \to F (x \to x\varphi)$ with $(x+y)\varphi = x\varphi + y\varphi$ and $(xy)\varphi = (x\varphi)(y\varphi)$ for all $x, y \in F$. Since maps $\varphi \in \operatorname{End} F$ act on the right and are injections, it follows that $\psi_1\varphi = \psi_2\varphi$ implies $\psi_1 = \psi_2$ for all $\psi_i, \varphi \in \operatorname{End} F$. Hence $\operatorname{End} F$ is a monoid with right cancellation containing the subgroup $\operatorname{Aut} F$ of all automorphisms of F. If F has characteristic $\chi F = p > 0$, then F has a very special endomorphism Φ, defined by $x\Phi = x^p$ for all $x \in F$. The Frobenius endomorphism Φ generates a central submonoid

$$\operatorname{Frob} F = \langle \Phi \rangle = \{\Phi^n : n \in \mathbb{N} \cup \{0\}\}$$

of $\operatorname{End} F$. If $\chi F = 0$, we define $\operatorname{Frob} F = \{1\}$, where 1 is the identity map on F. Clearly $\operatorname{End} F/\operatorname{Frob} F$ is a monoid with right cancellation as well. Elements in $\operatorname{Frob} F$ have a trivial, but remarkable property

(3.1) $\operatorname{Frob} F \subseteq \operatorname{Iness} F$, where $\operatorname{Iness} F$ is the set of all inessential endomorphisms of F which extend to any extension of F. It is a less trivial problem (answered below), that equality holds in (3.1). Again following the general conception, we will say that F is endo-rigid if $\operatorname{End} F = \operatorname{Frob} F$ which is justified only afterwards if equality is shown in (3.1). Similarly F is auto-rigid if $\operatorname{Aut} F = 1$. This is in accordance with the general definition in Section 2, after $\operatorname{Iness} F = 1$ is shown for automorphism groups. The general conception now leads to the following problems.

(3.2) If K is a field and G is a monoid with right cancellation, then we must (be able to) find an extension field $F \subset K$ such that $\operatorname{End} F/\operatorname{Frob} F \simeq G$.

(3.2′) Moreover, we want some fine tuning. The above isomorphism should be strengthened to

(a) $\operatorname{End} F \simeq \operatorname{Frob} F \rtimes G$, where $\rtimes$ denotes the usual (semi) group semidirect product.

(b) What is the cardinal spectrum of (K, G)? This is the class $\operatorname{spec}(K, G)$ of all cardinals $|F|$ as in (a).

(c) What can we say about the fixed field F_G? We define

$$F_G = \{x \in F : xg = x \text{ for all } g \in G\}$$

as in the case of groups G.

(3.3) If K is a field and G is a group, then we must (be able to) find an extension field $F \supset K$ with Aut $F \simeq G$.

(3.3′) Again, we want some fine tuning.

(a) What is spec(K, G)?

(b) What can we say about the fixed field F_G?

We have the surprising new result, that there is a big difference if we consider spec(K, G) for monoids G or for groups G respectively. In order to tackle these problems, we first transport some ideas from abelian groups to fields. Baer [4] characterized torsion-free abelian groups of rank 1 by introducing the notion of types. Types measure the degree of divisibility of elements by primes. If the torsion-free abelian group of rank 1 is also a ring, only special types with entries 0 or ∞ are needed and they may be viewed as subsets of all primes. Now we are ready to carry types over to fields. If K is a field, we consider the set $\mathbb{P}_K$ of all odd primes which are different from χK. The prime 2 is excluded in order to avoid difficulties with Abel's theorem for fields, see [51]. An (abstract) type will be subset of $\mathbb{P}_K$ and the type of $a \in K$ will be the set

$$\tau(a) = \{p \in \mathbb{P} : \forall n \in \mathbb{N}\ \exists x \in K \text{ with } x^{p^n} = a\}.$$

If Σ is a set of abstract types from $\mathbb{P}_K$ and $K \subset F$, then F is a Σ-field over K if F is generated by elements of type in Σ and K. Moreover F is a Σ-field if F is a Σ-field over its prime field. If K is algebraically closed, then K is a $\mathbb{P}_K$-field, where we omit set-brackets. Clearly $\mathbb{Q}$ is a $\emptyset$-field because all elements different from $1, -1$ and 0 have type $\emptyset$. Any function-field is an $\emptyset$-field as well.

Prüfer [60] investigated abelian p-groups by introducing the notion of a "Servanzuntergruppe". Such subgroups of abelian groups are now commonly (except in Russian literature and their (bad) translation by the AMS) called pure subgroups. Elements in pure subgroups have the same degree of divisibility if they are viewed as elements of the overgroup. We say that a subfield K is pure in the field F if any element $a \in K$ has the same type $\tau_K(a) = \tau_F(a)$ evaluated in K and F respectively.

The following trivial observation is the key for a solution of (3.2) and (3.3)

Observation 3.4. *Let K be a field and $a \in K$.*

(a) *If $\alpha \in$ End K, then $\tau(a\alpha) \supseteq \tau(a)$.*

(b) *If $\alpha \in$ Aut K, then $\tau(a\alpha) = \tau(a)$.*

We see that types (like in abelian groups) restrict the "activity" of endomorphisms.

Using the Black Box discussed in Section 1 we have the following

Theorem 3.5 ([23]). *If G is a monoid with right cancellation, K is a field and $\lambda > |G| \cdot |K|$ is any infinite cardinal, then we can find an extension field F of K with the following properties*

(a) $\operatorname{End} F \simeq \operatorname{Frob} F \rtimes G$

(b) $|F| = \lambda^{\aleph_0}$, *provided* $\lambda^{|K|\cdot\aleph_0} = \lambda^{\aleph_0}$

(c) $F_G = K$, *if G is infinite.*

Special cases (with much larger extension fields or $K = \mathbb{Q}$) are due to E. Fried [33], Fried and Kollar [34], Kuyk [48] and Pröhle [58, 59]. Theorem 3.5 answers a question of Baer's (1967) and of C. Jensen (1985), see also [23]. Recall that all cardinals $\kappa^{|K|}$ with $\kappa \geq |G|$ belong to $\operatorname{spec}(K, G)$ for infinite fields K. We also would like to draw attention to the following (new) special case of (3.5).

Corollary 3.6. *Any algebraically closed field K has an endo-rigid extension field F with $|F| = |K|^+$, the successor of $|K|$.*

This applies to the complex numbers $\mathbb{C}$ which can be embedded into an endo-rigid field F of cardinality $(2^{\aleph_0})^+$. Conversely, Krull proved the following proposition, showing that a cardinal jump is necessary.

Proposition 3.7. *If F is an extension of the algebraically closed field K and $|F| = |K|$, then* $|\operatorname{End} F| = 2^{|F|}$.

Hence F is far away from endo-rigid. The 1994-proof of (3.7) is very easy: Take any transcendency base B_K and B_F of K and F respectively. Clearly $|B_K| = |K| = |F| = |B_F|$ and any bijection $B_F \to B_K$ extends to an endomorphism of F.

Refining the arguments in [23] it should not be complicated to determine $\operatorname{spec}(K, G)$ completely.

Surprisingly, the cardinal spectrum is different in case of automorphism groups of fields. The reason is hidden in the definition of a group and is not a defect of the tools which are at disposal. Proposition 3.7 can be modified to give the following

Proposition 3.8. *If $K \subseteq F = K(Y)$ are fields with $|Y| < |K|$ and K is algebraically closed, then also* $|\operatorname{Aut} F| = 2^{|F|}$.

Hence F is never auto-rigid if F is a "small" extension of an algebraically closed field. Many of the $2^{|K|}$ automorphisms of K survive the extension. We indicate an argument. Let B be a transcendency base of K and choose a maximal transcendental set T of F over K, hence $K(T) \subseteq F \subseteq \widehat{K(T)}$, where $\widehat{\cdots}$ denotes the algebraic closure. Next we define support $[y] \subset B$ of any $y \in Y$ which is some (minimal) finite subset $[y]$ of B with $y \in P(\widehat{[y], T})$, where P is the prime field of K. It follows that $C = \bigcup_{y \in Y} [y]$ has $|C| < |F|$. Take any permutation α of $B \backslash C$ and put $\alpha|_{\widehat{F'}} = 1$ where $F' = P(C \cup T)$, then α extends to an endomorphism of $\widehat{F'}(B \backslash C)$ which can be extended further to $\widehat{\alpha} \in \operatorname{End} \widehat{F'(B \backslash C)}$. However $\widehat{F'(B \backslash C)} = \widehat{F}$ and if $\alpha = \widehat{\alpha}|_F$, then α is an automorphism of F because $B\alpha = (B \backslash C)\alpha \cup C\alpha = B$ which is a transcendency base of the algebraically closed field K. From $|B \backslash C| = |B| = |F|$ follows $|\operatorname{Aut} F| = 2^{|F|}$.

We see from (3.8) that equipotent auto-rigid extensions are not yet excluded. This however could only be possible by adjoining an equipotent set of new elements.

The combinatorial tools used for (3.5) do not provide an answer. However we are able to apply a module theoretic result concerning endomorphism rings. The required theorem will be discussed in Section 4 (see Theorem 4.1). It will appear in [24] and [25]. We will use Observation 3.4 to code modules into fields. First results of this kind, namely coding graphs into fields, are due to E. Fried [33] and Fried and Kollar [34].

We will indicate the construction of the desired extension fields F in (3.9) and (3.10) using (4.1), a recent theorem in module theory, which is applied in the proof of

Lemma 3.9 ([24] and [25]). *Let G be a subgroup of the automorphism group of a field K and Σ be a set of types $\neq \mathbb{P}_K$ such that the following hold.*

(1) *K is algebraic over a $\Sigma \cup \{\mathbb{P}_K\}$-field L.*

(2) *There are seven distinct primes $p, q_0, \ldots, q_5 \in \mathbb{P}_K \backslash \bigcup \Sigma$. Then there is an extension field F of cardinality λ for any $\lambda \geq \aleph_0 \cdot |K| \cdot |G|$ such that*
 - (i) *K is algebraically closed in F,*
 - (ii) $\operatorname{Aut} F|_K = G$,
 - (iii) *there is a field L' between K and F such that F is algebraic over L' and L' is a $\{\{q_0\}, \ldots, \{q_5\}\}$-field over K,*
 - (iv) *if $b \in F$ and $\tau(b) \in \{\mathbb{P}_K\} \cup \Sigma$, then $b \in K$,*
 - (v) *K is pure in F.*

If P is the prime field of K and $A = P[G]$ is the group ring, then $\lambda \geq |A| \cdot \aleph_0$ and we may apply (4.1) and find a free A-module $\mathbf{V} = (V, V_1, \ldots, V_4)$ of rank λ (and $\dim_P V = \lambda$) such that $\operatorname{End} \mathbf{V} = A$. Let X be an A-basis of V and X_i be an A-basis of V_i (for $i = 1, \ldots, 4$) and recall that $XG = \{xg : x \in X, g \in G\}$ is a P-basis of V and similarly $X_i G$ is a P-basis of V_i. Let $Y = XG \cup \{t_1, \ldots, t_5\}$ be XG adjoined five new elements, all transcendental over K.

Now we want to apply Observation 3.4 and mark field elements by specific type. If $z \in K(Y)$, and q is some prime, then $\langle z \rangle_q = \{{}_i z : i \in \omega\}$ denotes elements ${}_i z \in \widehat{K(Y)}$ with ${}_{i+1}z^q = {}_i z$ and ${}_0 z = z$ for all $i \in \omega$, commonly called a root chain of z. Set $K_0 = K(\langle xg \rangle_{q_0}, xg \in XG)$ and observe that K is pure in K_0 but $\tau_{K_0}(xg) = \{q_0\}$. It is trivial to extend $G \subseteq \operatorname{Aut} K_0$ to $G \subseteq \operatorname{Aut} K$ canonically. So V is marked inside K_0. Similarly, we define K_i for $1 \leq i \leq 4$ such that

$$K_i = K_{i-1}(\langle t_i \rangle_{q_i}, a_i, b_{xg} : xg \in X_i G)$$

where $a_i^p = t_i - 1$, $(b_{xg})^p = t_i - xg$ for all $xg \in X_i G$ and the V_i's are marked. Finally, if $(x \to a_x)$ denotes a surjection of X onto K, then $F = K_5 = K_4(\langle t_5 \rangle_{q_5}, a_5, {}^5 b_{xg} : xg \in XG)$ where $a_5^p = t_5 - 1$, $({}^5 b_{xg})^p = t_5 - xg - a_x g$. If $h \in G$ and ${}_j t_i h = {}_j t_i$ and $b_{xg} h = g_{xgh}$ then G extends to $G \subseteq \operatorname{Aut} F$.

Repeated application of Observation 3.4 shows (ii) and the remaining condition in (3.9) are obvious by construction, cf. [25].

It needs a few more steps to derive

Theorem 3.10 ([25]). *Let K be a field, G be a group and λ be any infinite cardinal $\geq |K| \cdot |G|$. Then there exists an extension field F with* $\operatorname{Aut} F = G$ *and* $|F| = \lambda$.

This answers problems (3.3) and (3.3′)(a) completely and we can say more about (3.3′)(b). The required field F will be the union of a countable chain of intermediate fields $\bigcup_{n\in\omega} K_n$ arising by repeated application of (3.9). We may assume that $K \subset K_0'$ and K_0' is the algebraic closure of K. This reduces the number of used types to one, which is $\mathbb{P}_k$. If K_{n+1} is derived from K_n by (3.9), then it is necessary to ensure $K_{n+1}\alpha \subseteq K_{\alpha+1}$ for any $\alpha \in \operatorname{Aut} F$ in order to apply the implication of (3.9). Hence $\mathbb{P}_K \backslash \{p, q\}$ is enumerated by $\{q_i : i \in \omega\}$, where p, q are two distinct primes in $\mathbb{P}_K$. First replace K_0' by $K_0'' = K_0'(\langle x_g \rangle_q : g \in G)$ with $\langle x_g \rangle_q$ as in the Lemma and apply (3.9) to derive K_0. At stage K_n we apply the Lemma using p, q and the primes $\{q_{6n+i} : 0 \leq i < 6\}$. Now type arguments ensure $K_{n+1}\alpha \subseteq K_{n+1}$, hence $\alpha|_{K_{n+1}} = g_{n+1} \in G$ and it is easy to conclude $g_n = g_{n+1}$, hence $\operatorname{Aut} K = G$, see [25].

Remarks. The fixed field F_G in (3.10) is strictly larger then K which is caused by our definition of the action of G on intermediate extensions with fixed points $t_i g = t_i$ for all $g \in G$ given in (3.9). However if G is infinite, an easy modification of the group action on those transcendental variables would reduce the fixed field to $F_G = K$.

The methods presented in the second part of this section could be used to fill cardinal gaps in (3.2′)(b) due to Shelah's Black Box, hence of (3.5). It remains the following

Problem. Can algebraically closed field of cardinality $\kappa < 2^{\aleph_0}$ be embedded into endo-rigid fields of cardinality κ^+?

This problem has obvious variations, which are unanswered. The answer is yes if we replace "endo" by "auto", see above.

We will conclude this section dealing shortly with formally real fields. The problem behind the mentioned results is due to A. MacIntyre. We are most grateful to him, drawing our attention to these questions at a Colloquium at Oxford University.

Any ordered field has characteristic 0. Moreover any automorphism group of an ordered field (by definition) preserves the ordering of the field and must be right orderable by a result of Conrad [10]. We now have the converse result, which is

Theorem 3.11 ([24]). *If K is an ordered field and G is a right orderable group, then we can find an extension field F with* $\operatorname{Aut} F \simeq G$.

The results concerning cardinal spectrum and fixed fields are very similar to the field case. The results using Shelah's Black Box for formally real field are given in [24]. The results using module theory can be derived similarly as in the field case. The latter proofs are left to the reader as some advanced seminar work. Answers to the above problem for particular formally real fields are given by Baer [6] and Scott [67]. Particular cases of (3.11) are also due to E. Fried and Kollar [34].

4. Representations of P.O. Sets (and More)

In order to complete an argument in constructing groups of automorphisms of field (Section 3, proof of Lemma 3.9) we need results on module theory presented in this section. Let R be any commutative ring with $1 \neq 0$. Then we consider the category of R_n-modules for some positive integer n. Its objects are $\mathbf{M} = (M, M_1, \ldots, M_n)$, where M is an R-module and $M_1, \ldots M_n$ are n distinguished submodules. Morphisms σ between two R_n-modules $\mathbf{M}$ and $\mathbf{M}'$ are the R-homomorphisms $\sigma \in \operatorname{Hom}_R(M, M')$ with $M_i\sigma \subseteq M_i$ for all $i \leq n$. If $\{1, \ldots, n\}$ is replaced by a partially ordered set I such that $M_i \subseteq N_j$ for all $i \leq j \in I$, then the category of R_I-modules is a refinement of R_n-modules where $I = \{1, \ldots, n\}$ may be given the trivial ordering of an antichain. Using ideas from Kronecker [46], Ringel [62], Corner [11], Fuchs [37] and Shelah [71] we derive [40] the following theorem incorporating many older partial results.

Theorem 4.1 ([40]). *Let A be an R-algebra with $\leq \lambda$ generators (as algebra), where λ is an infinite cardinal. Then we can find an R_4-module $\mathbf{M} = (M, M_1, \ldots, M_4)$ such that M and M_i are isomorphic to $\bigoplus_\lambda A$ and M_i is an A-summand of M for each $i \leq 4$. Moreover,* $\operatorname{End} \mathbf{M} \simeq A$.

This answers a problem in Corner [11], which deals with R_5-modules. The non-trivial proof needs some tickly Shelah-combinatorics which fortunately is valid in ZFC set theory and puts no restriction on λ. The number 4 in (4.1) is minimal for fields R as shown by Ringel, Tachikawa [63] and independently by Simson [73], see also [74]. If R_n-modules are replaced by R_I-modules with $(I, <)$ an partially ordered set, then a similar result involving Kleiner's five minimal poset was derived in [7]. More recently 110 posets are characterized in [39], if the notion of R_I-modules is extended in the sense of Simson [75], see also [41]. If fields are excluded, then 4 is no longer minimal. In particular we may replace 4 by 2 if R in (4.1) is the ring $\mathbb{Z}$ of integers, see [29].

Theorem 4.1 and its variants have many applications in module theory. It can be used to construct indecomposable torsion-free abelian groups, see [32]. Moreover it can be used to construct Butler groups, which is a new result in [26]. We applied (4.1) in Section 3 for constructing fields with prescribed automorphism group. We close this section, which provides auxiliary tools by giving an interesting application of the results mentioned above. (4.2) has many applications. It serves for coding groups into algebras.

Corollary 4.2 ([29]). *If G is any group and λ is an infinite cardinal $\geq |G|$, then we can find a free abelian group F with two distinguished (pure) subgroups F_1, F_2 such that*

$$\mathrm{End}(F, F_1, F_2) = \mathbb{Z}G,$$

the group ring of G.

5. Group Theory

A. All results of this section have a general method in common. In order to construct particular non-commutative groups we will provide specially designed abelian groups which are used as building blocks for their non-commutative relatives. The closest relatives of abelian groups are nilpotent groups of class 2. In fact we will deal with a particular class of these groups G first and some calculation [21], [20] show that in this category the inessential morphisms in the sense of Section 2 constitute a well-known normal subgroup of Aut G, which is the stabilizer of G. Recall that

$$\mathrm{Stab}\ G = \{\alpha \in \mathrm{Aut}\, G,\ \alpha|ZG = 1 \text{ and the “induced } \alpha\text{” on } G/ZG \text{ is } 1\}.$$

Hence we have Iness G = Stab G in the category. Moreover there is a well-known "Ext-theory" in this case, which is due to Baer [5], see also Warfield [79]. These groups G are central extensions

$$1 \to A \to G \to B \to 1$$

of abelian groups A, B, and by Hall's commutator formulae they lead to an alternating bilinear map

$$f : B \times B \to A.$$

Conversely the Baer–Lazard–Theorem (see [79, p.41, Theorem 5.14]) gives rise to groups which are nilpotent of class 2. Suppose G^+ is abelian, 2-divisible and 2-torsion-free with alternating bilinear map $f : G \times G \to G$ such that $f(f(x, y), z) = 0$ for all $x, y, z \in G$. If $xy = x + y + \frac{1}{2} f(x, y)$, then $G^\bullet = G_f^\bullet$ is nilpotent of class 2 and $(G^+, f) \to G_f^\bullet$ is a category isomorphism. Moreover $\alpha \in \operatorname{Aut} G_f^\bullet$ if and only $\alpha \in \operatorname{Aut} G^+$ and f is α-invariant. This is a key for constructing nilpotent groups. Now we want to prescribe their automorphisms modulo inessential maps. Using model theoretic techniques (Shelah's Black Box, cf. [70]) we can find the desired abelian group: If the group H is essentially the wanted automorphism group, then we choose the group ring $R = \mathbb{Z}H$.

Observe that R^+ is a free abelian group, hence cotorsion-free and topological methods from [19] apply. Beginning with a free R-module F, we may define an alternating, bilinear map on F which can be extended (as shown in [21]) to an R-submodule G of the $\mathbb{Z}$-adic completion of F such that — and here the Black Box is used —

$$\operatorname{End} G = R \oplus f_G \qquad (*)$$

where f_G is an ideal of $\operatorname{End} G$ and $(*)$ is a split extension. The elements of the ideal f_G are the endomorphisms coming from the bilinear form: If $g \in G$, then $f(g, \_) : G \to G$ is in f_G. By the Baer–Lazard–Theorem we trade (G, f) with $(*)$ for a nilpotent group.

Corollary 5.1. *If H is any group and $\lambda = \lambda^{\aleph_0} > |H|$ then we find a torsion-free, nilpotent group G of class 2 with $|G| = \lambda$ and* $\operatorname{Aut} G = \operatorname{Stab} G \rtimes H$ *where* $\operatorname{Stab} G / \operatorname{Inn} G = \oplus Z_2$.

Changing the ring leads to many results of similar type which are discussed in [21]. Observe that Stab G is abelian by a general theorem of Hall and Hartley [42].

Another method for constructing torsion-free nilpotent groups comes from linear algebra and was exploited by Zalesskii [83]. As matrix multiplication becomes too complicated if the dimension is > 3, we (unfortunately) like [83] restrict to 3×3-matrices. However we will work over an arbitrary commutative ring R with $1 \neq 0$ and R^+ torsion free and p-reduced for infinitely many primes p. Decompose this set of primes into two infinite subsets P_1, P_2 and write

$$(a, b, c) = \begin{pmatrix} 1 & b & a \\ 0 & 1 & c \\ 0 & 0 & 1 \end{pmatrix} \in R^{3\times 3}$$

for the upper triangular matrix. Then we consider the group

$$G = G(R) = \left\langle \left(\frac{r}{p_2^2}, 0, \frac{r}{p_2}\right), \left(\frac{r}{p_1^2}, \frac{r}{p_1^2}, 0\right), \left(0, \frac{r}{p_0}, \frac{r}{p_0}\right), \; r \in R, \; p_i \in P_i \right\rangle$$

Elementary group calculation carried out in [20] show that

(5.2)(a) *G is torsion-free, nilpotent of class 2 and*

$$\operatorname{Aut} G = \operatorname{Stab} G \rtimes \operatorname{Aut} R \qquad (**)$$

The group theoretic problem is reduced to the question about realizing groups as automorphism groups of rings with the above prime-property. This is connected with E-rings whose existence (a question of R. S. Pierce and P. Schultz) for many cardinals was established only recently [30] using methods from [14]. Recall that a ring R is an E-ring if $R^{+\bullet} \simeq \operatorname{End}_{\mathbb{Z}} R^+$ canonically, e.g. $\mathbb{Z}$ is an E-ring. Again, using the Black Box we can sharpen this result, taking care of a group H: *If $\lambda > |H|$ is infinite and I is a set of $\lambda^{|H|}$ commuting variables, then we can find a ring R between $\mathbb{Z}[I] \subseteq R \subseteq \widehat{\mathbb{Z}[I]}$ such that* $\operatorname{End}_{\mathbb{Z}} R^+ = R[H]$ *is the group ring,* $\operatorname{Aut} R = H$ *and* $H = R[H] \cap \operatorname{Aut} R$.

Using $(**)$ we utilize the ring R to obtain:

(5.2)(b) *A torsion-free nilpotent group of class 2 with* $\operatorname{Aut} G = \operatorname{Stab} G \rtimes H$ *as desired.*

There is another (more complicated) construction where $\operatorname{Stab} G$ can be replaced by $\operatorname{Inn} G$. Moreover G is torsion-free and locally nilpotent, see [28].

B. In this second part of Section 5 we discuss automorphism groups of locally finite p-groups.

Abelian p-groups are locally finite p-groups and we must strengthen a recent result on abelian p-groups in order to handle their non-commutative relatives. First more complicated abelian p-groups were constructed by Corner [12] and later by Shelah [68]. These results are unified and extended in [18] and [14] using heavily that inessential endomorphisms in Ines G are Pierce's "small" homomorphisms discovered more then thirty years ago: If A is any torsion-free and reduced algebra over the p-adic integers J_p, then $\operatorname{End} G/\operatorname{Small} G \simeq A$ for some abelian p-group G if and only if A is complete in the p-adic topology. The Black Box constructions allows more space for additional requirements (see [27]): Let λ be an infinite cardinal with $\lambda = \lambda^{\aleph_0}$. Let C be a direct sum of λ cyclic p-groups and A as above with $|A| < \lambda$, then there exist a separable abelian p-group S of cardinality λ such that

$$\operatorname{Hom}(S, S') = A \oplus \operatorname{Small}(S, S')$$

where $S' = \bigoplus_{\lambda} S \oplus C$ and A is the completion of $\bigoplus_{\lambda} J_p$ mapping S in S' by $(s \to (s\pi_\alpha)_\alpha \in \bigoplus_{\alpha<\lambda} S \subseteq S')$ where $\pi = (\pi_\alpha) \in A$ and π_α is the α-th component of π in $A = \widehat{\bigoplus_{\alpha<\lambda} J_p}$. Now the construction of locally finite p-groups G is based on a transfinite induction on $\alpha < \lambda^+$, and the resulting group will belong to the class $\{E_s, L\}\{H, B, Z_{p^n} : n < \omega\}$, where B is any infinite group, H is a given countable group, E_s and L are obvious Hall-closure operators: E_s for split extensions and L for "local". Moreover $\operatorname{Aut} G = \operatorname{Inn} G \rtimes H$ and $|G| = \lambda^+$. We will give the construction and indicate the use of S. We want $G = \bigcup_{\alpha<\lambda^+} G_\alpha$ a smooth chain of subgroups. Let

$$E = \{\alpha < \lambda^+ : cf\ \alpha = \omega\}$$

which is stationary in λ^+ and suppose G_α is constructed such that $H \subseteq \operatorname{Aut} G_\alpha$ and H is controlled by $Y \subseteq G_0$, which means $G_0 = H \wr B$, $|Y| = \aleph_0$, $Y \subseteq G_0$ such that

$$\text{If } h \in H,\ Y^* \subseteq Y \text{ of finite index and } h|_{Y^*} \in \operatorname{Inn}(H|_B)|Y^*, \text{ then } h = 1. \qquad (+)$$

We must define $G_{\alpha+1}$. If $\alpha \notin E$, then $G_{\alpha+1} = \bigoplus_n Z_{p^n} \wr G_\alpha$ is the ordinary wreath product. If $\alpha \in E$, then we choose $G_{\alpha+1}$ between $A_\alpha \wr G_\alpha \subseteq G_{\alpha+1} \subseteq \widehat{A_\alpha \wr G_\alpha}$ the p-adic closure of the wreath product (which may be viewed inside the "complete" cartesian) wreath product. The abelian group A_α is a copy of S in the above theorem. This completes the construction and somewhat lengthy calculations show the following results [27].

Theorem 5.3. *If $\lambda = \lambda^{\aleph_0}$ and H is a countable group and B is any group of cardinality $\leq \lambda$, then we can find a group*

$$G \in \{E_s, L\}\{B, H, Z_{p^n} : n \in \omega\} \text{ such that}$$

$\operatorname{Aut} G = \operatorname{Inn} G \rtimes H$ *and* $B \subseteq G$.

By the same token — using ideas due to S. Thomas [77] — we also derive a theorem on ULF-groups.

Theorem 5.4. *If $\lambda = \lambda^{\aleph_0}$ and B is locally finite of cardinality $\leq \lambda$, then there exists a family of 2^{λ^+} pairwise non-isomorphic, complete ULF-groups G of cardinality λ^+ containing B.*

C. Some historical notes are in order. Theorems that inessential automorphisms are the inner automorphisms for various classes of groups are shown in some resent papers. Schupp [66] proved this for the class of all groups and Pettet [56] established this for the class $\mathcal{C}$ of finite groups. More recently Pettet [56] investigated independently a class $\mathcal{C}$ closely related to results discussed in A of this section. He considered nilpotent groups G of class 2 with $G^{p^2} = 1$ and derived for any group H the existence of such a G with $\operatorname{Aut} G = \operatorname{Aut} G|_{(G/ZG)} \rtimes H$. The question about the existence of p-groups G with prescribed outer automorphism group $\operatorname{Out} G = \operatorname{Aut} G/\operatorname{Inn} G$ has a relatively long history. After some partial results, Gaschütz [38] showed in 1962 that 1 can never be realized by a finite p-group $\neq 1$. Later Webb [80] gave a more elementary proof of Gaschütz's result, avoiding homology. A. E. Zalesskii [82] extended Gaschütz result to arbitrary nilpotent p-groups. Recall however that 1 can be realized by torsion-free nilpotent groups of class 2 by Zalesskii [83] or the results above. Hence "p" is crucial! Using model theoretic arguments by Kueker, Thomas [76] showed that any infinite, countable p-group must satisfy $|\operatorname{Out} G| = 2^{\aleph_0}$, hence $\operatorname{Out} G = 1$ is obviously impossible. Another negative result is due to Buckley and Wiegold [9], [8] who determined $|\operatorname{Out} G|$ for nilpotent p-groups. Moreover, Menegazzo and Stonehewer [53] observed that $t_p \operatorname{Out} G \neq 1$ for nilpotent p-groups G up to a few exceptions. Here $t_p X$ denotes the elements of p-power order in X. In 1985 Thomas [76] established the existence of a p-group of cardinality $\aleph_1$ with $\operatorname{Out} G = 1$. His result was shown in Gödel's constructible universe, which shows that a positive answer to Ph. Hall's problem about the existence of complete, locally finite, p-groups ($\operatorname{Out} G = ZG = 1 \neq G$) is relatively consistent with ZFC set theory. Thomas uses a clever self-embedding of the unique countable ULF p-group. The existence of locally finite, complete p-groups was shown in [27] and is discussed above. We extract our examples from (5.3) for $H = B = 1$:

(5.3*). *There is a locally finite p-group G with $ZG = \operatorname{Out} G = 1$ and $|G| = (2^{\aleph_0})^+$.*

Our basic method are applications of nasty abelian p-groups and a very useful p-adic wreath product which may play a role in future constructions of groups. The advantage of p-adic wreath products is the fact that "many" automorphisms become continuous and topology applies naturally. It may be interesting to answer the following stronger version of Hall's problem.

Problem. Can we find a locally finite, complete p-group of cardinality $\aleph_1$? There is a less ambitious program: Replace $\aleph_1$ by $2^{\aleph_0}$.

D. Another approach of realizing groups as automorphism groups is based on extensions of McLain's group. Here the power of results discussed in Section 1–4 with new results in order theory provides interesting applications. Recall the classical definition of McLain's group from excellent text books as D. J. S. Robinson [64] or from articles [52], [65] or [81]. Inspired from Section 4, the linear ordering — e.g. $(\mathbb{Q}, <)$ — used for defining classical McLain groups may be replaced by an unbounded partially ordered set S. Moreover the field $\mathbb{Z}/p\mathbb{Z}$ may be replaced by a ring R with $1 \neq 0$, which is not necessarily commutative. Elements of the generalized McLain group $G = G(R, S)$ are of the form $g = 1 + \sum_{\alpha<\beta} a_{\alpha\beta} e_{\alpha\beta}$ where $\alpha, \beta \in S$ and $e_{\alpha\beta}$ is a $S \times S$-matrix with all entries 0 except 1 at $\alpha\beta$. The triangular form of the matrix representation of g ensures that G is locally nilpotent and G is a p-group if and only if the characteristic χR of R is p. It can be shown [15] that "McLain Theory" holds if and only if S is locally linear and R has no zero-divisors $\neq 0$. By "McLain Theory" we mean an order theoretic description of maximal abelian normal subgroups of G, which are a basis for describing automorphisms in terms of order -and ring-automorphism. Under this assumption on (R, S) we have the following

Theorem 5.5 ([15]). (a) *$G(R, S) \simeq G(R', S')$ if and only if $R \simeq R'$, and $S \simeq S'$.*

(b) $\operatorname{Aut} G \simeq (((\operatorname{Lin} G \rtimes U(R)^{|S|}) \rtimes \operatorname{Aut} R) \rtimes \operatorname{Aut} S) \rtimes \langle\tau\rangle$, *where* $\operatorname{Lin} G$ *is the group* *of locally inner automorphisms, $U(R)$ is the group of units of the ring R,* Aut R *are the ring-automorphisms of R and* Aut S *denotes the order preserving automorphisms of S. The element τ is either the identity on G or $\tau^2 = 1$ is induced by a reflection on S.*

Recall that an automorphism is locally inner if its restriction to finitely generated subgroups is inner.

We refer to [15] for the lengthy proof of Theorem 5.5. A special case is due to [65] and extensions of the latter to particular linear orderings S are obtained by [81]. It is also interesting to note that Aut G determines G as shown in [16]. We now may specify S and R to obtain particular groups G. The reader can easily apply results from section 4. To have a definite example, we choose a rigid countable ordering S — as constructed in [17] — and conclude Aut $S = 1$ and $\tau = 1$. Moreover let $R = \mathbb{Z}/2\mathbb{Z}$. Then we find a countable, infinite, locally nilpotent locally finite 2-group G with Aut $G =$ Linn G. Clearly $|$ Linn $G| = 2^{\aleph_0}$ according to an earlier remark. We derive at the question about realizing groups as Aut $G/$ Linn G. Hence 1 is realized and further results are in [15].

6. Finite Rank Torsion-Free Modules over Valuation Domains

The final section is a survey of recent results concerning the complexity of modules of finite rank over valuation domains.

Let R be a discrete valuation domain with prime element p, $\widehat{R}$ the p-adic completion of R, $Q(\widehat{Q})$ the field of quotients of $R(\widehat{R})$. Let M be a torsion-free R-module of finite rank n. Then M contains a basic submodule $B \simeq R^r$, $r \leq n$, such that $M/B \simeq Q^k$, $n = r + k$. It is easy to see [1] that there exists a $k \times r$ matrix Γ over $\widehat{R}$ such that

$$M \simeq A[\Gamma] = [(\widehat{R})^r \oplus Q^k] \cap \{(x + y\Gamma, y) : y \in Q^k,\ x \in Q^r\}.$$

Unfortunately, Γ is not at all uniquely determined by M. There exists an isomorphism $\varphi : A[\Delta] \to A[\Gamma]$ if and only if

(a) Δ and Γ are both $k \times r$ matrices on R;

(b) there exists a non-singular Q-matrix $M = \begin{pmatrix} \Pi & P \\ \Sigma & T \end{pmatrix}$ where Π is $k \times k$, P is $k \times r$, Σ is $r \times k$ and T is an $r \times r$ matrix over Q, such that

  (i) $\Pi\Gamma + P = \Delta(\Sigma\Gamma + T)$,

  (ii) $\Sigma\Gamma + T$ is a matrix with entries in $\widehat{R}$ whose determinant is a unit in $\widehat{R}$.

Note that $\Sigma\Gamma + T$ is the matrix induced by φ acting on the basic submodule of $A[\Delta]$. If φ is not required to be an isomorphism, one can relax (2) by eliminating the conditions placed on the determinates.

Let $K \supseteq Q$ be the field generated by the entries of Γ. Then the splitting field of $A[\Gamma]$ (cf. [50]) is contained in K, i.e. $A[\Gamma] \otimes_R K \simeq K^r \oplus Q^k$. Lady in [50] investigated modules with a fixed splitting field K. Recall that a module M is strongly indecomposable if M is indecomposable in the quasi-isomorphism category. Lady used fairly deep results from representation theory to show:

If $[K : Q] = 2$, then R, Q, K are the only strongly indecomposable modules with splitting field contained in K.

If $[K : Q] = 3$ and $\{1, u, v\}$ is a basis of K over Q, then $R, Q, K, A[u]$ and $A[u, v]$ are the only strongly indecomposable modules with splitting field contained in K. Here we view $[u]$ ($[u, v]$) as (1×1) or (1×2) matrices.

If $[K : Q] \geq 4$, then there are strongly indecomposable modules of arbitrarily large (finite) rank where the case $[K : Q] = 4$ corresponds to tame representation type.

In [3] one finds easy examples of indecomposable modules of arbitrarily large ranks in the case $[K : Q] \geq 4$.

For any domain R, we define $fr(R) = \infty$ if there are arbitrarily large indecomposable torsion-free finite rank modules and $fr(R) = n$ if there is such a module of rank n but all modules of larger rank split. (All modules are understood to be torsion-free of finite rank.)

Vamos [78] investigates $fr(R)$ for valuation domains R: He characterizes valuation domains R that are $\mathbb{Q}$-algebras with $fr(R) < \infty$. In this case $fr(R) < \infty$ implies $fr(R) \in \{1, 2\}$ with both cases occurring. This lead Vamos' to conjecture that $fr(R) \in \{1, 2, \infty\}$ for all valuation domains R. Actually, he was able to show this for many *non*-discrete valuation domains.

From now on, let R be a ***discrete*** valuation domain. It is well known that $fr(R) = \infty$ if only $[\widehat{Q} : Q]$ is infinite. Since Nagata [55, Ex E33 p. 207] constructed examples of such rings and we follow Zanardo [84] in calling discrete valuation domains R with $[\widehat{Q} : Q] < \infty$ ***Nagata valuation domains***. Ribenboim [61] first investigated such rings and showed that they have positive characteristic. The examples in [3] mentioned above show that $fr(R) = \infty$ if $[\widehat{Q} : Q] \geq 4$. If $[\widehat{Q} : Q] = 1$, then $fr(R) = 1$ since $\widehat{R} = R$ in this case. Zanardo showed that if $[\widehat{Q} : Q] = 2$, then $fr(R) = 2$ and $R, Q, \widehat{Q}, A[u]$ are the only indecomposable R-modules where $\widehat{Q} = Q(u)$. This leaves us with the case $[\widehat{Q} : Q] = 3$. Lady settled the question for ***strongly*** indecomposable modules up to quasi-isomorphism with a very short list. We will see that, up to isomorphism, the list is infinite but $fr(R) = 3$. So let $[\widehat{Q} : Q] = 3$. Then $\chi\ R = 3$ and there is a unit in R with $\widehat{Q} = Q(u)$.

Theorem 6.1 ([3]). *Let R be as above. Then R, $\widehat{R}$, $\widehat{Q}$, Q, $A[u]$, $A[u, p^i u_2]$, $i = 1, 2, \ldots$, is a complete list (without repetitions) of indecomposable R-modules up to isomorphism.*

We would like to outline the proof.

First one shows that all indecomposable R-modules of rank ≤ 3 are in the list given in the theorem. Thus, by Lady's result, all finite rank R-modules are quasi-isomorphic to direct sums of modules in that list. (Zanardo's results [84] may be used to settle the case of rank 2 modules)

Recall that for R-modules A and X, $S_A(X) = \Sigma\{f(A) | f \in \text{Hom}(A, X)\}$ is the A-socle of X. The following is crucial: If $i \geq j$, then, for $A = A[u, p_i u_2]$ and $X = A[u, p_j u_2]$ we have $S_A(X) = X$. The same holds if $A = A[u]$. Moreover $\text{End}(A[u])$ and $\text{End}(A[u, p^i u_2])$ are isomorphic to subrings of $\widehat{Q}$ and hence commutative.

This allows us to use Baer's Lemma to show that each finite rank torsion-free R-module, which, as we know thanks to Lady, is quasi-isomorphic to a direct sum of modules in our list, is actually isomorphic to such a sum.

The modules of the form $A[u, p^i u_2]$ have rank $=$ p-rank $+1$, i.e. they are co-purely indecomposable. The following may be found in [1]: (R is now any discrete valuation domain)

Lemma 6.2. *Let X be a finite rank torsion-free R-module with no free summand. The following are equivalent:*

(a) *X is co-purely indecomposable.*

(b) *X is reduced and* rank $X = 1 + rk_p X$, *where $rk_p X$ denotes the p-rank of X.*

(c) *If Y is a submodule of X of lesser rank, then Y is free.*

(d) *Each torsion-free epimorphic image of X is strongly indecomposable.*

Let X, Y be such modules and B a (free) basic submodule of X. Then there is a short exact sequence $0 \to B \to X \to Q \to 0$ which might raise some hope that the class of co-purely indecomposable modules might be accessible to classification up to isomorphism. (A classification up to quasi-isomorphism in terms of arithmetic conditions in $\widehat{Q}$ is not hard).

Unfortunately, this seems to be a hard problem, solved only if $[\widehat{Q} : Q] \leq 3$. Define $X \leq Y$ if $S_X(Y) = Y$. This turns out to be a partial order on the set of co-purely indecomposables. If $[\widehat{Q} : Q] = 3$, then this order is linear, a crucial fact in the proof of the Theorem above.

If $[\widehat{Q} : Q] = 4$ this order is no longer linear. We want to conclude this section posing the following

Question. *If $[\widehat{Q} : Q] \geq 4$, determine the structure of the poset of co-purely indecomposable R-module with respect to the order defined above.*

References

[1] Arnold, D., A duality for torsion-free modules of finite rank over a discrete valuation ring. Proc. London Math. Soc. (3) 24 (1972), 204–216.

[2] —, Finite Rank Torsion Free Abelian Groups and Rings. Lecture Notes in Math. 931, Springer, New York 1982.

[3] —, Dugas, M., Indecomposable modules over Nagata valuation domains. Submitted to Acta. Math.

[4] Baer, R., Abelian groups without elements of finite order. Duke Math. J. 3 (1937), 68–122.

[5] —, Groups with abelian central quotient group. Trans. Amer. Math. Soc. 44 (1938), 357–386.

[6] —, Dichte, Archimedizität und Starrheit geordneter Körper. Math. Ann. 188 (1970), 165–205.

[7] Böttinger, C., Göbel, R., Endomorphism algebras of modules with distinguished partially ordered submodules over commutative rings. J. Pure Appl. Algebra 76 (1991), 121–141.

[8] Buckley, J., Wiegold, J., On the number of outer automorphisms of an infinite nilpotent p-group. Arch. Math. 31 (1978), 321–328.

[9] —, —, On the number of outer automorphisms of an infinite nilpotent p-group II. Arch. Math. 36 (1981), 1–5.

[10] Conrad, P., Right-ordered groups, Michigan Math. J. 6 (1959), 267–275.

[11] Corner, A. L. S., Endomorphism algebras of large modules with distinguished submodules. J. Algebra 11 (1969), 155–185.

[12] —, On endomorphism rings of primary abelian groups. Quart. J. Math. Oxford 20 (1969), 277–296.

[13] —, A class of pure subgroups of the Specker group. Talk given at Montreal 1967.

[14] —, Göbel, R., Prescribing endomorphism algebras - A unified treatment. Proc. London Math. Soc. (3) 50 (1985), 447–479.

[15] Droste, M., Göbel, R., McLain's groups over arbitrary rings and orderings. Math. Proc. Cambridge Phil. Soc. 117 (1995), 439–467.

[16] —, —, The automorphism groups of generalized McLain-groups. Submitted.

[17] Droste, M., Holland, W. C., Macpherson, H. D., Automorphism groups of homogeneous semilinear orders: normal subgroups and commutators. Canad. J. Math. 43 (1991), 721–737.

[18] Dugas, M., Göbel, R., On endomorphism rings of primary abelian groups. Math. Ann. 261 (1982), 359–385.

[19] —, —, Every cotorsion-free algebra is an endomorphism algebra. Math. Z. 181 (1982), 451–470.

[20] —, —, Torsion-free nilpotent groups and E-modules. Arch. Math. 54 (1990), 340–351.

[21] —, —, Automorphisms of torsion-free nilpotent groups of class two. Trans. Amer. Math. Soc. 332 (1992), 633–646.

[22] —, —, Field extensions in L - A solution of C. U. Jensen's $25-problem. Proc. Oberwolfach, 1985, Abelian Group Theory, R. Göbel and E. A. Walker eds. Gordon and Breach, London 1986.

[23] —, —, All infinite groups are Galois groups over any field. Trans. Amer. Math. Soc. 304 (1987), 355–384.

[24] —, —, Automorphism groups of fields. Manuscripta Math. 85 (1994), 227–242.

[25] —, —, Automorphism groups of fields II. Submitted.

[26] —, —, Realizing rings as endomorphism rings of Butler groups. Submitted to Proc. Lond. Math. Soc.

[27] —, —, On locally finite p-groups and a problem of Philip Hall's. J. Algebra 159 (1993), 115–138.

[28] —, —, Outer automorphism groups. Illinois J. Math. 35 (1991), 27–46.

[29] —, —, May, W., Free modules with two distinguished submodules. Submitted.

[30] —, Mader, A., Vinsonhaler, C., Large E-rings exist. J. Algebra 108 (1987), 88–101.

[31] Eklof, P., Mekler, A., Almost Free Modules. Set-Theoretic Methods. North Holland, Amsterdam 1990.

[32] Franzen, B., Göbel, R., The Brenner–Butler–Corner theorem and its application to modules. In: Abelian Group Theory, Gordon and Breach, London 1987, 209–227.

[33] Fried, E., A comment on automorphism groups of fields. Studia Sci. Math. Hungar. 14 (1979), 315–317.

[34] —, Kollar, J., Automorphism groups of fields. Colloq. Math. Soc. Janos Bolyai 29 (1977), 293–303.

[35] Fuchs, L., Infinite Abelian Groups, Vol.I. Academic Press, New York 1970.

[36] —, Infinite Abelian Groups, Vol II. Academic Press, New York 1973.

[37] —, Large indecomposable modules in torsion theories. Aequationes Math. 34 (1987), 106–111.

[38] Gaschütz, W., Nicht abelsche p-Gruppen besitzen äussere p-Automorphismen. J. Algebra 4 (1966), 1–2.

[39] Göbel, R., May, W., Endomorphism algebras of peak I-spaces over posets of infinite prinjective type. Submitted to Trans Amer. Math. Soc.

[40] —, —, Four submodules suffice for realizing algebras over commutative rings. J. Pure Appl. Algebra 65 (1990), 29–43.

[41] Göbel, R., Simson, D., Endomorphism ring problem via embeddings of Kronecker modules into the category of prenjective modules. In preparation.

[42] Hall, P., Hartley, B., The stability group of a series of subgroups. Proc. Lond. Math. Soc. 16 (1966), 1–39.

[43] Hasse, H., Zahlentheorie. Akademie-Verlag, Berlin 1969.

[44] Jensen, R., The fine structure of the constructible hierarchy. Ann. Math. Logic 4 (1972), 229–308.

[45] Jonsson, B., Topics in Universal Algebra, Springer, Berlin 1972.

[46] L. Kronecker, Algebraische Reduktion der Schaaren bilinearer Formen. Sitzungsberichte der Königl. preuss. Akademie der Wiss. Berlin (1890), 1225–1237.

[47] Kegel, O. H, Wehrfritz, B.A.F., Locally Finite Groups. North Holland, Amsterdam-London 1973.

[48] Kuik, W., The construction of fields with infinite cyclic automorphism group. Canad. J. Math. 17 (1965), 665–668.

[49] Lady, L., On classifying torsion-free modules over discrete valuation rings. In: Abelian Group Theory, Lectures Notes in Math. 616, Springer, Berlin 1977, 168–172.

[50] —, A seminar on splitting rings for torsion-free modules over Dedekind domains. In: Abelian Group Theory, Lectures Notes in Math. 1006, Springer, Berlin 1983, 1–48.

[51] Lang, S., Algebra. Addison-Wesley, Reading 1965.

[52] McLain, D. H., A characteristically-simple group. Math. Proc. Cambridge Philos. Soc. 50 (1954), 641–642.

[53] Menegazzo, F., Stonehewer, S. E., On the automorphism groups of a nilpotent p-group. J. London Math. Soc. (2) 3 (1985), 272–276.

[54] Mitchel, B., Theory of Categories. Academic Press, New York 1965.

[55] M. Nagata, Local Rings. Wiley Interscience, New York 1962.

[56] Pettet, M. R., On inner automorphisms of finite groups. Proc. Amer. Math. Soc. 105 (1989), 1–4.

[57] —, Characterizing inner automorphisms of groups. Arch. Math. 55 (1990), 422–428.

[58] Pröhle, P., Does a given subfield of characteristic zero imply any restrictions to the endomorphism monoids of fields? Acta Sci. Math. 50 (1986), 15–38.

[59] —, Does the Frobenius endomorphism always generate a direct summand in the monoid of prime characteristic? Bull. Austral. Math. Soc. 30 (1984), 335–356.

[60] Prüfer, H., Untersuchungen über die Zerlegbarkeit der abzählbaren primären abelschen Gruppen. Math. Z. 17 (1923), 35–61.

[61] Ribenboim, P., On the completion of a valuation ring. Math. Ann. 155 (1964), 392–396.

[62] Ringel, C. M., Infinite-dimensional representations of finite-dimensional hereditary algebras. Sympos. Math. 23 (1979), 321–412.

[63] —, Tachikawa, H., QF-3 rings, J. Reine Angew. Math. 272 (1975), 49–72.

[64] Robinson, D. J. S., A Course in the Theory of Groups. Springer, Berlin 1982.

[65] Roseblade, J. E., The automorphism group of McLain's characteristically simple group. Math. Z. 82 (1963), 267–287.

[66] Schupp, P. E., A characterization of inner automorphisms. Proc. Amer. Math. Soc. 101 (1987), 226–228.

[67] Scott, D., On completing ordered fields. In: Applications of model theory to algebra, analysis and probability, Proc. Intern. Sympos. of the Institute of Technology, Holt-Rinehart and Winston, New York 1969, 274–278.

[68] Shelah, S., Existence of rigid-like families of abelian p-groups. In: Model theory and algebra, Lecture Notes in Math. 498, Springer, Berlin 1975, 384–402.

[69] —, Classification Theory and the Number of Non-Isomorphic Models. North Holland, Amsterdam 1978.

[70] —, A combinatorial theorem and endomorphism rings of abelian groups II. In: Abelian Groups and Modules, CISM Courses and Lectures 287, Springer, Wien 1984, 37–86.

[71] —, Infinite abelian groups, the Whitehead problem and some constructions. Israel J. Math. 18 (1974), 243–256.

[72] —, Existence of endo-rigid boolean algebras. In: Around Classification Theory of Models, Lecture Notes in Mathem. 1182, Springer, Berlin 1980, 91–119.

[73] Simson, D., Functor categories in which every flat object is projective. Bull. Acad. Polon. Ser. Math. 22 (1974), 375–380.

[74] —, Linear representation of partially ordered sets and vector space categories. Algebra, Logic and Appl. 4, Gordon and Breach, London 1992.

[75] —, Posets of finite prinjective type and a class of orders, J. Pure Appl. Algebra 90 (1993), 73–103.

[76] Thomas, S., Complete existentially closed locally finite groups, Arch. Math. 44 (1985), 97–109.

[77] —, Complete universal locally finite groups of large cardinality. In: Logic Colloquium, North Holland, Amsterdam 1986, 277–301.

[78] Vámos, P., Decomposition problems for modules over valuation domains. J. London Math. Soc. 41 (1990), 10–26.

[79] Warfield, R. B., Nilpotent Groups. Lecture Notes in Math. 513, Springer, Berlin 1976.

[80] Webb, U. H., An elementary proof of Gaschütz's theorem. Arch. Math. 35 (1980), 23–26.

[81] Wilson, J. S., Groups with many characteristically simple subgroups. Math. Proc. Cambridge Philos. Soc. 86 (1979), 193–197.

[82] Zalesskii, A. E., A nilpotent p-group has an outer automorphism. Dokl. Akad. Nauk SSSR 196 (1971), 751–754 = Soviet Math. Dokl. 12 (1971), 227–230.

[83] —, An example of a torsion-free nilpotent group having no outer automorphisms. Mat. Zametki 11 (1972), 21–26 = Math. Notes 11 (1972), 16–19.

[84] Zanardo, P., Kurosch invariants for torsion-free modules over Nagata valuation domains. J. Pure Appl. Algebra 82 (1992), 195–209.

Groups Satisfying the Minimal Condition on Non-Subnormal Subgroups

Silvana Franciosi and Francesco de Giovanni

1991 Mathematics Subject Classification: 20E15

1. Introduction

Let χ be a property pertaining to subgroups. The investigation of groups satisfying the minimal condition on subgroups which do not have the property χ was started by S. N. Chernikov in a series of articles (for a general reference see [3]). In particular, he studied groups satisfying the minimal condition on non-abelian subgroups or the minimal condition on non-normal subgroups, and proved that, if such groups have a series with finite factors, then either they satisfy the minimal condition on subgroups or all their subgroups have the prescribed property. Similar problems were considered by Phillips and Wilson in [13], where, among several other choices for the property χ, they considered groups satisfying the minimal condition on non-serial subgroups or the minimal condition on subgroups which are not locally nilpotent. Moreover, groups satisfying the minimal condition on non-pronormal subgroups have recently been characterized in [6]. In order to avoid Tarski groups and other similar difficulties, Phillips and Wilson worked with the class $\mathfrak{W}$ of groups whose finitely generated subgroups either are nilpotent or have a finite non-nilpotent homomorphic image. Clearly every locally finite group belongs to $\mathfrak{W}$, and it follows from a result of Robinson (see [14]) that $\mathfrak{W}$ contains the class of locally hyper-(abelian or finite) groups, and in particular that of locally soluble groups. Moreover, it is known that all linear groups are $\mathfrak{W}$-groups (see [19]). In the first part of this article we shall prove that, if G is a $\mathfrak{W}$-group satisfying the minimal condition on non-subnormal subgroups, then either G is a Chernikov group or all subgroups of G are subnormal. To prove this theorem we will make large use of the results of Möhres on infinite groups whose subgroups are all subnormal (see [10], [11], [12]).

The above theorem and other results from [13] will be applied in the last section to study groups with finitely many conjugacy classes of subgroups with a prescribed property. In [2] and [16] it was proved that, if k is a positive integer and G is an infinite

Offprint from
Infinite Groups 94, Eds.: de Giovanni/Newell

locally graded group with finitely many conjugacy classes of subgroups which are not subnormal of defect at most k, then G is nilpotent. Here we shall consider groups with finitely many conjugacy classes of non-subnormal subgroups. Also, groups with finitely many conjugacy classes of subgroups which are not locally nilpotent will be studied.

Most of our notation is standard and can be found in [15].

2. The Minimal Condition

The ***Baer radical*** of a group G is the subgroup generated by all abelian subnormal subgroups of G, and a group is said to be a ***Baer group*** if coincides with its Baer radical. It is well-known that a group is a Baer group if and only if its finitely generated subgroups are all subnormal.

Lemma 2.1. *Let G be a soluble Baer p-group which does not satisfy the minimal condition on subgroups. Then G contains an infinite characteristic subgroup of finite exponent.*

Proof. Since G is a soluble group and does not satisfy the minimal condition on subgroups, it contains characteristic subgroups H and K such that $K < H$, K is a Chernikov group and H/K is infinite abelian of exponent p. The finite residual J of K is a radicable abelian normal subgroup of the Baer p-group G, and hence it is contained in $Z(G)$ (see [15] Part 1, Lemma 3.13). In particular, $K/Z(K)$ is finite, so that K' is finite and K contains a finite characteristic subgroup L such that K/L is a radicable abelian group. As L is characteristic in G, it can be assumed without loss of generality that K is radicable abelian, and so is contained in $Z(G)$. Thus H is nilpotent of class ≤ 2, and $H/Z(H)$ has exponent at most p, so that H' has exponent at most p. Moreover, either H' is infinite or the socle S/H' of H/H' is infinite. In both cases S is an infinite characteristic subgroup of G of finite exponent. □

We will first consider Baer groups satisfying the minimal condition on non-subnormal subgroups. In our proofs we will use the important result of Möhres [12] that a group whose subgroups are all subnormal is soluble.

Lemma 2.2. *Let G be a Baer group satisfying the minimal condition on non-subnormal subgroups. Then every subgroup of G is subnormal.*

Proof. Suppose first that G is hyperabelian, and assume by contradiction that G contains non-subnormal subgroups. Let K be a minimal non-subnormal subgroup of G. Then every proper subgroup of K is subnormal in G, and in particular K is a soluble group (see [12]). Thus $H = K'$ is properly contained in K, and so is subnormal in G.

If

$$H = H_0 \lhd H_1 \lhd \cdots \lhd H_n = G$$

is the standard series of H in G, there exists $i \leq n-1$ such that KH_i is not subnormal in KH_{i+1}. Since H is normal in K, the subgroup H_i is normal in KH_{i+1}, and KH_i/H_i is an abelian subgroup of KH_{i+1}/H_i. Let K_0H_i/H_i be a minimal non-subnormal subgroup of KH_{i+1}/H_i contained in KH_i/H_i. Replacing G by KH_{i+1}/H_i and K by K_0H_i/H_i, it can be assumed without loss of generality that K is abelian. Clearly K is not generated by two proper subgroups, so that it is locally cyclic, and so of type p^∞ for some prime p, since G is a Baer group. Let

$$1 = G_0 \leq G_1 \leq \cdots \leq G_\tau = G$$

be an ascending normal series of G with abelian factors. The subgroup K is not ascendant in G (see [15] Part 1, Corollary to Lemma 4.46), and so there exists an ordinal $\alpha < \tau$ such that KG_α is not subnormal in $KG_{\alpha+1}$. Clearly KG_α/G_α is also of type p^∞, so that it is a minimal non-subnormal subgroup of $KG_{\alpha+1}/G_\alpha$. Replacing G by $KG_{\alpha+1}/G_\alpha$ and K by KG_α/G_α, it can also be assumed that $G = KA$, where A is an abelian normal subgroup of G. If $\bar{Z}(G)$ is the hypercentre of G, we have that K is ascendant in $K\bar{Z}(G)$, so that $K\bar{Z}(G)$ is not subnormal in G and $K\bar{Z}(G)/\bar{Z}(G)$ is a minimal non-subnormal subgroup of $G/\bar{Z}(G)$. Thus we may suppose that $Z(G) = 1$, so that in particular $A \cap K = C_A(K) = 1$, and hence $N_G(K) = K$.

Let $\mathcal{L}$ be the set of all subgroups of G which properly contain K and are not subnormal in G. If $\mathcal{L}$ is not empty, let L be a minimal element of $\mathcal{L}$, and put $L = G$ if $\mathcal{L} = \emptyset$. Clearly K is contained in the subgroup T consisting of all elements of finite order of L, so that $T = K(A \cap T)$, and $K < T$ since $N_G(K) = K$. Let $a \neq 1$ be an element of $A \cap T$. Then $[K, a] = K[K, a] \cap A$ is a normal subgroup of $G = KA$. Moreover $[K, a]$ is not contained in K and a does not belong to $K[K, a]$, so that $K < K[K, a] < T \leq L$ and $K[K, a]$ is a subnormal subgroup of G. The normal closure of $K[K, a]/[K, a]$ in $T/[K, a]$ is a periodic radicable abelian group (see [15] Part 1, Lemma 4.46), so that it is contained in the centre of $T/[K, a]$ (see [15] Part 1, Lemma 3.13). It follows that the group $T/[K, a]$ is abelian, so that T' is contained in $[K, a]$, and in particular a does not belong to T'. Thus $A \cap T' = (A \cap T) \cap T' = 1$, and $T' = 1$ since G/A is abelian. In particular K is a normal subgroup of T. This contradiction proves the lemma when G is hyperabelian. It follows that all hyperabelian subgroups of G are soluble (see [12]), so that also G is soluble (see [4], Lemma 4), and every subgroup of G is subnormal. □

Theorem 2.3. *Let G be a $\mathfrak{W}$-group satisfying the minimal condition on non-subnormal subgroups. Then either G is a Chernikov group or every subgroup of G is subnormal.*

Proof. Suppose that G is not a Chernikov group. Since G satisfies the minimal condition on non-serial subgroups, it follows from Theorem C(ii) of [13] that G is locally nilpotent. Suppose that G contains non-subnormal subgroups. Then G is not a Baer group by Lemma 2.2, and so its Baer radical B is a proper subgroup. Assume that the set $G \setminus B$ contains elements of infinite order, and let $\langle g \rangle$ be an infinite cyclic minimal non-subnormal subgroup of G. If p and q are distinct prime numbers, the subgroups $\langle g^p \rangle$ and $\langle g^q \rangle$ are both subnormal in G, so that also $\langle g \rangle = \langle g^p, g^q \rangle$ is subnormal. This contradiction shows that B contains all elements of infinite order of G. In particular, G is not generated by its elements of infinite order, and so is periodic, since it is locally nilpotent. Suppose first that G is a p-group for some prime p. As G does not satisfy the minimal condition on abelian subgroups (see [18]), it contains an infinite abelian subgroup A of exponent p; then A has a subgroup of finite index A_0 which is subnormal in G, and hence B does not satisfy the minimal condition on subgroups. Every subgroup of B is subnormal by Lemma 2.2, so that B is soluble (see [12]), and by Lemma 2.1 it contains an infinite characteristic subgroup N of finite exponent. Moreover the subgroup N is nilpotent (see [11]). Let x be an element of G such that $\langle x \rangle$ is not subnormal. The subgroup $K = \langle x, N \rangle$ does not satisfy the minimal condition on normal subgroups (see [15] Part 1, Corollary 2 to Theorem 5.27), and N has finite index in K, so that N has an infinite strictly descending chain of K-invariant subgroups

$$N_1 > N_2 > \cdots > N_n > \cdots$$

Assume that there exists a positive integer m such that

$$\langle x, N_m \rangle = \langle x, N_{m+1} \rangle = \cdots$$

Then $N_m = N_n(\langle x \rangle \cap N_m)$, and so $|N_m : N_n| \leq |\langle x \rangle|$ for every $n \geq m$. This contradiction shows that the chain

$$\langle x, N_1 \rangle \geq \langle x, N_2 \rangle \geq \cdots \geq \langle x, N_n \rangle \geq \cdots$$

is infinite, and hence there exists a positive integer k such that $\langle x, N_k \rangle$ is a subnormal subgroup of G. On the other hand $\langle x, N_k \rangle$ is nilpotent (see [15] Part 2, Lemma 6.34), and so $\langle x \rangle$ is subnormal in G. This contradiction shows that G is not a p-group. Let q be a prime such that the Sylow q-subgroup G_q of G is not a Baer group, and let y be an element of G_q such that $\langle y \rangle$ is not subnormal. It follows from the above argument that G_q is a Chernikov group, so that the Sylow q'-subgroup $G_{q'}$ of G has an infinite strictly descending chain of subgroups

$$L_1 > L_2 > \cdots > L_n > \cdots$$

Then

$$\langle y, L_1 \rangle > \langle y, L_2 \rangle > \cdots > \langle y, L_n \rangle > \cdots$$

is an infinite strictly descending chain of non-subnormal subgroups of G. This last contradiction completes the proof. □

Our next result shows that in the statement of Theorem 2.3 the hypothesis that G is a $\mathfrak{W}$-group can be omitted if G is not periodic.

Theorem 2.4. *Let G be a non-periodic group satisfying the minimal condition on non-subnormal subgroups. Then every subgroup of G is subnormal.*

Proof. Assume by contradiction that G contains non-subnormal subgroups, so that G properly contains its Baer radical B by Lemma 2.2. As in the proof of Theorem 2.3 it can be proved that all elements of infinite order of G belong to B, so that G is not generated by its elements of infinite order. Let T be the largest periodic normal subgroup of G. Clearly the factor group G/T is also a counterexample, and hence without loss of generality it can be assumed that G has no non-trivial periodic normal subgroups. By Lemma 2.2 every subgroup of the torsion-free group B is subnormal, and so B is hypercentral (see [10]). Let a be a non-trivial element of $Z(B)$, and let x be an element of $G \setminus B$. Then x has finite order, and so the normal closure $A = \langle a\rangle^{\langle x,B\rangle}$ is a finitely generated torsion-free abelian group. Let p be a prime which does not divide the order of x. Since $B \cap \langle x\rangle = 1$, the chain of subgroups

$$\langle x, A\rangle > \langle x, A^p\rangle > \langle x, A^{p^2}\rangle > \cdots$$

is strictly descending, and hence there exists a positive integer m such that $\langle x, A^{p^n}\rangle$ is subnormal in G for every $n \geq m$. Then A/A^{p^n} and $\langle x, A^{p^n}\rangle/A^{p^n}$ are subnormal subgroups with coprime orders of the finite group $\langle x, A\rangle/A^{p^n}$, and so $[A, x] \leq A^{p^n}$ for every $n \geq m$. Thus

$$[A, x] \leq \bigcap_{n\geq m} A^{p^n} = 1,$$

and in particular $ax = xa$. It follows that ax has infinite order, and so belongs to B, so that also x is in B. This contradiction completes the proof. □

Corollary 2.5. *Let G be an infinite $\mathfrak{W}$-group whose infinite subgroups are subnormal. Then G is soluble.*

Proof. Clearly G satisfies the minimal condition on non-subnormal subgroups, and it follows from Theorem 2.3 that either G is a Chernikov group or every subgroup of G is subnormal. In the latter case G is soluble by the result of Möhres [12]. Suppose that G is a Chernikov group, and let J be the finite residual of J. Then every subgroup of the finite group G/J is subnormal, so that G/J is nilpotent and G is soluble. □

3. Conjugacy Classes of Subgroups

Groups satisfying the minimal condition on subgroups with a given property χ are naturally involved in the investigations concerning groups with a finite number of conjugacy classes of χ-subgroups. To see this, the following result is needed.

Lemma 3.1 ([1], Lemma 4.6.3). *Let G be a group locally satisfying the maximal condition on subgroups. If H is a subgroup of G such that $H^x \leq H$ for some element x of G, then $H^x = H$.*

In our next lemma χ will denote a property pertaining to subgroups, such that, if H is a χ-subgroup of a group G and α is an automorphism of G, then also H^α is a χ-subgroup of G.

Lemma 3.2. *Let G be a group locally satisfying the maximal condition on subgroups. If G has finitely many conjugacy classes of χ-subgroups, then G satisfies the minimal condition on χ-subgroups.*

Proof. By Lemma 3.1 the elements of any chain of subgroups of G are pairwise non-conjugate. Thus every chain of χ-subgroups of G is finite, and in particular G satisfies the minimal condition on χ-subgroups. □

A group G is called ***locally graded*** if every finitely generated non-trivial subgroup of G has a proper subgroup of finite index. It is obvious that all $\mathfrak{W}$-groups are locally graded. Moreover, every free group is locally graded, and hence homomorphic images of locally graded groups need not be locally graded. However, if N is a soluble normal subgroup of a locally graded group G, it is easy to show that also the factor group G/N is locally graded (see [9]). It is now possible to characterize locally graded groups having a finite number of conjugacy classes of non-subnormal subgroups.

Theorem 3.3. *Let G be an infinite locally graded group with finitely many conjugacy classes of non-subnormal subgroups. Then every subgroup of G is subnormal. In particular, G is soluble.*

Proof. Let B be the Baer radical of G. The factor group G/B has finitely many conjugacy classes of cyclic subgroups, so that in particular it contains only finitely many normal subgroups. Let

$$1 = V_0 \leq V_1 = B \leq V_2 \leq \cdots \leq V_t = V$$

be the upper Baer series of G. Then G/V contains no non-trivial abelian subnormal subgroups, and hence it has only finitely many conjugacy classes of abelian subgroups, so that G/V has finite exponent (see [16], Lemma 1). Assume that G/B is not periodic, and let $i < t$ be the largest positive integer such that V_{i+1}/V_i contains an element of infinite order xV_i. Clearly the subgroup $\langle x^n \rangle$ is not subnormal in G for each positive integer n, and so there exists a subgroup H of $\langle x \rangle$ and an element g of G such that H properly contains H^g. Since G/V_{i+1} is periodic, the group $\langle g, V_{i+1} \rangle / V_i$ locally satisfies the maximal condition on subgroups, so that $HV_i = H^g V_i$ by Lemma 3.1 and hence $H = H^g$. This contradiction shows that G/B is periodic. In particular for every element g of G the group $\langle V, g \rangle$ is (locally nilpotent)-by-(locally finite), and so locally satisfies the maximal condition on subgroups. It follows from Lemma 3.1 that every chain of non-subnormal subgroups of V is finite, and hence Theorem 2.3 yields that either V is a Chernikov group or every subgroup of V is subnormal. In both cases V is soluble, and G/V is locally graded. Then G/V is locally finite (see [16], Lemma 3), and so also G locally satisfies the maximal condition on subgroups. Then G satisfies the minimal condition on non-subnormal subgroups by Lemma 3.2, and hence either it is a Chernikov group or all its subgroups are subnormal by Theorem 2.3. On the other hand, if G is a Chernikov group, as in the first part of the proof of Theorem 2.4 of [2] it can be shown that every subgroup of G is subnormal. □

It is known that a finitely generated group is finite-by-nilpotent if and only if it has a finite number of self-normalizing subgroups (see [8], Corollary 6.3.4). Our next two results deal with certain (generalized) soluble groups having finitely many conjugacy classes of self-normalizing subgroups.

Theorem 3.4. *Let G be a finitely generated hyper-(abelian or finite) group. Then G has finitely many conjugacy classes of self-normalizing subgroups if and only if it is finite-by-nilpotent.*

Proof. Assume that G has finitely many conjugacy classes of self-normalizing subgroups but is not finite-by-nilpotent. Since finitely generated finite-by-nilpotent groups are finitely presented, the group G contains a normal subgroup M such that G/M is not finite-by-nilpotent, but G/N is finite-by-nilpotent for every normal subgroup N of G properly containing M (see [15] Part 2, Lemma 6.17). Replacing G by G/M, it can be assumed that all proper homomorphic images of G are finite-by-nilpotent. Then G has no non-trivial finite normal subgroups, and so it contains a non-trivial abelian normal subgroup. It follows that G is abelian-by-finite-by-nilpotent, and so also abelian-by-nilpotent-by-finite. Then every maximal subgroup of G has finite index (see [15] Part 2, Theorem 9.57), and hence has only finitely many conjugates, so that G has finitely many maximal subgroups which are not normal. Therefore the intersection of all non-normal maximal subgroups of G has finite index, and so the Frattini factor group $G/\Phi(G)$ is finite-by-abelian. Thus G is finite-by-nilpotent (see [7], Theorem A), and this contradiction proves the theorem. □

Theorem 3.5. *Let G be a soluble residually finite minimax group. Then G has finitely many conjugacy classes of self-normalizing subgroups if and only if it is finite-by-nilpotent.*

Proof. Suppose that G has finitely many conjugacy classes of self-normalizing subgroups. Since every maximal subgroup of a soluble minimax group has finite index, the group G contains only finitely many maximal subgroups which are not normal, and hence G is finite-by-nilpotent (see [5]). □

In the last part of the article we shall consider groups with a finite number of conjugacy classes of subgroups which are not locally nilpotent.

Lemma 3.6. *Let $\mathfrak{X}$ be a subgroup-closed class of groups, and let G be an infinite locally finite group with finitely many conjugacy classes of subgroups which are not locally $\mathfrak{X}$. Then G is locally $\mathfrak{X}$.*

Proof. Assume that the set $\mathcal{L}$ of all finite subgroups of G which are not in $\mathfrak{X}$ is not empty. Since G has finitely many conjugacy classes of such subgroups, $\mathcal{L}$ contains an element L of maximal order. If x is an element of $G \setminus L$, the subgroup $\langle x, L \rangle$ does not belong to $\mathcal{L}$, and hence is an $\mathfrak{X}$-group. Then also L is an $\mathfrak{X}$-subgroup of G. This contradiction proves that G is locally $\mathfrak{X}$. □

Our next lemma is very easy and certainly well-known.

Lemma 3.7. *Let G be a locally soluble group containing finitely many normal subgroups. Then G is finite.*

Proof. Since every chief factor of a locally soluble group is abelian, and the chief series of G are finite, it follows that G is soluble. It is now clear that G is finite. □

Lemma 3.8. *Let G be a hyper-(locally finite or locally soluble) group with finitely many conjugacy classes of subgroups which are not locally nilpotent, and let H be the Hirsch–Plotkin radical of G. Then the factor group G/H is finite.*

Proof. Since every normal subgroup of G properly containing H is not locally nilpotent, it follows from the hypothesis that G/H has finitely many normal subgroups. Let

$$H = G_0 < G_1 < \cdots < G_t = G$$

be a normal series of minimal length whose factors either are locally finite or locally soluble, and put $K = G_{t-1}$. Since G/K either is locally finite or locally soluble, it follows from Lemma 3.6 and Lemma 3.7 that G/K is finite. Then K has finitely many conjugacy classes of subgroups which are not locally nilpotent (see [2], Lemma 2.2) and by induction on t we obtain that K/H is finite, so that also G/H is finite. □

Theorem 3.9. *Let G be an infinite group whose chief factors either are locally finite or locally soluble. If G has finitely many conjugacy classes of subgroups which are not locally nilpotent, then G is locally nilpotent.*

Proof. Let H be the Hirsch–Plotkin radical of G. Clearly G/H has finitely many normal subgroups, and so G is hyper-(locally finite or locally soluble). Then G/H is finite by Lemma 3.8, and G is (locally nilpotent)-by-finite. In particular G locally satisfies the maximal condition on subgroups, and hence it satisfies the minimal condition on non-(locally nilpotent) subgroups by Lemma 3.2. It follows that either G is locally nilpotent or a Chernikov group (see [13], Theorem B(vi)), so that G is locally nilpotent by Lemma 3.6. □

We finally note that Theorem 3.9 has the following consequence.

Corollary 3.10 (H. Smith [17]). *Let c be a positive integer, and let G be an infinite group whose chief factors either are locally finite or locally soluble. If G has finitely many conjugacy classes of subgroups which are not nilpotent of class $\leq c$, then G is nilpotent of class at most c.*

Proof. The group G is locally nilpotent by Theorem 3.9, so that it locally satisfies the maximal condition on subgroups. It follows from Lemma 3.2 that G satisfies the minimal condition on subgroups which are not nilpotent of class $\leq c$. Application of Theorem B(ix) of [13] to the class $\mathfrak{N}_c$ of nilpotent groups of class $\leq c$ yields that G either is nilpotent of class at most c or is a Chernikov group. In the latter case, we also obtain that G belongs to $\mathfrak{N}_c$ by Lemma 3.6. □

References

[1] Amberg, B., Franciosi, S., de Giovanni, F., Products of Groups. Oxford Mathematical Monographs, Clarendon Press, Oxford 1992.

[2] Brandl, R., Franciosi, S., de Giovanni, F., Groups with finitely many conjugacy classes of non-normal subgroups. Proc. Roy. Irish Acad. To appear.

[3] Chernikov, S. N., Investigation of groups with given properties of the subgroups. Ukrain. Math. J. 21 (1969), 160–172.

[4] Franciosi, S., de Giovanni, F., On groups with many subnormal subgroups. Note Mat. 13 (1993), 99–105.

[5] —, —, On maximal subgroups of minimax groups. Atti Accad. Naz. Lincei Rend. Cl. Sci. Fis. Mat. Natur. To appear.

[6] de Giovanni, F., Vincenzi, G., Groups satisfying the minimal condition on non-pronormal subgroups. Boll. Un. Mat. Ital. To appear.

[7] Lennox, J. C., Finite by nilpotent Frattini factors in finitely generated hyper-(abelian by finite) groups. Arch. Math. (Basel) 25 (1974), 463–465.

[8] —, Stonehewer, S. E., Subnormal Subgroups of Groups. Oxford Mathematical Monographs, Clarendon Press, Oxford 1987.

[9] Longobardi, P., Maj, M., Smith, H., A note on locally graded groups. To appear.

[10] Möhres, W., Torsionfreie Gruppen deren Untergruppen alle subnormal sind. Math. Ann. 284 (1989), 245–249.

[11] —, Auflösbare Gruppen mit endlichem Exponenten deren Untergruppen alle subnormal sind II. Rend. Sem. Mat. Univ. Padova 81 (1989), 269–287.

[12] —, Auflösbarkeit von Gruppen deren Untergruppen alle subnormal sind. Arch. Math. (Basel) 54 (1990), 232–235.

[13] Phillips, R. E., Wilson, J. S., On certain minimal conditions for infinite groups. J. Algebra 51 (1978), 41–68.

[14] Robinson, D. J. S., A theorem on finitely generated hyperabelian groups. Invent. Math. 10 (1970), 38–43.

[15] —, Finiteness Conditions and Generalized Soluble Groups. Springer, New York–Berlin–Heidelberg 1972.

[16] Smith, H., Groups with finitely many conjugacy classes of subgroups with large subnormal defect. Glasgow Math. J. To appear.

[17] —, Groups with finitely many conjugacy classes of subgroups of large derived length. To appear.

[18] Shunkov, V. P., On the minimality problem for locally finite groups. Algebra and Logic 9 (1970), 137–151.

[19] Wehrfritz, B. A. F., Frattini subgroups in finitely generated linear groups. J. London Math. Soc. 43 (1968), 619–622.

Metabelian Groups with All Cyclic Subgroups Subnormal of Bounded Defect

David J. Garrison and Luise-Charlotte Kappe

Abstract. In this paper we investigate metabelian groups with all cyclic subgroups subnormal of bounded defect. Specifically, we give bounds in terms of the defect for the nilpotency class of such groups whenever they exist, resulting in estimates for the nilpotency class of metabelian groups with all subgroups subnormal of bounded defect.

1991 Mathematics Subject Classification: 20E15

1. Introduction

A subgroup H of a group G is called n-subnormal, or subnormal of defect n, denoted by $H \lhd_n G$, if there exists a normal series of subgroups of length $n+1$, starting at G and ending at H. In a group of nilpotency class n, all subgroups are subnormal of defect at most n. Conversely, Roseblade in [9] has shown that a group G with all subgroups n-subnormal is nilpotent, where the nilpotency class of G is a function of n, say $\mu(n)$. However, $\mu(n)$ is not explicitly given in [9]. Recent direct calculations by F. Menegazzo using Roseblade's method yield $\mu(n) \leq 2^{2^{2^{2^n}}}$ in the general case, and the exact values are only known in the cases $n = 1$ and 2. In the case of metabelian groups, Menegazzo's estimate is $\mu(n) \leq n^{n^2}$, or $\leq 2^{n-1}$ if $n \geq 3$ and only "large primes" are involved in the element orders (oral communication). In this paper we will show that for metabelian groups, $\mu(n)$ essentially grows linearly with n (Theorem 3.1 and Theorem 4.1).

Groups in which all subgroups are 1-subnormal, in other words normal, are exactly the Dedekind groups (see e.g. Theorem 6.1.1 in [6]). In the case $n = 1$ all subgroups being n-subnormal is equivalent to all cyclic subgroups being n-subnormal. This is no longer the case if $n \geq 2$. Heineken obtained the following results on groups with all cyclic subgroups 2-subnormal:

Theorem 1.1 [2; Theorem 1]. *Let G be a non-torsion group. Then G has all cyclic subgroups 2-subnormal if and only if G is 2-Engel.*

Offprint from
Infinite Groups 94, Eds.: de Giovanni/Newell

Theorem 1.2 [2; Theorem 2]. *Let G be a torsion group with all cyclic subgroups 2-subnormal, then $G/Z(G)$ is 2-Engel and $G_5 = (G_4)^3 = 1$.*

Mahdavianary in [7] showed that the prime 3 does not play an exceptional role as Theorem 1.2 suggests. We have instead:

Theorem 1.3 [7; Theorem 1]. *Let G be a torsion group with all cyclic subgroups 2-subnormal, then G is nilpotent of class 3.*

Theorem 1.1 together with Theorem 1.3 yield $\mu(2) \leq 3$, and it is easy to see that this bound is sharp. A bound for the nilpotency class of groups with all cyclic subgroups n-subnormal gives an estimate for $\mu(n)$. But we observe that such groups do not have to be nilpotent under certain conditions. A group with all cyclic subgroups n-subnormal is an $(n+1)$-Engel group, and a metabelian n-Engel group has all cyclic subgroups n-subnormal (Lemma 2.6). By a result of Gruenberg [1], for every prime $p < n$, there exists a metabelian p-group of arbitrary large nilpotency class which is n-Engel. From this discussion it becomes evident that to succeed in our investigations we need detailed knowledge about Engel-groups, which for general n is only available in the class of metabelian groups (see [4]).

As for $n = 2$, the case for metabelian groups for general n splits into two subcases: non-torsion and torsion. In the absence of elements of small order, the non-torsion case for metabelian groups and $n \geq 3$ (Theorem 3.1) is an analogue of Theorem 1.1. Our result for metabelian groups and $n \geq 3$ in the torsion case (Theorem 4.1) does not resemble Theorem 1.3, the corresponding result for $n = 2$, rather it has similarities with Theorem 1.2, Heineken's earlier result. This raises the question on whether groups with elements of prime order $p \leq n$ play an exceptional role in this context. In the case $n = 3$ (Theorem 5.1) we have answered the question completely. What came as a surprise in view of Theorem 1.3, is that the primes 2 and 3 indeed play an exceptional role. Example 5.2 and 5.3 are examples of a 2-group and a 3-group, respectively, for which $G/Z(G)$ is not 3-Engel, however all cyclic subgroups are 3-subnormal. It should be mentioned here that the two groups have been constructed with the aid of GAP (see [10]). It is still an open question on whether for $n \geq 4$, primes $p \leq n$ play indeed an exceptional role in Theorem 4.1. But judging from the outcome for $n = 3$, we conjecture that the exclusion of such primes is not only sufficient but also necessary.

In conclusion, we want to mention that there are indications to the effect that the estimates for $\mu(n)$ in the metabelian case resulting from our discussions here are sharp bounds. This issue will be dealt with in a forthcoming paper by the first author.

2. Some Preparatory Steps

For the convenience of the reader, this section contains several formulas and preparatory results to be applied in the rest of the paper. We make use of the following standard notation for commutators: $x^{-1}y^{-1}xy = [x, y] = [x, {}_1y]$, and recursively, $[x, {}_ny] = [[x, {}_{n-1}y], y]$ for $x, y \in G$ and $n \in \mathbb{N}$. In addition, we observe that for a metabelian group G we always have $[a, b, c] = [a, c, b]$, if $a \in G'$ and $b, c \in G$.

Our first lemma is an expansion formula for Engel commutators in metabelian groups and is a more explicit version of Lemma 1 in [3]. In this lemma and the following two we will use additive notation for elements in the commutator subgroup.

Lemma 2.1. *Let G be a metabelian group, $m \in \mathbb{N}$, $u \in G'$, $v, w \in G$. Then*

$$[u, {}_mvw] = \sum_{i=0}^{m}\binom{m}{i}\sum_{j=0}^{i}\binom{i}{j}[u, {}_iv, {}_{m-j}w] = \sum_{i=0}^{m}\sum_{j=0}^{i}\binom{m}{i}\binom{i}{j}[u, {}_iv, {}_{m-j}w].$$

The proof follows by induction on m and the usual commutator expansion, and thus is omitted here. The next lemma is an immediate consequence of Lemma 2.1 and can be found as Lemma 2 in [3] together with its proof.

Lemma 2.2. *Let G be a metabelian group and g an m-right-Engel element in G. Then for each $k \geq m$ there exists an integer δ_k, divisible only by primes $p \leq m-1$, such that for all $h \in G$ we have $\delta_k[g, {}_ih, {}_jg] = 0$, whenever $i + j \geq k$.*

Lemma 2.3. *Let G be a metabelian group with $[y, {}_sx, {}_{t+1}y] = 0$. Then:*

(i) $[y, {}_sx, y^{\alpha_1}, \ldots, y^{\alpha_{t+1}}] = 0$ *for all* $\alpha_i \in \mathbb{N}$, $1 \leq i \leq t+1$;

(ii) $[y, {}_sx, y^{\alpha_1}, \ldots, y^{\alpha_t}] = (\prod_{i=1}^{t}\alpha_i)[y, {}_sx, {}_ty]$ *for all* $\alpha_i \in \mathbb{N}$, $1 \leq i \leq t$.

Proof. The proof follows by induction on $\sum_{i=1}^{k}\alpha_i$, $k = t+1$ or t, and by standard commutator expansion. □

The next three results discuss the connection between Engel length and the bound for the subnormality defect of cyclic subgroups.

Lemma 2.4. *If $\langle x\rangle \lhd_n G$, then $[y, {}_nx] \in \langle x\rangle$ for all $y \in G$. Conversely, if G is metabelian and $[y, {}_nx] \in \langle x\rangle$ for all $y \in G$, then $\langle x\rangle \lhd_n G$.*

Proof. From (13.1.3) in [8] we obtain that $\langle x\rangle \lhd_n G$ is equivalent to $[y, x^{\alpha_1}, \ldots, x^{\alpha_n}] \in \langle x\rangle$ for all $\alpha_i \in \mathbb{Z}, i = 1, \ldots, n$. Hence, in particular, $[y, {}_nx] \in \langle x\rangle$. Conversely, assume $[y, {}_nx] \in \langle x\rangle$. Thus $[y, {}_nx]^k \in \langle x\rangle$ for all $k \in \mathbb{Z}$. If G is metabelian we have by Lemma 2.3 that $[y, x^{\alpha_1}, \ldots, x^{\alpha_n}] = [y, {}_nx]^{\alpha}$, where $\alpha = \prod_{i=1}^{n}\alpha_i$, hence $[y, x^{\alpha_1}, \ldots, x^{\alpha_n}] \in \langle x\rangle$ for all $\alpha_i \in \mathbb{Z}$. Thus, by the above, this implies $\langle x\rangle \lhd_n G$. □

The following corollary is now an immediate consequence.

Corollary 2.5. *Let G be a group with all cyclic subgroups n-subnormal. Then G is an $(n+1)$-Engel group, i.e. $[y, {}_{n+1}x] = 1$ for all $x, y \in G$.*

Conversely, in the case of metabelian groups we can give a bound for the subnormality defect of a cyclic subgroup if the Engel length of the group is given.

Lemma 2.6. *Let G be a metabelian n-Engel group. Then $\langle x \rangle \lhd_n G$ for all $x \in G$.*

Proof. By Corollary 1 in [5], we have that for a metabelian group G the conditions G being n-Engel and the normal closure x^G of every element x in G having nilpotency class $n-1$ are equivalent. Hence $\langle x \rangle \lhd_{n-1} x^G$. Since $x^G \lhd G$, it follows $\langle x \rangle \lhd_n G$, the desired result. □

Theorem 3 of [4] gives estimates for the embedding of right n-Engel elements in the upper central series of a metabelian group. In this context we will make frequent use of a special case of this theorem, namely the case that the group in question is an n-Engel group. Theorem 3 of [4] then provides us with a bound for the nilpotency class of the group. For the convenience of the reader we formulate this result as a corollary.

Corollary 2.7. *Let $n \geq 2$ be an integer and G a metabelian n-Engel group. Further, let p be a prime.*

(i) *If $n+1$ is not prime and G contains no elements of order $p \leq n-1$, or if $n+1$ is prime and G contains no elements of order $p \leq n+1$, then G has nilpotency class not exceeding n.*

(ii) *If $n+1$ is prime and G contains no elements of order $p \leq n-1$, then G has nilpotency class not exceeding $n+1$.*

(iii) *If G contains elements of order $p \leq n-1$, no upper bound for the nilpotency class of G exists.*

The following familiar argument is often called the Vandermonde argument, since the determinant of the system in question is a Vandermonde determinant. To facilitate its application, we will state it here as a lemma without proof.

Lemma 2.8. *Let W be a module over the integers without elements of order $\leq m$. Then the system of equations*

$$0 = \sum_{i=1}^{k} j^i w_i, \quad j = 1, \ldots, k,$$

has only the trivial solution $w_i = 0$, $i = 1, \ldots, k$, provided $k \leq m$.

3. Non-Torsion Groups

In this section we first prove an analogue of Theorem 1.1 for general n in the case of metabelian non-torsion groups.

Theorem 3.1. *Let $n \geq 3$ be an integer and G a metabelian non-torsion group without elements of order $\leq n-1$. Then G is n-Engel if and only if all cyclic subgroups are n-subnormal. Furthermore, G is nilpotent of class at most $n+1$ or n, depending on whether $n+1$ is or is not a prime.*

Proof. Suppose first that G is a metabelian n-Engel group. By Lemma 2.6, it follows that $\langle x \rangle \lhd_n G$ for all $x \in G$. We observe that no restrictions on element orders are needed to prove this direction of our claim.

Conversely, assume that G is a metabelian non-torsion group having all cyclic subgroups n-subnormal and without elements of order $\leq n-1$. By Corollary 2.5 we know that G is an $(n+1)$-Engel group. We will prove that G is n-Engel by discussing the following 3 cases: Let $x, y, z \in G$, then:

(i) $[x,_n y] = 1$, if $|y| = \infty$;

(ii) $[z,_n x] = 1$, if $|x| < \infty$ and $|z| < \infty$;

(iii) $[y,_n x] = 1$, if $|y| = \infty$ and $|x| < \infty$.

(i) Let $|y| = \infty$. By Lemma 2.4 it follows that there exists an integer m such that $[x,_n y] = y^m$, hence $[x,_{n-1} y]^{-1} y^{-1} [x,_{n-1} y] = y^{m-1}$. Now $y^m \in G'$, since G is metabelian, hence

$$1 = [x,_{n-1} y, y^m] = [x,_{n-1} y]^{-1} y^{-m} [x,_{n-1} y] y^m = ([x,_{n-1} y]^{-1} y^{-1} [x,_{n-1} y])^m y^m.$$

Thus, by the above, we obtain $1 = (y^{m-1})^m y^m = y^{m^2}$. Since $|y| = \infty$, it follows $m = 0$, and hence $[x,_n y] = 1$.

(ii) We will reduce this case to the next one. Since G is $(n+1)$-Engel, we have that G is locally nilpotent. Hence the elements of finite order form a characteristic subgroup of G. Thus, with $|y| = \infty$ and $|z| < \infty$ we have $|yz| = \infty$. By (iii) we obtain $1 = [yz,_n x]$. Straightforward expansion yields

$$1 = [y,_n x][z,_n x][y,_n x, z].$$

Since $[y,_n x] = 1$ by (iii), we obtain $[z,_n x] = 1$, the desired result.

(iii) In this case we will use additive notation for the elements of G'. By (i) we have $[y,_n y^\alpha x] = 0$ for any non-zero integer α, and $[y,_n y^\alpha x] = [y, x,_{n-1} y^\alpha x]$. By Lemma

2.3 we obtain $[y,_{n-j} x,_{n-1} y^\alpha] = 0$. Thus expansion by Lemma 2.1 yields

$$0 = [y, x,_{n-1} y^\alpha x] = \sum_{i=0}^{n-2} \sum_{j=0}^{i} \binom{n-1}{i} \binom{i}{j} [y,_{n-j} x,_i y^\alpha]. \tag{3.1.1}$$

We shall show now by induction on k, $k = 1, \dots, n-1$, that

$$0 = [y,_{n-j} x,_{(n-1)-k+i} y] \text{ for } 0 \le i \le n-2,\ 0 \le j \le i, \tag{3.1.2}$$

provided G has no elements of order $\le n-1$. Let $k = 1$. Commuting (3.1.1) $((n-1)-k)$-times, i.e. $(n-2)$-times, by y and setting $\alpha = 1$, we obtain

$$0 = \sum_{i=0}^{n-2} \sum_{j=0}^{i} \binom{n-1}{i} \binom{i}{j} [y,_{n-j} x,_{n-2+i} y]. \tag{3.1.3}$$

For $i > 0$ we have $n-2+i \ge n-1$. Thus $[y,_{n-j} x,_{n-2+i} y] = 0$ in this case. Hence (3.1.3) reduces to $\binom{n-1}{0}\binom{0}{0}[y,_n x,_{n-2} y] = 0$. Therefore $[y,_n x,_{n-2} y] = 0$ too, and we conclude that (3.1.2) holds for $k = 1$.

Suppose now there exists $k \in \mathbb{N}$, $1 \le k < n-1$ such that

$$0 = [y,_{n-j} x,_{n-1-l+i} y] \text{ for } 0 \le i \le n-1,\ 0 \le j \le i, \text{ and } 1 \le l \le k. \tag{3.1.4}$$

We will show next that our claim is also true for $k+1$. Commuting (3.1.1) $((n-1)-(k+1))$-times by y gives:

$$0 = \sum_{i=0}^{n-2} \sum_{j=0}^{i} [y,_{n-j} x,_i y^\alpha,_{n-2-k} y].$$

By our hypothesis and Lemma 2.3 this leads to

$$0 = \sum_{i=0}^{n-2} \binom{n-1}{i} \alpha^i [y,_{n-i} x,_{n-2-k+i} y].$$

If $i > k$ we have $n-2-k+i \ge n-1$. Hence in this case $0 = [y,_{n-i} x,_{n-2-k+i} y]$. Thus the above reduces to

$$0 = \sum_{i=0}^{k} \alpha^i \binom{n-1}{i} [y,_{n-i} x,_{n-2-k+i} y].$$

Setting $w_i = \binom{n-1}{i}[y,_{n-i} x,_{n-2-k+i} y]$, we obtain by Lemma 2.8 that $w_i = 0$ for $i = 0, \dots, k$. Since G contains no elements of order $\le n-1$, it follows $0 = [y,_{n-i} x,_{n-2-k+i} y]$ for $i = 0, \dots, k$. Thus (3.1.4) holds for $k+1$ instead of k, and we conclude that (3.1.2) is true as claimed. Setting $k = n-1$, $i = j = 0$, in (3.1.2) yields $0 = [y,_n x]$, the desired result. This concludes the proof of (iii), and we conclude that G is n-Engel.

Finally, we observe that the class restrictions are immediate consequences of Corollary 2.7. □

We want to mention here that we have calculations indicating that the above theorem remains valid if we only require the absence of elements of order $\leq n-2$. However, as already mentioned in the introduction, in the presence of elements of order $\leq n-1$, the nilpotency class of metabelian n-Engel groups, hence metabelian groups with all cyclic subgroups n-subnormal, is no longer bounded.

Our next result shows that for small values of n, i.e. $n \leq 5$, the equivalence of Theorem 3.1 holds without restrictions on element orders. For $n = 6$ our methods are no longer capable of proving this equivalence.

Theorem 3.2. *Let G be a metabelian non-torsion group and $n \in \mathbb{N}$, $3 \leq n \leq 5$. Then G is n-Engel if and only if G has all cyclic subgroups n-subnormal.*

Proof. We observe that in Theorem 3.1 the restriction on element orders was only used in (iii). Thus it suffices to show that for $n = 3$, 4 or 5, and $|y| = \infty$, $|x| < \infty$, we have $[y,_n x] = 1$ without restrictions on element orders in G. We will reach our goal in a case by case discussion, in which we use again additive notation for the elements of G'.

Setting $n = 3$ in (3.1.1) leads to

$$0 = [y,_3 x] + 2[y,_3 x, y^\alpha] + 2[y,_2 x, y^\alpha]. \tag{3.2.1}$$

Commuting the above by y and setting $\alpha = 1$ leads to $0 = [y,_3 x, y]$ by (i) in Theorem 3.1. Thus, by Lemma 2.3, (3.2.1) reduces to

$$0 = [y,_3 x] + 2\alpha[y,_2 x, y].$$

By setting $\alpha = 1, 2$ in the above and solving the resulting system of equations we obtain $0 = [y,_3 x]$, the desired result.

Now setting $n = 4$ in (3.1.1) yields

$$0 = \sum_{i=0}^{2} \sum_{j=0}^{i} \binom{3}{i}\binom{i}{j}[y,_{4-j} x,_i y^\alpha]. \tag{3.2.2}$$

Setting $\alpha = 1$ in the above and commuting twice by y leads to $0 = [y,_4 x,_2 y]$. Hence, by Lemma 2.3, (3.2.2) reduces to a sum of commutators of weight not exceeding 6. Commuting this reduced version of (3.2.2) by y once and applying Lemma 2.3 yield

$$0 = [y,_4 x, y] + 3\alpha[y,_3 x,_2 y].$$

Setting $\alpha = 1, 2$ in the above and solving the resulting system lead to

$$0 = [y,_4 x, y] = 3[y,_2 x,_2 y].$$

Thus, with the help of Lemma 2.3, we obtain the following reduction of (3.2.2):

$$0 = [y,_4 x] + 3\alpha[y,_3 x, y] + 3\alpha^2[y,_2 x,_2 y].$$

Setting $\alpha = 1, 2, 3$ in the above and solving the resulting system leads to $[y,_4 x] = 0$, the desired result.

Finally, setting $n = 5$ in (3.1.1) yields

$$0 = \sum_{i=0}^{3} \sum_{j=0}^{i} \binom{4}{i} \binom{i}{j} [y,_{5-j} x,_i y^\alpha]. \tag{3.2.3}$$

Commuting (3.2.3) three times by y and setting $\alpha = 1$ leads to

$$0 = [y,_5 x,_3 y]. \tag{3.2.4}$$

Subsequently commuting (3.2.3) twice by y and observing (3.2.4) and Lemma 2.3, we obtain

$$0 = [y,_5 x,_2 y] + 4\alpha[y,_4 x,_3 y].$$

Setting $\alpha = 1, 2$ in the above and solving the resulting system leads to

$$0 = [y,_5 x,_2 y] = 4[y,_4 x,_3 y].$$

Now (3.2.4) together with the above imply that (3.2.3) reduces to a sum of commutators of weight not exceeding 7. Commuting this reduced version of (3.2.3) by y and applying Lemma 2.3 leads to

$$0 = [y,_5 x, y] + 4\alpha[y,_4 x,_2 y] + 6\alpha^2[y,_3 x,_3 y]. \tag{3.2.5}$$

By setting $\alpha = 1, 2, 3$ in (3.2.5) and solving the resulting system, we obtain

$$0 = [y,_5 x, y] = 8[y,_4 x,_2 y] = 12[y,_3 x,_3 y]. \tag{3.2.6}$$

However, (3.2.6) is not sufficient to reduce (3.2.3) to a sum of commutators of weight not exceeding 6, since the coefficients of $[y,_4 x,_2 y]$ and $[y,_3 x,_3 y]$ in (3.2.3) are 12 and 6, respectively. Therefore we commute (3.2.3) by x and set $\alpha = 1$. Thus we obtain

$$0 = 6[y,_4 x,_2 y] + 4[y,_3 x,_3 y].$$

Multiplying the above by 3 leads to $0 = 18[y,_4 x, 2y]$. This together with (3.2.6) yields $0 = 2[y,_4 x,_2 y]$. Now (3.2.5) for $\alpha = 1$ implies $0 = 6[y,_3 x,_3 y]$. Hence, by Lemma 2.3 and the above, (3.2.3) reduces now to

$$0 = \sum_{i=0}^{3} \alpha^i \binom{4}{i} [y,_{5-i} x,_i y]. \tag{3.2.7}$$

Setting $\alpha = 1, 2, 3, 4$ in (3.2.7) and solving the resulting system we obtain $[y,_5 x] = 0$, the desired result. □

4. Torsion Groups

In this section we give a sufficient condition for a metabelian torsion group with all cyclic subgroups n-subnormal to be nilpotent and a bound for its nilpotency class. Our result resembles Theorem 1.2 in the case $n = 2$. In view of Theorem 1.3 this might suggest that the restrictions on element orders are not necessary. However, we have indications to the contrary. In the next section we will show that the restrictions are indeed necessary in case $n = 3$. At present it is not possible to extend the GAP-aided methods to construct examples of groups of the size necessary for larger n. But the examples strongly suggest that the restrictions on element orders in our next theorem are indeed necessary.

Theorem 4.1. *Let $n \geq 3$ be an integer and G a metabelian torsion group with all cyclic subgroups n-subnormal and having no elements of order $\leq n$. Then $G/Z(G)$ is n-Engel and G has nilpotency class not exceeding $n+1$, if $n+1$ is not prime, and $n+2$, if $n+1$ is prime.*

Proof. Since all cyclic subgroups of G are n-subnormal, it follows by Corollary 2.5 that G is an $(n+1)$-Engel group. Thus, as a solvable group, G is locally nilpotent, hence the direct sum of its primary components. Therefore we may assume that G is a p-group, p a prime $> n$.

From now on we will use multiplicative notation for the elements of G'. By Lemma 2.2 we observe that

$$[y, {}_nx, y] = 1 \text{ for all } x, y \in G. \tag{4.1.1}$$

Let x, y, z be arbitrary elements in G, then $[xz, {}_ny, xz] = 1$. By straightforward expansion of this commutator and use of (4.1.1) we obtain

$$[x, {}_ny, z][z, {}_ny, x][x, {}_ny, z, z] = 1. \tag{4.1.2}$$

Commuting (4.1.2) by x and observing (4.1.1) leads to $[z, {}_ny, {}_2x] = 1$. Since x, y, z are arbitrary we have $[x, {}_ny, z, z] = 1$, and thus (4.1.2) becomes

$$1 = [x, {}_ny, z][z, {}_ny, x]. \tag{4.1.3}$$

Since $\langle y \rangle \lhd_n G$, Lemma 2.4 implies that $[x, {}_ny]$ and $[z, {}_ny]$ are in $\langle y \rangle$. Hence there exist nonnegative integers i, j, and integers α, β, relatively prime to p, such that

$$[x, {}_ny] = y^{\alpha p^i} \text{ and } [z, {}_ny] = y^{\beta p^j}.$$

Without loss of generality we may assume $j \geq i$. Thus there exists an integer t such that $[z, {}_ny] = [x, {}_ny]^t$, and we obtain by (4.1.1)

$$[z, {}_ny, x] = [[x, {}_ny]^t, x] = [x, {}_ny, x]^t = 1.$$

This together with (4.1.3) yields $1 = [x, {}_ny, z]$ for all $x, y, z \in G$. Thus $G/Z(G)$ is n-Engel, the desired result.

To obtain the class bounds, we observe that in the case $n+1$ is not a prime, $G/Z(G)$ has class not exceeding n by Corollary 2.7.i, and in case $n+1$ is a prime, $G/Z(G)$ has class not exceeding $n+1$ by Corollary 2.7.ii. Thus we obtain the desired class bounds for G as claimed. □

5. Three-Subnormal Cyclic Subgroups

In this section we will give a complete account on the structure and bound for the nilpotency class of a metabelian group with all cyclic subgroups 3-subnormal. The first two parts of the theorem are a summary on what was shown in Section 3 and 4 in the case $n = 3$. The third part shows that the restriction to torsion groups without elements of order 2 and 3 in the second part are indeed necessary. The two examples of groups constructed with the help of GAP (see [10]) are discussed in detail at the end of this section.

Theorem 5.1. *Let G be a metabelian group.*

(i) *If G is a non-torsion group, then all cyclic subgroups are 3-subnormal if and only if G is 3-Engel. Furthermore, if G has no elements of even order, then the class of G does not exceed 3.*

(ii) *If G is a torsion group with all cyclic subgroups 3-subnormal and without elements of order 2 or 3, then $G/Z(G)$ is 3-Engel and the class of G does not exceed 4.*

(iii) *There exist metabelian 2-groups and 3-groups having all cyclic subgroups 3-subnormal and $G/Z(G)$ is not 3-Engel.*

Proof. The cases (i) and (ii) are just a rephrasing of Theorem 3.2 and 4.1, respectively, in the case $n = 3$. For (iii) we refer to the groups of Example 5.2 and 5.3 which provide a 2-group and a 3-group with the properties claimed in (iii). □

Gruenberg in [1] has shown that there exist metabelian 3-Engel 2-groups of arbitrary large nilpotency class. By Lemma 2.6 all their cyclic subgroups are 3-subnormal. Hence there is no class restriction on metabelian groups with all cyclic subgroups 3-subnormal having elements of even order, regardless of whether they are torsion or non-torsion groups.

By [1], there also exist metabelian 4-Engel 3-groups of arbitrary large nilpotency class. On the other hand, any group with all cyclic subgroups 3-subnormal is 4-Engel by Corollary 2.5. It is an open question whether the 3-subnormality condition on cyclic subgroups imposes a class restriction on metabelian 3-groups. If such a bound exists it must be at least 5, as can be seen from Example 5.3.

We conclude this section with the two examples of groups which show that the restrictions imposed on the element orders in Theorem 5.1.ii are not only sufficient but also necessary. These groups have been constructed and the claims verified with the help of GAP on a SUN 10/30, and in case of Example 5.3, simultaneously on several of them. After stating the examples together with their relevant properties we will give a detailed description, so that the interested reader in sufficient command of GAP and with comparable computing facilities can reproduce these results.

Example 5.2. Let $K = G/M^G$, where

$$\begin{aligned} G =\langle x, y\ ;\ & x^{64} = y^{64} = [x, y]^{32} = [x, y, x]^{16} = [x, y, y]^{16} = [y, {}_3x]^8 = \\ &= [x, {}_3y]^8 = 1,\ [x, {}_2y, x]^8 = [x, {}_3y, x]^2 = [y, {}_3x, y]^2 = [x, {}_4y] = \\ &= [y, {}_4x] = 1,\ G'' = 1\rangle, \end{aligned}$$

and M^G is the normal closure of

$$\begin{aligned} M =\langle & [x, {}_3y]y^{-8},\ [y, {}_3x]x^{-8},\ [x, {}_3xy](xy)^{-16},\ x^{32}y^{32}, \\ & [x, y, x]^2[x, {}_2y]^2[x, {}_2y, x][x, {}_3y]\rangle. \end{aligned}$$

Then K has all cyclic subgroups 3-subnormal, hence is 4-Engel, and $K/Z(K)$ is not 3-Engel. Furthermore, K is nilpotent of class 5 and $|K| = 2^{17}$.

Proof. We first describe the construction of the group G, and then give an outline of how this construction can be implemented with GAP. We start from a finite abelian 2-group, corresponding to the commutator subgroup of G, from which we reach G by two cyclic split extensions.

Let $A = \langle u\rangle \times X \times Y \times Z$, where $\langle u\rangle \cong C_{32}$, $X = \langle x_1\rangle \times \langle x_2\rangle \cong C_{16} \times C_{16}$, $Y = \langle y_1\rangle \times \langle y_2\rangle \times \langle y_3\rangle \cong C_8 \times C_8 \times C_8$, $Z = \langle z_1\rangle \times \langle z_2\rangle \cong C_2 \times C_2$.

Next, let $B = [A]\langle x\rangle$, the semidirect product of A with a cyclic group $\langle x\rangle$, where $\langle x\rangle \cong C_{64}$. The automorphism induced by x on A has order 2^k for some $k \leq 6$. The action of x on the generators of A is given as follows:

$$\begin{cases} [u, x] = x_1,\ [x_1, x] = y_1,\ [x_2, x] = y_2, \\ [y_2, x] = z_2,\ [y_3, x] = z_1,\ [y_1, x] = [z_1, x] = [z_2, x] = 1. \end{cases} \tag{5.2.1}$$

The defining relations of B are those of A, (5.2.1) and $x^{64} = 1$.

Now let $G = [B]\langle y\rangle$ with $\langle y\rangle \cong C_{64}$, where y induces an automorphism of order 2^j on B for some $j \leq 6$. The action of y on the generators of B is given as follows:

$$\begin{cases} [x, y] = u,\ [u, y] = x_2,\ [x_1, y] = y_2,\ [x_2, y] = y_3, \\ [y_1, y] = z_2,\ [y_2, y] = z_1,\ [y_3, y] = [z_1, y] = [z_2, y] = 1. \end{cases} \tag{5.2.2}$$

The defining relations of G are those of B, (5.2.2) and $y^{64} = 1$.

The implementation of the construction of G can in principle be given manually in the familiar and cumbersome way. We have used here the method provided by GAP for finite polycyclic groups, which has at its core the GAP-functions ***AgGroupFpGroup*** and ***RefinedAgSeries***, and produces as the end-result a presentation of the finite polycyclic group as an ***ag-group*** with all cyclic factors of prime order. The starting point of the construction is a list of generators and a list of relators. In our case we have 10 generators, namely the 8 generators of A together with x and y. The list of relations for G in these generators consists of the relations of G as given above. We obtain an *ag-group* presentation for G as a polycyclic group with a refined cyclic series, i.e. every step has order 2. It can be easily verified that the resulting group G has order 2^{36} and has nilpotency class 5. Identifying the generators of A with appropriate commutators in x and y leads to the presentation of G given in this example.

We will proceed now to a suitable factor group of G for which the desired properties can be verified. Among the list of generators of G in its final presentation as an *ag-group*, there are exactly 2 of order 64, call them x and y, and the subgroup generated by them is all of G. We construct now the subgroup M, as $M = \langle [x, {}_3y]y^{-8}, [y, {}_3x]x^{-8}, [x, {}_3xy](xy)^{-16}, x^{32}y^{32}, [x, y, x]^2[x, {}_2y]^2[x, {}_2y, x][x, {}_3y]\rangle$ and M^G, its normal closure. The factor group $K = G/M^G$ is presented as an *ag-group* with a refined series. The group K has order 2^{17} and has nilpotency class 5. Among the generators of K there are again exactly 2 generators, call them a and b, having order 64, and $\langle a, b\rangle = K$. For these elements we find $[a, {}_3b, a] \neq 1$, hence $K/Z(K)$ is not 3-Engel.

The final step, verifying that each cyclic subgroup of K is 3-subnormal, was done in a straightforward manner by using the GAP-function ***SubnormalSeries*** to check that $\langle g\rangle \lhd_j K$ for some $1 \leq j \leq 3$ for all $g \in K$. The running time for this program on the SUN 10/30 was approximately 10 hours. □

Example 5.3. Let $H = F/L^F$, where

$$\begin{aligned} F = \langle x, y\ ;\ & x^{729} = y^{81} = [x, y]^{81} = [y, {}_2x]^{81} = [y, {}_2x, y]^{81} = [y, {}_3x]^{27} = \\ & = [x, {}_3y]^{27} = 1,\ [x, {}_2y]^{81} = [y, {}_2x, {}_2y]^{27} = [y, {}_3x, y]^3 = [x, {}_4y] = \\ & = [y, {}_4x] = 1,\ F'' = 1\rangle, \end{aligned}$$

and let L^F be the normal closure of

$$\begin{aligned} L = \langle & [x, {}_3y],\ [y, {}_3x]x^{27},\ [x, {}_3xy](xy)^{8\cdot 27}, [y, {}_3xy](xy)^{25\cdot 27}, [x, {}_3xy^{-1}](xy^{-1})^{25\cdot 27}, \\ & [y, {}_3xy^{-1}](xy^{-1})^{25\cdot 27}, [x, y, x]^3[x, {}_2y, x],\ [x, y]^9[x, y, {}_2x]\rangle. \end{aligned}$$

Then $H = \langle a, b\rangle$ has all cyclic subgroups 3-subnormal, hence is 4-Engel, $H/Z(H)$ is not 3-Engel. Furthermore, H is nilpotent of class 5, and $|H| = 3^{14}$.

Proof. The construction of F together with its quotient H are similar to the one described in Example 5.2. Thus we omit the details.

It can be easily verified that the resulting group H has order 3^{14} and nilpotency class 5. Among the generators of H there exist 2 elements, a and b, of order 243 and 81, respectively, such that $H = \langle a, b\rangle$. Since $[a, {}_2b, {}_2a] \neq 1$, we have $H/Z(H)$ is not 3-Engel.

The verification that each cyclic subgroup of H is 3-subnormal was done in a similar manner as in Example 5.2. To facilitate these computations, we make use of the fact that every $g \in H$ can be represented as $g = a^i b^j c$, where $c \in H'$ and $0 \le i, j \le 26$, since $a^{27}, b^{27} \in H'$. This was done simultaneously on several SUN's of equal or lesser speed, translating into a total running time of approximately 200 hours for the single machine used in Example 5.2. □

References

[1] Gruenberg, K. W., The Engel Elements of a Soluble Group, Illinois J. Math. 3 (1959), 151–168.

[2] Heineken, H., A Class of Three-Engel Groups. J. Algebra 17 (1971), 341–345.

[3] Kappe, L. C., Right and Left Engel Elements in Metabelian Groups. Comm. Algebra 9 (1981), 1295–1306.

[4] —, Engel Margins in Metabelian Groups. Comm. Algebra 11 (1983), 1965–1987.

[5] —, Morse, R. F., Levi-Properties in Metabelian Groups. Contemp. Math. 109 (1990), 59–72.

[6] Lennox, J. C., Stonehewer, S. E., Subnormal Subgroups of Groups. Clarendon Press, Oxford 1987.

[7] Mahdavianary, S. K., A special class of three-Engel groups. Arch. Math. (Basel) 40 (1983), 193–199.

[8] Robinson, D. J. S., A Course in the Theory of Groups. Springer, Berlin–New York–Heidelberg 1982.

[9] Roseblade, J. E., On groups in which every subgroup is subnormal. J. Algebra 2 (1965), 402–412.

[10] Schönert, M., et. al., GAP-Groups, Algorithms, and Programming (3rd edn). Lehrstuhl D für Mathematik, RWTH, Aachen 1993.

Locally Finite Groups Whose Sylow Subgroups Are Chernikov

Brian Hartley

1991 Mathematics Subject Classification: 20F50

1. Introduction

Let G be a group, let $\mathbb{P}$ denote the set of all primes, and for each prime p, let $\mathrm{Syl}_p\,(G)$ be the set of all Sylow (that is, maximal) p-subgroups of G. Of course, if G is finite and S_p is any member of $\mathrm{Syl}_p\,(G)$, one for each prime p (we speak of this as a ***complete set of Sylow subgroups***), then

$$G = \langle S_p : p \in \mathbb{P} \rangle. \tag{1.1}$$

This is proved by noting that the two sides have the same order. In this situation, $S_p = 1$ for all but a finite number of primes p.

It seems reasonable to ask to what extent (1.1) remains true if we weaken the condition that G be finite, perhaps replacing it by some other condition guaranteeing good behaviour of Sylow subgroups. This question was at least implicitly raised in a paper of Baer [1] in 1970. There, he considered periodic locally soluble groups in which all Sylow subgroups are finite. If P and Q are Sylow p-subgroups of such a group G, they are clearly conjugate in the finite group $\langle P, Q \rangle$, and so we have a good analogue of Sylow's Theorem in these groups. Nevertheless, it turned out that in the infinite wreath product

$$W = \cdots \wr (C_{p_3} \wr (C_{p_2} \wr C_{p_1})), \tag{1.2}$$

where $p_1, p_2, \ldots$ is an infinite sequence of distinct primes, it is possible to choose a complete set of Sylow subgroups that generate a proper subgroup. More remarkably, it is even possible to embed W in an uncountable periodic locally soluble group $\tilde{W}$, in such a way that each Sylow subgroup of W is also a Sylow subgroup of $\tilde{W}$. Thus, the countable group W is generated by a complete set of (finite) Sylow subgroups of the uncountable group $\tilde{W}$. These examples can be found in Baer's paper *loc. cit.*, and we will outline the constructions shortly.

Offprint from
Infinite Groups 94, Eds.: de Giovanni/Newell

Now let G be a countable periodic locally soluble group with finite Sylow subgroups. Baer asked, though in somewhat different terms, under what circumstances can we be sure that every complete set of Sylow subgroups of G generates the whole group G. Perhaps it may be worth considering that question in more general locally finite groups. However, Baer suggested that for the groups we are considering, every complete set of Sylow subgroups generates G if and only if G is hyperfinite. This was proved in 1988 by Belyaev, in the more general class of periodic locally soluble groups whose Sylow subgroups are Chernikov groups [5]. It was also proved at about the same time, and by quite different methods, by Bell. Bell's work does not assume local solubility and involves some interesting methods from the character theory of finite groups. Further, it sheds light on other aspects of the subgroup structure of locally finite groups in which every Sylow subgroup is a Chernikov group, and on uncountable groups of this type. The main aim of this article is to draw attention to Bell's work, which has so far only appeared in preprint form [2]. The reader requiring more details than we shall give, is referred to this preprint.

Now we give the details of Baer's construction and enlarge a little more on Baer's paper. First we note the following elementary lemma.

Lemma 1.1. *Let A and H be non-trivial finite groups, and let K be a subgroup of H. Let $L = A \wr H$, where the wreath product is formed with respect to the regular permutation representation of H. Thus, L is the semidirect product $L = \overline{A}H$, where $\overline{A}$ is the base group. Then there exists an element $a \in \overline{A}$ such that $H^a \cap H = K$.*

To see this, we choose a non-identity element $b \in A$, and thinking of the base group as consisting of functions on H with values in A, we let a be the function whose value is b on elements of K, and 1 elsewhere. Now let $p_1, p_2, \dots$ be an infinite sequence of distinct primes. Let A_i be a cyclic group of order p_i. Put $B_1 = A_1$, and let $B_i = A_i \wr B_{i-1}$ for $i > 1$. These wreath products are formed with respect to the regular representation of the top group. Considering B_i as a subgroup of B_{i+1} in the obvious way, we have a tower

$$B_1 < B_2 < \cdots \tag{1.3}$$

whose union W is a countable periodic locally soluble group with finite Sylow subgroups. Applying Lemma 1.1 to $B_3 = A_3 \wr B_2$, we find that this group has a subgroup B_2^*, conjugate to B_2, such that $B_2^* \cap B_2 = B_1$. Clearly, B_2^* contains Sylow subgroups of B_3 corresponding to the primes p_1 and p_2. If B_n^* is a conjugate of B_n in B_{n+1} such that $B_n^* \cap B_2 = B_1$, we can extend it to a conjugate B_{n+1}^* of B_{n+1} in B_{n+2} such that $B_{n+1}^* \cap B_{n+1} = B_n^*$. Hence $B_{n+1}^* \cap B_2 = B_1$. It follows from this equation that if we put $W^* = \cup_{i=2}^{\infty} B_i^*$, then $W^* \cap B_2 = B_1$. In particular, W^* is a proper subgroup of W. It is clear that W^* contains a complete set of Sylow subgroups of W.

Now it follows from Baer's work (see Lemma 2.3) that if G is a countable periodic locally soluble group with finite Sylow subgroups, and if H is a subgroup of G generated by a complete set of Sylow subgroups, then $G \simeq H$. In particular, $W^* \simeq W$.

This means that we can build a tower of groups

$$W = W_1 < W_2 < \cdots,$$

indexed by the natural numbers, each isomorphic to W. Let $W_\omega = \cup W_i$. Then each finite p -subgroup of W_ω is contained in some W_i, and so is isomorphic to a subgroup of W. Therefore the finite p-subgroups of W_ω have order bounded by those of W, and it follows easily that each Sylow subgroup of W is a Sylow subgroup of W_ω. By the result of Baer just stated, it follows that $W_\omega \simeq W$, and so W_ω is a proper subgroup of a group $W_{\omega+1} \simeq W$. The construction can be continued through the countable ordinals to produce a strictly increasing tower of groups, each isomorphic to W, indexed by the countable ordinals. The union of this tower is a group $\tilde{W}$ of cardinal $\aleph_1$, the first uncountable ordinal, of which W contains a complete set of Sylow subgroups.

This construction is taken from Baer [1]

Hickin has shown [11] that there are $2^{\aleph_0}$ variations on $\tilde{W}$, each of cardinal $2^{\aleph_0}$, such that no uncountable subgroup of one is isomorphic to a subgroup of another. Bell has extended Hickin's result to all countable locally finite non-hyperfinite groups whose Sylow subgroups are Chernikov, and has also shown that from any such group, $2^{2^{\aleph_0}}$ uncountable variations can be constructed. For more precise statements of these results, see Section 5.

2. Basis Subgroups

Let $\mathcal{K}$ denote the class of all locally finite groups G such that, for each prime p, every p-subgroup of G is Chernikov. In the Russian literature, these groups are often called SF-groups. Of course, a locally finite group G belongs to this class, if and only if G satisfies the minimal condition on p-subgroups for each prime p.

First we note the following result.

Theorem 2.1. *Every $\mathcal{K}$-group is almost locally soluble.*

As usual, a group is said almost to have a certain property, if it has a subgroup of finite index with the property.

This result was first proved by Belyaev [4] using a number of deep results from finite group theory. In particular, a result of Glauberman [8] on soluble signalizer functors plays an important part. It can also be easily deduced from the classification of finite simple groups. By [12, 3.17], it follows from the fact that every simple $\mathcal{K}$-group is finite, and that follows from [12, 4.8, 1.L.5].

The following much stronger structural result for $\mathcal{K}$-groups then follows [12, 3.18, 3.17].

Theorem 2.2. *Let $G \in \mathcal{K}$. Then G contains a normal abelian subgroup D such that each Sylow subgroup of D is a divisible Chernikov group, and G/D is a residually* *finite group whose Sylow subgroups are all finite. Further, if π is any finite set of primes, then $G/O_{\pi'}(G)$ is Chernikov.*

It is easy to deduce from this that, for each prime p, the Sylow p-subgroups of a $\mathcal{K}$-group are conjugate. It is interesting to note that this was earlier proved by Shunkov [14] without using the deep structural information provided by Theorem 2.1.

The next result is designed to pave the way for the definition of basis subgroups of $\mathcal{K}$-groups. In it, an endomorphism ϕ of a group G is called ***locally inner***, if given any finite subset X of G, there exists $g \in G$ such that $x^{\phi} = x^{g}$ for all $x \in X$.

Lemma 2.3. *Let G be a countable $\mathcal{K}$-group and H be a subgroup of G. Then the following conditions are equivalent.*

1. *There is a locally inner endomorphism of G onto H.*
2. *$G \simeq H$.*
3. *H contains an isomorphic copy of each finite subgroup of G.*
4. *H contains an isomorphic copy of each finite subgroup of prime power order of G.*
5. *For each prime p, H contains a Sylow p-subgroup of G.*
6. *H is generated by a complete set of Sylow p-subgroups of G.*

Proof. Trivially, (1) $\Rightarrow$ (2) $\Rightarrow$ (3) $\Rightarrow$ (4).

Assuming (4), choose a tower of finite subgroups $H_1 \leq H_2 \leq \cdots$ such that $H = \cup_{i=1}^{\infty} H_i$, and choose a Sylow p-subgroup Q_i of H_i such that $Q_1 \leq Q_2 \leq \cdots$ Let $Q = \cup_i Q_i$. Then Q is a Sylow p-subgroup of H that contains a conjugate of every finite p-subgroup of H. Let P be a Sylow p-subgroup of G containing Q. Then Q contains an isomorphic copy of every finite subgroup of P, and so as P and Q are Chernikov, Q contains an isomorphic copy of P [12, 3.6]. But a Chernikov group cannot contain a proper subgroup isomorphic to itself. Hence $Q = P$. This gives (5).

Now every countable locally finite group is generated by some complete set of Sylow p-subgroups (we can construct such Sylow subgroups as above), and so H is generated by a complete set of Sylow subgroups of itself. If we assume (5) and use the conjugacy of the Sylow subgroups, we find that every Sylow subgroup of H is a Sylow subgroup of G, and obtain (6).

Finally, let us assume (6). Perhaps the most direct way to obtain (1) is to use the approach of Dixon and Tomkinson [7]; in the case when the Sylow subgroups are finite the result can be found in [1]. First recall that if π is a set of primes, then a finite group L is called ***π-separable,*** if L has a series in which each factor is either a π-group or a π'-group. It is well known that every finite π-separable group L satisfies the condition D_{π}, meaning that L has a unique conjugacy class of Hall π-subgroups, and each π-subgroup of L is contained in a Hall π-subgroup of L [9, Chapter 6].

Returning to our group G, let S be a normal locally soluble subgroup of finite index of G (see Theorem 2.1), and choose a finite set π of primes such that G/S is a π-group. Then G is locally π-separable, and using this, it is not hard to show that in

G, we can find a Hall π-subgroup G_π and Sylow p-subgroups G_p for $p \notin \pi$, such that any two of them permute, and they generate G. One may argue using a tower of finite subgroups as above. In the locally soluble case these would constitute a Sylow basis of G; let us call them a Sylow basis in this case also. Similarly, H is generated by the members of a Sylow basis consisting of a Hall π-subgroup H_π and Sylow subgroups H_p for $p \notin \pi$. Using (6) and the fact that H_π and G_π are Chernikov, it is not hard to see that H_π is a Hall π-subgroup of G. Now we can argue exactly as in [7] to show that there is a locally inner endomorphism ϕ of G such that $G_\pi^\phi = H_\pi$, and $G_p^\phi = H_p$ for each $p \notin \pi$. Thus, $G^\phi = H$. In fact, ϕ turns out to be inner on any subgroup generated by G_π and finitely many of the G_p. □

Definition 2.4. If G is a countable $\mathcal{K}$-group, a subgroup H satisfying any of the above equivalent conditions is called a basis subgroup of G. A basis subgroup of an uncountable $\mathcal{K}$-group is defined to be a countable subgroup H of G satisfying any of the conditions (3)–(6), which are easily seen to be equivalent. If G_1 and G_2 are $\mathcal{K}$-groups and some basis subgroup of G_1 is isomorphic to a basis subgroup of G_2, then G_1 and G_2 are said to have the same basis type.

It follows easily from Lemma 2.3 that if $G \in \mathcal{K}$ and B is a basis subgroup of G, then every countable subgroup of G containing B is a basis subgroup of G and so is isomorphic to B.

We can now state precisely the theorem proved by Belyaev in the locally soluble case and by Bell in general.

Theorem 2.5. *Let G be a countable $\mathcal{K}$-group. Then G has a proper basis subgroup if and only if G is not hyperfinite.*

In one direction this result is easy (see Lemma 2.9). Before we explain why, a few preliminary remarks are in order.

We recall the definition of the FC-hypercentre of a group. Let $FC(G)$ be the FC-centre of a group G, which in the locally finite case is the product of the finite normal subgroups of G. The upper FC-central series of G is defined for ordinals α by $FC_0(G) = 1$; $FC_{\alpha+1}(G)/FC_\alpha(G) = FC(G/FC_\alpha(G))$; and $FC_\beta(G) = \bigcup_{\alpha<\beta} FC_\alpha(G)$ for limit ordinals β. Let $FC^\infty(G)$ be the terminus of this series, the FC-hypercentre of G. Then $G/FC^\infty(G)$ contains no non-trivial finite normal subgroup.

Lemma 2.6. *Let $G \in \mathcal{K}$. Then every normal hyperfinite subgroup of G is contained in $FC^\infty(G)$.*

Proof. It suffices to show that if $FC(G) = 1$, then G contains no non-trivial hyperfinite normal subgroup. So suppose that N is a non-trivial hyperfinite normal subgroup of G. Then N contains a minimal normal subgroup M (of itself). Now M is either an elementary abelian p-group for some prime p, or a direct power of some non-abelian finite simple group S. In the first case, the normal closure $\langle M^G \rangle$ of M in G is also elementary abelian, and must be finite as every p-subgroup of G is Chernikov. In the second, $\langle M^G \rangle$ is also a direct power of S, and must again be finite. In either case we have a contradiction to the fact that $FC(G)$ is trivial. □

Since the FC-hypercentre is clearly hyperfinite, it follows that in the case of a $\mathcal{K}$-group, the FC-hypercentre is also the unique maximal hyperfinite normal subgroup. We will therefore sometimes refer to it as the hyperfinite radical.

Definition 2.7. The class of hyperfinite $\mathcal{K}$-groups will be denoted by $\mathcal{H}$ and the hyperfinite radical of a $\mathcal{K}$-group G will be denoted by $\mathcal{H}(G)$.

The following statements are then trivial and will be used without mention.

Lemma 2.8. *Let $G \in \mathcal{K}$.*

1. *If $N \lhd G$, then $\mathcal{H}(N) \leq \mathcal{H}(G)$.*
2. $\mathcal{H}(G/\mathcal{H}(G)) = 1$.

Now it is an easy exercise to verify that a basis subgroup of a $\mathcal{K}$-group G contains every finite normal subgroup of G. Hence it contains the FC-centre of G, and by a simple induction, it also contains the FC-hypercentre. This gives one implication of Theorem 2.5.

Lemma 2.9. *If $G \in \mathcal{K}$ and B is a basis subgroup of G, then $B \geq \mathcal{H}(G)$.*

The argument given for the group W in the Introduction shows that if G is any countable $\mathcal{K}$-group having a proper basis subgroup, then G can be embedded as a basis subgroup of an uncountable $\mathcal{K}$-group. In due course we shall give some results of Bell stating in how many ways this can be done.

We note in passing that the class of $\mathcal{H}$ is more complicated than one might suspect.

Lemma 2.10. *For any countable ordinal λ, there exists an $\mathcal{H}$-group whose upper FC-central series has length exactly λ.*

Proof. We prove by transfinite induction that if α is any countable ordinal and π is any infinite set of primes, then there exists a locally finite group $G(\pi, \alpha)$ with the following properties.

1. $G(\pi, \alpha)$ has finite Sylow subgroups.
2. $G(\pi, \alpha)$ is a π-group.
3. The upper FC-central series of $G(\pi, \alpha)$ reaches G in precisely α steps.

We begin with $G(\pi, 0) = 1$.

Suppose α is a countable ordinal and we have the groups $G(\sigma, \beta)$ for all infinite sets σ of primes and all ordinals $\beta < \alpha$. First suppose that α is a limit. Partition π as the union of infinitely many pairwise disjoint subsets π_i ($i = 1, 2, \ldots$), and let (β_i)

be a sequence of ordinals less than α, indexed by the natural numbers, converging to α. Put $G(\pi, \alpha) = \mathrm{Dr}_{i=1}^{\infty} G(\pi_i, \beta_i)$, the restricted direct product of the $G(\pi_i, \beta_i)$. It is trivial to verify that $G(\pi, \alpha)$ satisfies the required conditions.

Now let $\alpha = \beta + 1$ for some ordinal β. Choose $p \in \pi$, and partition $\pi \setminus \{p\}$ into an infinite number of pairwise disjoint infinite subsets π_i $(i = 1, 2, \ldots)$. Let $H = \mathrm{Dr}_{i=1}^{\infty} G(\pi_i, \alpha)$ and $G(\pi, \alpha) = H \wr < x >$, where x is an element of order p. Let $H_\gamma = FC_\gamma(H)$ for ordinals $\gamma \leq \beta$ and write $G(\pi, \alpha) = G$. Suppose that, for some ordinal $\gamma < \beta$, we know that

$$FC_\gamma(G) = \prod_{i=0}^{p-1} H_\gamma^{x^i}. \tag{2.4}$$

Then $G / FC_\gamma(G) \simeq (H/H_\gamma) \wr \langle x \rangle$. The group H/H_γ is infinite, and it is easy to see that every element outside its base group has infinitely many conjugates. It follows that (2.4) continues to hold when γ is replaced by $\gamma + 1$. We deduce that the upper FC-series of G reaches the base group of G at precisely the term β, as required. □

3. Representation Theoretic Methods

In this section we shall indicate how Theorem 2.5 can be proved.

It is perhaps natural to try to mimic the argument used for the wreath product in the Introduction, though as we shall see, Belyaev used a quite different approach. The wreath product W is a repeated semidirect product

$$W = \ldots V_3 V_2 V_1,$$

where V_i is a finite elementary abelian p_i-subgroup normalized by $G_{i-1} = V_{i-1} \ldots V_2 V_1$. We can think of V_i as the regular $\mathbb{F}_{p_i} G_{i-1}$-module, and since it is the regular module, distinct subgroups of G_{i-1} have distinct centralizers in V_i. This was the crucial ingredient in showing that W has a proper basis subgroup.

Suppose now that G is an infinite $\mathcal{K}$-group with $\mathcal{H}(G) = 1$. Let H be a finite subgroup of G. Now G contains no non-trivial divisible abelian normal subgroup and so the Sylow subgroups of G are finite. From Lemma 2.2, if π is the set of prime divisors of $|H|$ and $P = O_{\pi'}(G)$, we have that $|G : P| < \infty$. Hence H operates faithfully on P. Routine arguments show that H normalizes a Sylow subgroup of P for each prime not in π, and thus we can begin to build up semidirect products

$$\ldots P_3 P_2 P_1 H$$

inside G, where each P_i is a finite p-subgroup normalized by the subgroups to its right. One might hope to force a regular H-module to appear somewhere in the action of H on a section of some P_i, and hence to approach the wreath product configuration. However this hope is naive, as was pointed out by Bell. To see this, we consider a different wreath product from that above.

For the general theory of the wreath product of (possibly infinitely many) permutation groups, we refer the reader to Hall [10]. We give just what we need. Let $\Sigma = (p_1, p_2, \ldots)$ be an infinite sequence of distinct primes, and let A_i be a group of order p_i. Let Ω be the restricted direct product of the A_i, considered simply as sets with a distinguished element.

$$\Omega = \cdots \times A_3 \times A_2 \times A_1. \tag{3.5}$$

We think of the elements of Ω as sequences

$$\omega = (\ldots, \omega_3, \omega_2, \omega_1) \tag{3.6}$$

with $\omega_i \in A_i$, and where $\omega_i = 1$ for all but finitely many i. We define an action of A_i on the right as follows. Let $\omega \in \Omega$ and $a \in A_i$. Then

$$\begin{aligned} (\omega a)_j &= \omega_j \text{ for all } j \neq i. \\ (\omega a)_i &= \omega_i \text{ if } \omega_j \neq 1 \text{ for some } j < i. \\ (\omega a)_i &= \omega_i a \text{ otherwise} \end{aligned}$$

Let $G =< \ldots, A_3, A_2, A_1 >$, which we denote by

$$W_{\text{perm}}(\Sigma). \tag{3.7}$$

This is the permutational wreath product of the A_i considered as permutation groups. Let $G_i = \langle A_i, \ldots, A_1 \rangle$, and let $\Omega_i = A_i \times \cdots \times A_1$. Then G_i also acts naturally on Ω_i. Let $V_i = \langle A_i^{G_i} \rangle$. Then V_i is the direct product of conjugates of A_i indexed by the elements of Ω_{i-1}, if $i > 1$, and G_{i-1} permutes these conjugates in the same way as it permutes Ω_{i-1}. The group G is the repeated semidirect product

$$G = \ldots V_3 V_2 V_1, \tag{3.8}$$

and $G_i = V_i V_{i-1} \ldots V_1$. Regarded as a module for G_{i-1}, V_i is the permutation module

$$V_i = \mathbb{F}_{p_i} \Omega_{i-1}. \tag{3.9}$$

It is not too difficult to verify the following.

Lemma 3.1. *If $j > i > 1$, then V_j does not contain the regular $\mathbb{F}_{p_j} G_i$-module, and in fact there are irreducible $\mathbb{F}_{p_j} G_i$-modules that do not appear as constituents of $V_j \mid_{G_i}$.*

For more details, see [2, Proposition 4.4].

The main representation theoretic result of Bell produces certain permutation modules, instead of regular modules, in a context where they can be used to prove Theorem 2.5. Before stating it, we need a little terminology. The Fitting height of a finite soluble group X is denoted by $h(X)$. The Fitting subgroup of X is denoted by $F(X)$, and the Fitting series of X is defined by $F_0(X) = 1$, and $F_{i+1}(X)/F_i(X) = F(X/F_i(X))$

for $i \geq 0$. Then $h(X)$ is the least integer h such that $F_h(X) = X$. Also, for an arbitrary finite group X, we denote by $\ell(X)$ the length of the longest chain of subgroups of X.

Proposition 3.2 (Bell). *Let H be a finite group containing a Hall subgroup K with complement E. Let H act on a finite soluble group L such that $L = [L, K]$. Suppose that*

$$h(L) > \ell(H),$$

and also

$$q > 2|H| \tag{3.10}$$

for all prime divisors q of $|L|$.

Then there exist a prime p dividing $|L|$, an H-invariant elementary abelian section P/P_0 of p-power order of L, and a subgroup C of H not containing K, such that as $\mathbb{F}_p H$-module, P/P_0 is isomorphic to the permutation module 1_C^H, the permutation module on the cosets of C.

Bell suggests that the hypotheses here are much stronger than necessary, and in particular that the "large prime divisor" assumption (3.10) is superfluous, but the above is good enough for the present purposes. We shall not say anything about the proof, except that it uses the H-towers introduced by Turull, and ideas from Turull's papers [15], [16], [17].

Theorem 3.3 (Bell). *Let H be a finite group containing a Hall subgroup K with complement E, and let H act on a finite soluble group L such that $L = [L, K]$. Assume (3.10), and that either*

$$C_L(H) = 1, \tag{3.11}$$

or

$$C_L(E) = C_L(H). \tag{3.12}$$

Then

$$h(L) \leq \ell(H).$$

The case when (3.11) holds may be compared with Turull [16], where the same result is obtained with a condition less stringent than (3.10).

Proof. Suppose that $h(L) > \ell(H)$, and let V be the elementary abelian section of L given by Proposition 3.2. By (3.10), we have $(|H|, |L|) = 1$. Hence if (3.11) holds, we have $C_V(H) = 0$, and if (3.12) holds, then $C_V(E) = C_V(H)$. Now since V is a permutation module corresponding to a transitive permutation representation of H, we have $\dim C_V(H) = 1$. This is a contradiction to (3.11). The dimension of $C_V(E)$

is the number of orbits of E on the cosets of C, that is, the number of double cosets CxE. Since C does not contain K, the order of CE is not divisible by $|K|$, and so $CE < H$. Therefore E has at least two orbits on the cosets of C, and $\dim C_V(E) \geq 2$. This contradicts (3.12). □

Theorem 3.3 is used by Bell in constructing the locally inner endomorphisms needed in the proof of Theorem 2.5.

A result very similar to Theorem 3.3, but in which H is a cyclic group whose order is the product of two distinct primes, is used by Belyaev. We state it below for comparison. We will then feed Bell's Theorem into Belyaev's method to obtain Theorem 2.5, since this yields some intermediate results of independent interest.

Lemma 3.4 (Belyaev). *Let $H = E \times K$, where E and K are cyclic groups of distinct prime orders. Let H act on a finite soluble group L, and assume that $C_L(E) = C_L(H)$. Further, assume that $(|K|, |L|) = 1$. Then $(|E|, |[L, K]|) = 1$, and either*

$$[L, K] \leq F(G),$$

or $|K| = 2$, $2|E| - 1$ is a power of a prime q, and $[L, K]/F([L.K])$ is a non-abelian q-group.

Now we consider fixed points of finite groups acting on $\mathcal{K}$-groups. The next few results are taken from Belyaev's paper [5], slightly modified to incorporate Bell's result. The first is in any case well known.

Lemma 3.5. *Let H be a finite subgroup of a locally finite group G, and N be a normal $\mathcal{K}$-subgroup of G. Then*

$$|C_{G/N}(H) : C_G(H)N/N| < \infty.$$

Proof. Let π be the set of prime divisors of $|H|$. Then $N/O_{\pi'}(N)$ is a Chernikov group. This reduces the proof to two cases. In the first, N is a π'-group and in the second, N is Chernikov. If N is a π'-group, then $C_{G/N}(H) = C_G(H)N/N$. If N is Chernikov, and $C/N = C_{G/N}(H)$, then $HN = L$ is a normal Chernikov subgroup of $M = HC$. Since every periodic group of outer automorphisms of a Chernikov group is finite [13, 3.29.2], we have $|M : LC_M(L)| < \infty$. Clearly, $C_M(L) \leq C_M(H) = C_G(H)$, from which the result follows. □

Let $\pi(X)$ denote the set of primes p such that the group X has an element of order p.

Lemma 3.6. *Let the finite group A act on a $\mathcal{K}$-group G.*

1. *If $\pi(A) \cap \pi(G) = \emptyset$, then*

$$\mathcal{H}(C_G(A)) = C_G(A) \cap \mathcal{H}(G).$$

2. *In any case, if $C_G(A) \in \mathcal{H}$, then $G \in \mathcal{H}$.*

Proof. (1). Of course, $C_G(A) \cap \mathcal{H}(G) \leq \mathcal{H}(C_G(A))$. For the converse, it suffices to assume that $\mathcal{H}(G) = 1$ and prove that $\mathcal{H}(C_G(A)) = 1$. For having done that, let $H = \mathcal{H}(G)$ and $C = C_G(A)$. We have $1 = \mathcal{H}(C_{G/H}(A)) = \mathcal{H}(CH/H) \geq \mathcal{H}(C)H/H$, whence $\mathcal{H}(C) \leq H$.

So now assume that $\mathcal{H}(G) = 1$, in which case, by Lemma 2.2, the Sylow subgroups of G are finite. For a contradiction, assume that $\mathcal{H}(C) \neq 1$. Then C contains a non-trivial finite normal subgroup M. Let π be the set of primes q such that $q \leq 2|MA|$, and let $P = O_{\pi'}(G)$. Then $|G : P| < \infty$ by Lemma 2.2 and P is locally soluble by the Feit–Thompson Theorem. Now $[P \cap C, M] \leq P \cap M = 1$. In other words, $P \cap C = C_P(A) \leq C_P(M)$, and so $C_P(A) = C_P(A \times M)$. Noting that $\pi(A \times M) \cap \pi(P) = \emptyset$, we have from Theorem 3.3 that $P_0 = [P, M]$ has a finite series with locally nilpotent factors. If $P_0 \neq 1$, then there is a prime p such that $O_p(P_0) \neq 1$. Since P_0 is subnormal in G, it follows that $O_p(G) \neq 1$, and so G contains a non-trivial finite normal p-subgroup. Since this is contrary to assumption, we find that $[P, M] = 1$, and so $|G : C_G(M)| < \infty$. But then $|G : N_G(M)| < \infty$, and so the normal closure of M in G is finite, a contradiction.

(2). Again let $C = C_G(A)$, and suppose that $C \in \mathcal{H}$. Let π be the set of primes not exceeding $2|A|$, and let $P = O_{\pi'}(G)$. Since G/P is Chernikov, it suffices to prove that $P \in \mathcal{H}$. Applying (1), we find that $C_P(A) \leq \mathcal{H}(P)$. Let $H = \mathcal{H}(P)$. Then $C_{P/H}(A) = 1$, since $\pi(A) \cap \pi(P) = 1$. It follows from Theorem 3.3, taking $E = 1$, that P/H has a finite series with locally nilpotent factors, and hence, since its Sylow subgroups are all Chernikov, that P/H is hyperfinite. Hence $P = H$, as required.

Lemma 3.7. *Let $G \in \mathcal{K}$, and let H be a finite subgroup of G. Let $C = C_G(H)$. Let $\mathbf{X}$ be the set of all $\mathcal{H}$-subgroups X of G such that (i) $H \leq N_G(X)$, and (ii) X is normalized by a subgroup of finite index of C. Then $\langle X : X \in \mathbf{X}\rangle \in \mathcal{H}$.*

Corollary 3.8. *If G and H are as in Lemma 3.7, then among the $\mathcal{H}$-subgroups of G normalized by H and $C_G(H)$, there is a unique maximal one, namely the join of all such subgroups.*

Proof of Lemma 3.7. Let $L = \langle X : X \in \mathbf{X}\rangle$. We may assume that $G = LH$. Let $J = \mathcal{H}(G)$. If $X \in \mathbf{X}$, then XJ/J is normalized by a subgroup of finite index in CJ/J, and hence, by Lemma 3.5, by a subgroup of finite index in $C_{G/J}(HJ/J)$. It follows that LJ/J is contained in the subgroup of G/J constructed in the same way from HJ/J. Therefore, we may assume that $J = 1$. By Lemma 3.6, the result will follow if we can show that $C_L(H) \in \mathcal{H}$.

Put $D = C_L(H)$, let $X \in \mathbf{X}$, and let $x \in X$. We claim that

$$|D : C_D(x)| < \infty. \tag{3.13}$$

To see this, let E be a normal subgroup of finite index in D that normalizes X, and consider $< X, H, E >= M$. Let $\pi = \pi(H)$, and let $P = O_{\pi'}(L)$. Then $\mathcal{H}(P) = 1$, and so by Lemma 3.6, $\mathcal{H}(C_P(H)) = 1$. Hence $\mathcal{H}(D)$ is finite, and it follows, since $E \cap X$ is a normal $\mathcal{H}$-subgroup of E, that $E \cap X$ is finite. Put $E_1 = C_E(x)$. Now

$E \leq D \cap M \leq C \cap M = C_M(H)$, and each is of finite index in the next. It follows that E_1 is a subgroup of finite index of $C_M(\langle H, x\rangle)$. Now $X < H, x >= XH \lhd M$, and it follows from Lemma 3.5 that $|M/X : C_M(\langle H, x\rangle)X/X| < \infty$, and so $|EX : E_1X| < \infty$. Hence $|E : E_1(E \cap X)| < \infty$. But we have noted that $E \cap X$ is finite. Hence $|E : E_1| < \infty$, which, since $|D : E| < \infty$, establishes (3.13).

Now each element of L is a product of finitely many elements like x, and so it follows from (3.13) that if $g \in L$, then $|D : C_D(g)| < \infty$. Since $D \leq L$, this tells us that D is an FC-group, and so belongs to $\mathcal{H}$, as required. □

Now let $G \in \mathcal{H}$. For each $H \leq G$, let $\theta(H)$ be the join of all the $\mathcal{H}$-subgroups of G normalized by H and $C_G(H)$. In particular, $\theta(1) = \theta(G) = \mathcal{H}(G)$. Lemma 3.7 tells us that *if H is finite, then $\theta(H) \in \mathcal{H}$*. In Belyaev's terminology [5], *the functor θ is complete.* This may be compared with the signalizer functor terminology used in finite group theory, see [8]. He deduces from this, by a quite straightforward argument, the following, which immediately implies the locally soluble case of Theorem 2.5.

Proposition 3.9. *If $G \in \mathcal{K}$ and $g \in G \setminus \mathcal{H}(G)$, then there exists a locally inner endomorphism ϕ of G such that $g \notin \phi(G)$.*

We refer to his paper for the details.

We mention two other results of Bell, the second of which uses Theorem 3.3, that describe certain kinds on $\mathcal{K}$-groups that are "minimal" among $\mathcal{K}$-groups with trivial $\mathcal{H}$-radical.

Theorem 3.10 (Bell). *Let G be a $\mathcal{K}$-group such that $\mathcal{H}(G) = 1$. Then*

1. *G contains an infinite semidirect product*

$$\ldots A_3A_2A_1,$$

in which each A_i is an elementary abelian p_i-group and is the unique minimal normal subgroup of $A_iA_{i-1}\ldots A_1$, and the primes $p_1, p_2, \ldots$ are all distinct;

2. *G contains a wreath product of the form $W_{\text{perm}}(\Sigma)$ (see (3.7)).*

4. Uncountable Groups and Pro-Chernikov Groups

Definition 4.1. A $\mathcal{K}$-group G is called maximal, if whenever G is a subgroup of a $\mathcal{K}$-group G_1, and G contains a basis subgroup of G_1, we have $G = G_1$.

This concept was introduced by Baer [1]. Since a $\mathcal{K}$-group can have cardinal at most $2^{\aleph_0}$, (this follows from Theorem 2.2), it is easy to see that each $\mathcal{K}$-group is a subgroup of a maximal one. Furthermore, Theorem 2.5 shows that a countable $\mathcal{K}$-group is maximal if and only if it is not hyperfinite. In that case we have

Theorem 4.2 (Bell). *Let G be a countable non-hyperfinite $\mathcal{K}$-group. Assume that $2^{\aleph_1} > 2^{\aleph_0}$. Then G is a basis subgroup of at least $2^{\aleph_1}$ pairwise non-isomorphic maximal $\mathcal{K}$-groups.*

This result is proved by making use of the ***pro-Chernikov completion*** of a $\mathcal{K}$-group G, which was introduced by Dixon [6]. Since G is a residually Chernikov group, we can embed G canonically in the inverse limit $\hat{G}$ of its Chernikov factor groups. When G has finite Sylow subgroups, as for instance when $\mathcal{H}(G) = 1$, then G is residually finite and $\hat{G}$ is just the profinite completion of G. In that case, G carries a canonical compact topology, and G is dense in $\hat{G}$. This is true in general; $\hat{G}$ can be equipped with a canonical compact topology, with respect to which G is dense. Now it is not hard to see that $\hat{G}$ has cardinal at most $2^{\aleph_0}$, and that any continuous automorphism of $\hat{G}$ is determined by its restriction to G. Hence we have

Lemma 4.3. *If G is a countable $\mathcal{K}$-group, then the group of continuous automorphisms of $\hat{G}$ has cardinal at most $2^{\aleph_0}$.*

Further, it turns out that the maximal $\mathcal{K}$-groups containing G are, up to isomorphism, precisely the maximal locally finite subgroups of $\hat{G}$ containing G, and two such maximal locally finite subgroups are isomorphic if and only if they are conjugate under the group of continuous automorphisms of $\hat{G}$. Thus, each isomorphism class of maximal locally finite subgroups of $\hat{G}$ that contain G has size at most $2^{\aleph_0}$. Now the key step is the following.

Lemma 4.4. *Let G be a countable non-hyperfinite $\mathcal{K}$-group. Then there exists a set of $2^{\aleph_1}$ subgroups of $\hat{G}$, each of which is a $\mathcal{K}$-group containing G, such that any two of them generate a group containing an element of infinite order.*

Each of these locally finite subgroups of $\hat{G}$ is contained in a maximal one, and thus we obtain $2^{\aleph_1}$ distinct maximal locally finite subgroups of $\hat{G}$, each containing G. The set-theoretic hypothesis of Theorem 4.2, combined with Lemma 4.3, shows that these fall into $2^{\aleph_1}$ conjugacy classes under the group of continuous automorphisms of $\hat{G}$, and hence into $2^{\aleph_1}$ isomorphism classes.

It is unknown whether the set theoretic hypothesis is necessary.

It is trivial to see that Theorem 4.2 implies that every countable non-hyperfinite $\mathcal{K}$-group has a proper basis subgroup, that is, Theorem 2.5. That theorem is stated separately for clarity, and is proved separately for the same reason, but it is not used in the proof of Theorem 4.2. Rather, the representation-theoretic result Proposition 3.2, together with the techniques used in its proof, is used again in a more delicate way.

Let G be any countable non-hyperfinite $\mathcal{K}$-group. Bell constructs in G two sequences of finite subgroups $(P_n)_{n\geq 0}$ and $(F_n)_{n\geq 0}$ such that the following hold. We use bars to denote images modulo $\mathcal{H}(G)$.

1. P_n is a p_n-group normalized by F_n, where the prime p_n is greater than $2|\overline{F}_n|$, and does not divide $|F_n|$.
2. $\overline{P}_0$ is F_0-simple, and for $n > 0$, the Frattini factor group of $\overline{P}_n$ is F_n-simple.

3. For $n > 0$, $F_n \geq F_{n-1}P_{n-1}$.
4. For $n > 0$, P_{n-1} does not centralize P_n.
5. $\bigcup_{n=0}^{\infty} F_n = G$.

Any finite subgroup H of G is contained in some F_n and so normalizes all but finitely many of the P_n. This enables many "infinite H-towers" to be extracted from the P_n, and the careful analysis of these towers is the central part of Bell's proof of Theorem 4.2, as well as of the results in the next section.

5. Uncountable Groups and Tree Limits

As stated in the Introduction, Hickin [11] has proved that there are $2^{\aleph_0}$ variations on the regular wreath products W mentioned there. His work also gives information about permutation representations of these groups. He constructs his groups by using the ***tree-limit***, which he describes as a model theoretic construction introduced by Shelah. He gives a rather clear description of the details of this construction. Bell has combined the tree limit construction with the towers mentioned at the end of the last section, to obtain the following two striking results.

Theorem 5.1. *Let G be any countable non-hyperfinite $\mathcal{K}$-group. Then there exists a $\mathcal{K}$-group $\tilde{G}$ containing G as a basis subgroup, and containing $2^{\aleph_0}$ subgroups G_α such that the following hold.*

1. *Each G_α contains G (as a basis subgroup).*
2. *Each G_α has cardinal $2^{\aleph_0}$.*
3. *If $\alpha \neq \beta$, then no uncountable subgroup of G_α is isomorphic to a subgroup of G_β.*

Theorem 5.2. *Let G be a countable non-hyperfinite $\mathcal{K}$-group and $\aleph$ an uncountable cardinal no bigger than $2^{\aleph_0}$. Then the number of isomorphism types of $\mathcal{K}$-group of cardinal $\aleph$, each having G as a basis subgroup, is precisely $2^{\aleph}$.*

The two results above do not depend on continuum hypothesis type assumptions. It does not seem easy to sum up the arguments in a few words, and we refer the reader to Bell's preprint [2], or his thesis [3].

References

[1] Baer, R., Lokal endlich-auflösbare Gruppen mit endlichen Sylowuntergruppen. J. Reine Angew. Math. 239/240 (1970), 109–144.

[2] Bell, S. D., Uncountable locally finite groups with Chernikov Sylow subgroups. Preprint 1993/15, Manchester Centre for Pure Mathematics, 1993.

[3] —, Locally Finite Groups with Chernikov Sylow subgroups. Ph.D.Thesis, University of Manchester 1994.

[4] Belyaev, V. V., Locally finite groups with Chernikov Sylow p-subgroups. Algebra and Logic 20 (1981), 393–402.

[5] —, Locally inner endomorphisms of SF-groups. Algebra and Logic 27 (1988), 1–11.

[6] Dixon, M. R., Some topological properties of residually Chernikov groups. Glasgow Math. J. 23 (1982), 65–82.

[7] —, Tomkinson, M. J., The local conjugacy of some Sylow bases in a class of locally finite groups. J. London Math. Soc. (2) 21 (1980), 225–228.

[8] Glauberman, G., On solvable signalizer functors in finite groups. Proc. London Math. Soc. (3) 33 (1976), 1–27.

[9] Gorenstein, D., Finite Groups. Harper and Row, New York 1968.

[10] Hall, P., Wreath products and characteristically simple groups. Proc. Cambridge Philos. Soc. 58 (1962), 170–184.

[11] Hickin, K. K., Some applications of tree limits to groups. Part 1. Trans. Amer. Math. Soc. 305 (1988), 797–839.

[12] Kegel, O. H., Wehrfritz, B. A. F., Locally Finite Groups. North Holland, Amsterdam 1980.

[13] Robinson, D. J. S., Finiteness Conditions and Generalized Soluble Groups. Part 1. Springer, Berlin 1972.

[14] Shunkov, V. P., Conjugacy of Sylow p-subgroups. Algebra i Logika 10 (1971), 587–597.

[15] Turull, A., Fitting height of groups and of fixed points. J. Algebra 86 (1984), 555–566.

[16] —, Generic fixed point free action of arbitrary finite groups. Math. Z. 197 (1984), 491–503.

[17] —, Fixed point free action with regular orbits. J. Reine Angew. Math. 371 (1986), 67–91.

Groups Which Are Dual to Layer-Finite

N. V. Kalashnikova and L. A. Kurdachenko

1991 Mathematics Subject Classification: 20E26

Let G be a group; write $G[n] = \{g \in G \mid g \text{ has order } n\}$ and $G^n = \langle g^n \mid g \in G\rangle$. The group G is ***layer-finite*** if $G[\infty] = \emptyset$ and the set $G[n]$ is finite for all $n \in \mathbb{N}$. Layer-finite groups have been studied by S. N. Chernikov, R. Baer, Ya. D. Polovitzki (see, for example, [7], 4.4). We shall say that G is ***co-layer-finite*** if the factor group G/G^n is finite for each $n \in \mathbb{N}$. Such groups may be considered as dual to layer-finite groups. Co-layer-finite groups arise in the study of groups with finiteness conditions (see, for example, [2], [10] and [11]) and in FC-groups [4]. On the other hand if $G = G^n$ for all $n \in \mathbb{N}$ (i.e. G is divisible in the sense of S. N. Chernikov), then G is clearly co-layer-finite. Other examples are locally soluble groups of finite rank (in the sense of Mal'cev) and locally finite groups with min-p for all primes p. This range of examples indicates the difficulty of studying co-layer-finite groups in general. In this article we consider nilpotent co-layer-finite groups and some generalized nilpotency conditions. Note that abelian co-layer-finite groups have already been described in [4].

We begin by observing some elementary closure properties of the class of co-layer-finite groups.

Lemma 1. *Let G be a group. Then:*

(i) *If $N \lhd G$ and G is co-layer-finite, then G/N is co-layer-finite.*

(ii) *If G is co-layer finite and H is a subgroup of finite index in G, then H is co-layer-finite.*

(iii) *If $N \lhd G$ and G/N and N are both co-layer-finite, then G is co-layer-finite.*

Proof. (i) This follows from $(G/N)^n = G^nN/N$.

(ii) If $|G : H| = k$ and H/H^n were infinite then G/G^{kn} would also be infinite.

(iii) $G/G^nN \simeq (G/N)/(G/N)^n$ is finite and $G^nN/G^n \simeq N/N \cap G^n$, a factor group of N/N^n. So G^nN/G^n is also finite and hence G/G^n is finite. □

Offprint from
Infinite Groups 94, Eds.: de Giovanni/Newell

If G is a locally nilpotent group then G/G^n, having finite exponent, is the direct product of its Sylow subgroups. It is then clear that G is co-layer-finite if and only if G/G^{p^k} is finite for all primes p and all $k \in \mathbb{N}$.

For the following results we introduce some notation. A group G is said to be $\mathfrak{F}$-***perfect*** if it has no non-trivial finite factor groups. G is ***residually-***$\mathfrak{F}_p$ if it has normal subgroups G_i $(i \in I)$ such that G/G_i is a finite p-group and $\bigcap_{i \in I} G_i = 1$.

Lemma 2. *Let G be locally nilpotent and residually-$\mathfrak{F}_p$. If G has only finitely many subgroups of index p, then G is nilpotent and its torsion subgroup $T(G)$ is finite.*

Proof. Let $R = G'G^p$ so that G/R is a finite elementary abelian p-group. Therefore $G = FR$ for some finitely generated (and hence nilpotent) subgroup F. Let L be any normal subgroup of finite p-power index in G and let $K/L = RL/L$ be the Frattini subgroup of G/L. Then $FK = G$ and hence $FL = G$. If c is the nilpotency class of F, then G/L has class at most c and, since G is residually-$\mathfrak{F}_p$, G is nilpotent of class c.

If H is any finite subgroup of G then there is an $L \lhd G$ with G/L a finite p-group and $H \cap L = 1$. Since $G/L \simeq F/F \cap L$, H is isomorphic to a subgroup of $F/F \cap L$. If r is the (special) rank of F, then H has rank at most r. Hence $T(G)$ is a nilpotent p-group of finite rank and so is a Chernikov group ([8], Corollary 2 to Theorem 6.36). Since G is residually finite, it follows that $T(G)$ is finite. □

Corollary 3. *Let G be a locally finite p-group with only finitely many subgroups of index p. Then G is a finite extension of an $\mathfrak{F}$-perfect group.*

Proof. Let M be the intersection of all subgroups of finite index of G. It follows from Lemma 2 that G/M is finite and hence M is $\mathfrak{F}$-perfect. □

Corollary 4. *Let G be a hypercentral p-group. If G/G^p is finite, then G is co-layer-finite.*

Proof. By Corollary 3, G/G^{p^k} is a finite extension of an $\mathfrak{F}$-perfect group. But a non-trivial hypercentral $\mathfrak{F}$-perfect group has infinite exponent. Hence G/G^{p^k} is finite for all $k \in \mathbb{N}$ and so G is co-layer-finite. □

Lemma 5. *Let G be a nilpotent co-layer-finite group containing a central subgroup N of finite exponent n such that G/N is torsion-free. Then there is a subgroup W of finite index in G such that $W \cap N = 1$ (and hence N is finite).*

Proof. By induction on the nilpotency class c of G/N. Let Y/N be the centre of G/N; then Y/N is torsion-free and also G/Y is torsion-free and nilpotent of class $c - 1$. For $y \in Y$ and $g \in G$, we have $[y, g^n] = [y, g]^n = 1$ and so $G^n \leq C_G(Y)$. Since

G is co-layer-finite so also is G^n. Now $Y \cap G^n \leq C = Z(Y)$. Also N is a bounded subgroup of C and C/N is torsion-free. Hence $C = N \times X$ for some torsion-free abelian subgroup X and so $C^n \cap N = 1$.

Now C/C^n has exponent n and $CG^n/C \simeq YG^n/Y$ is torsion-free and nilpotent of class at most $c - 1$. By induction, there is a subgroup W of finite index in CG^n such that $C \cap W = C^n$ and hence $W \cap N = C^n \cap N = 1$. □

Corollary 6. *If G is a nilpotent co-layer-finite group then its torsion subgroup $T(G)$ is co-layer-finite and each Sylow subgroup of $T(G)$ is a finite extension of a divisible abelian group.*

Proof. Consider $G/T(G)^p$. It follows from Lemma 5 that $T(G)/T(G)^p$ is finite and so $T(G)$ is co-layer-finite. Hence so is each Sylow subgroup of $T(G)$. By Corollary 3, a Sylow subgroup is a finite extension of an $\mathfrak{F}$-perfect group. But a nilpotent $\mathfrak{F}$-perfect p-group is abelian. □

Theorem 7. *Let G be a locally nilpotent co-layer-finite group. Let L be a central elementary abelian subgroup such that G/L is residually-$\mathfrak{F}_p$. Then there is a subgroup K of finite index in G such that $K \cap L = 1$ (and hence L is finite).*

Proof. By Lemma 2, G/L is nilpotent with finite torsion subgroup T/L. So there is a subgroup $H \lhd G$ such that G/H is a finite p-group and $H \cap T = L$. In particular H/L is torsion-free nilpotent. Applying Lemma 5 to the subgroup H it follows that there is a subgroup W of finite index in H such that $W \cap L = 1$. It is sufficient to take $K = \mathrm{Core}_G\, W$. □

Corollary 8. *Let G be a nilpotent co-layer-finite group and let*

$$M = \bigcap \{H \lhd G \mid G/H \text{ is a finite } p\text{-group}\}$$

be the $\mathfrak{F}_p$-residual of G. Then M is p-divisible.

Proof. It follows from Theorem 7 that M has no finite non-trivial p-image. Hence M is p-divisible (see, for example, [8] Corollary 1 to Theorem 9.23). □

This result raises the question of whether the $\mathfrak{F}_p$-residual is p-divisible in more general cases. A negative answer is given by the following example.

Let $A = \mathrm{Dr}_{n \in \mathbb{N}} \langle a_n \rangle$ be an elementary abelian p-group. The group A has an automorphism φ defined by

$$a_1\varphi = a_1, \quad a_{n+1}\varphi = a_n a_{n+1} \quad (n \in \mathbb{N}).$$

Put $G = A \rtimes \langle \varphi \rangle$; then G/A is infinite cyclic and so residually-$\mathfrak{F}_p$ and G is hypercentral with central height $\omega + 1$. It is easy to see that any G-admissible subgroup of A is finite and so A is the $\mathfrak{F}_p$-residual of G and G is co-layer-finite since $G^n \geq A$, for all $n \in \mathbb{N}$.

Note that in [5] there is a more complicated example in which the $\mathfrak{F}_p$-residual is bounded but nonabelian. Most of our remaining results concern hypercentral groups with central height ω. We refer to these as ***ω-hypercentral groups***.

Lemma 9. *Let G be a co-layer-finite group containing a central elementary abelian p-subgroup A such that G/A is torsion-free and ω-hypercentral. Then there is a subgroup $K \lhd G$ such that G/K is a finite p-group and $K \cap A = 1$ (and hence A is finite).*

Proof. Let $Y_n/A = Z_n(G/A)$; then Y_n/A and G/Y_n are torsion-free. Let $C_1 = C_G(Y_1)$; then G/C_1 is an elementary abelian p-group and so is finite. Hence C_1 is a co-layer-finite group. Let $B_1 = C_1 \cap Y_1 = Z(Y_1)$; then $B_1 = A \times X_1$ for some torsion-free abelian subgroup X_1. Hence $B_1^p \cap A = 1$.

We define the normal subgroups C_n, B_n inductively by

$$C_n = C_{n-1} \cap C_G(C_{n-1} \cap Y_n/B_{n-1}^p)$$

and

$$B_n = C_n \cap Y_n,$$

and prove, by induction, that they satisfy the following properties:

(1) $C_{n-1} \geq C_n$,

(2) $B_{n-1} \leq B_n$,

(3) C_{n-1}/C_n is a finite elementary abelian p-group,

(4) $[B_n, C_n] \leq B_{n-1}^p$,

(5) $B_n \cap Y_{n-1} = B_{n-1}$,

(6) B_n^p/B_{n-1}^p is torsion-free.

Property (1) is immediate from the definition. By (3), $|G/C_n|$ is finite and so C_n a co-layer-finite group. From (4) and (2), B_n centralizes $C_n \cap Y_{n+1}/B_n^p$ and so $B_n \leq C_{n+1}$. Hence $B_n \leq C_{n+1} \cap Y_{n+1} = B_{n+1}$ and (2) is proved.

Since B_n/B_n^p is a central elementary abelian subgroup of C_n/B_n^p and

$$[C_n \cap Y_{n+1}, C_n] \leq C_n \cap Y_n = B_n,$$

it follows that C_n/C_{n+1} is an elementary abelian p-group and so is finite. Now

$$[B_{n+1}, C_n] \leq [Y_{n+1}, C_n] \leq C_n \cap Y_n = B_n$$

and B_n/B_n^p is a central elementary abelian subgroup of C_n/B_n^p. Therefore

$$[B_{n+1}, C_{n+1}] \leq [B_{n+1}, C_n^p] \leq B_n^p$$

and (4) is proved. It is clear that $B_n \leq B_{n+1} \cap Y_n$ and, conversely,

$$B_{n+1} \cap Y_n = C_{n+1} \cap Y_n \leq C_n \cap Y_n = B_n$$

so that (5) is proved.

Finally $B_{n+1}/B_n \simeq B_{n+1}Y_n/Y_n$, which is torsion-free. Thus

$$B_{n+1}/B_n^p = B_n/B_n^p \times X_n/B_n^p$$

for some torsion-free abelian subgroup X_n/B_n^p, and so $B_{n+1}^p \cap B_n = B_n^p$. Hence $B_{n+1}^p/B_n^p \simeq B_{n+1}^p B_n/B_n$, which as a subgroup of B_{n+1}/B_n must be torsion-free. Having proved all the properties (1)–(6), we now consider $C = \bigcap_{n\in\mathbb{N}} C_n$ and $B = \bigcup_{n\in\mathbb{N}} B_n$. Since $G = \bigcup_{n\in\mathbb{N}} Y_n$, we have

$$C = \bigcup_{n\in\mathbb{N}}(C \cap Y_n) = \bigcup_{n\in\mathbb{N}}(C_n \cap Y_n) = \bigcup_{n\in\mathbb{N}} B_n = B.$$

Now $Y_nB/B \simeq Y_n/Yn \cap B$, a factor group of $Y_n/B_n \simeq Y_nC_n/C_n$, which is a finite p-group. So G/B is a p-group. Also since $B = \bigcap C_n$ and G/C_n is finite, G/B is residually finite and is co-layer-finite. By Lemma 2, G/B is finite. Put $K = BP$; then G/K is finite and $K \cap A = \bigcup_{n\in\mathbb{N}}(B_n^p \cap A) = B_1^p \cap A = 1$. □

Corollary 10. *Let G be a torsion-free ω-hypercentral co-layer-finite group. Then $Z_n(G)$ is co-layer-finite.*

Proof. The group $G/Z(G)$ is torsion-free and ω-hypercentral and so, by Lemma 9, Z/Z^p is finite for any prime p. Hence Z is co-layer-finite. The same argument shows that Z_i/Z_{i-1} is co-layer-finite for each $i \in \mathbb{N}$ and, by the extension closure (Lemma 1(iii)), Z_n is co-layer-finite. □

Corollary 11. *Let G be a torsion-free ω-hypercentral co-layer-finite group. For each prime p, there is a $k(p) \in \mathbb{N}$ such that Z_{n+1}/Z_n is p-divisible for all $n \geq k(p)$.*

Proof. In the proof of Lemma 9, take $A = 1$ and $Y_n = Z_n$ and again construct the subgroups C_n, B_n. We have $|B_n/B_n^p| \leq |B_{n+1}/B_{n+1}^p| \leq |B/B^p|$. Since $|G/B|$ and $|B/B^p|$ are finite, we can choose k so that $|B_k/B_k^p|$ is maximal and $G = Y_kB$. Then for all $n \geq k$, $B_nB_{n+1}^p = B_{n+1}$ and so B_{n+1}/B_n is p-divisible. But $Y_n \cap B = B_n$ and so $Y_{n+1}/Y_n = B_{n+1}/B_n$ is p-divisible. □

Lemma 12. *Let G be a torsion-free nilpotent co-layer-finite group. For each prime p, G has a subnormal series*

$$1 \lhd H_0 \lhd H_1 \lhd \cdots \lhd H_n = G$$

such that

(1) *H_0 has finite rank (in the Mal'cev–Prüfer sense),*

(2) *the groups H_i/H_{i-1} are torsion-free abelian and p-divisible, $1 \leq i \leq n$.*

Proof. By induction on the nilpotency class c of G. If $c = 1$ then the result follows from Lemma 10 of [4]. For $c > 1$, we may assume that there are subgroups

$$Z \lhd H_1 \lhd H_2 \lhd \cdots \lhd H_n = G$$

with H_1/Z having finite rank and H_i/H_{i-1} being torsion-free abelian and p-divisible $(2 \leq i \leq n)$.

Since Z is co-layer-finite (Corollary 10) it contains a finitely generated subgroup L, such that Z/L_1 is p-divisible. Since H_1/Z has finite rank there is a series

$$Z = B_0 \lhd B_1 \lhd \cdots \lhd B_r = H_1$$

with B_i/B_{i-1} having rank one, $1 \leq i \leq r$. Thus there is a finitely generated subgroup $L \geq L_1$ of H_1 such that $B_i/(B_i \cap L)B_{i-1}$ is torsion and p-divisible. Let H_0 be the isolator of L in H_1; then H_0 also has finite rank. Let $N = N_{H_1}(H_0)$; then N is also isolated in H_1 ([6], Section 67). But $Z \leq N$ and $B_i/(B_i \cap N)B_{i-1}$ is torsion for all i, $1 \leq i \leq r$. This implies that $N = H_1$ and so $H_0 < H_1$.

Since $B_i/(B_i \cap H_0)B_{i-1}$ is p-divisible for each i, it follows that H_1/ZH_0 is p-divisible. Also $ZH_0/H_0 \simeq Z/Z \cap H_0$ is p-divisible so that H_1/H_0 is p-divisible. Since H_0 is isolated in H_1, H_1/H_0 is torsion-free. Finally $ZH_0/H_0 \leq Z(H_1/H_0)$ and H_1/ZH_0 is torsion, so H_1/H_0 must be abelian. □

Theorem 13. *Let G be a torsion-free ω-hypercentral group. Then G is co-layer-finite if and only if for each prime p, there exists $k(p) \in \mathbb{N}$ such that*

(1) *Z_{n+1}/Z_n is p-divisible for all $n \geq k(p)$,*

(2) *the subgroup $Z_{k(p)}$ has a series*

$$1 \lhd H_0 \lhd H_1 \lhd \cdots \lhd H_n = Z_{k(p)}$$

where

(a) *H_0 has finite rank,*

(b) *H_i/H_{i-1}, is torsion-free abelian and p-divisible, $1 \leq i \leq n$.*

Proof. Suppose G is co-layer-finite. Then (1) follows from Corollary 11 and (2) follows from Corollary 10 and Lemma 12.

Conversely suppose that (1) and (2) are satisfied. We show that G/G^p is finite. Since Z_{n+1}/Z_n is p-divisible for each $n \geq k(p)$, we have $G^p Z_{k(p)} = G$ so that G/G^p is isomorphic to a factor group of $Z_{k(p)}/Z^p_{k(p)}$. Since H_i/H_{i-1} is p-divisible, Z_k/Z^p_k is isomorphic to a factor group of H_0/H^p_0 which is clearly finite. □

Note that Theorem 13 can be combined with Corollary 6 to give a complete description of nilpotent co-layer-finite groups.

Let G be a residually finite co-layer-finite group. Then we can consider the profinite topology on G in which the set of all subgroups of finite index forms a basis of neighbourhoods of the identity. We could also consider the $\mathbb{Z}$-adic topology in which the subgroups G^n ($n \in \mathbb{N}$) form a basis of neighbourhoods of the identity. It is easy to prove that these topologies coincide for co-layer-finite groups and the following theorem describes the completion of G in this topology.

Theorem 14. *Let G be a residually finite locally nilpotent co-layer-finite group. Then $G \leq \mathrm{Cr}_p\, G_p$, and each group G_p has a central series*

$$1 = L_0 \lhd L_1 \lhd \cdots \lhd L_n = G_p$$

in which the factors L_i/L_{i-1} are finite for $1 \leq i \leq k$ and L_i/L_{i-1} is isomorphic to the additive group of p-adic integers for $k+1 \leq i \leq n$.

Proof. Let R_p be the $\mathfrak{F}_p$-residual of G so that $G \leq \mathrm{Cr}_p\, G/R_p$. By Lemma 1, G/R_p is co-layer-finite and so it is nilpotent (Lemma 2) with finite torsion subgroup. Then the completion G_p of G/R_p has a series of the required form ([9], Theorem 7.7). □

References

[1] Fuchs, L., Infinite Abelian Groups, Vol. I, Academic Press, New York, 1970.

[2] Hall, P., A note on SI-groups. J. London Math. Soc. 39 (1964), 338–344.

[3] Kargapolov, M. I., Merzljakov, Y. I., Foundations of Group Theory, Nauka, Moscow, 1982.

[4] Kurdachenko, L. A., Some conditions for embeddability of an FC-group in a direct product of finite groups and a torsion free abelian group, Mat. Sbornik. 42 (1982), 499–514.

[5] —, Locally nilpotent groups with a condition Min-∞-n, Ukrain. Mat. Zh. 42 (1990), 340–346.

[6] Kurosh, A. G., The Theory of Groups, Nauka, Moscow, 1967.

[7] Robinson, D. J. S., Finiteness Conditions and Generalized Soluble Groups, Part 1, Springer, Berlin 1972.

[8] —, Finiteness Conditions and Generalized Soluble Groups, Part 2, Springer, Berlin 1972.

[9] Warfield, R. B., Nilpotent Groups, Lecture Notes in Math. 513, Springer, Berlin, 1976.

[10] Wilson, J. S., Groups satisfying the maximal condition for normal subgroups, Math. Z. 118 (1970), 107–114.

[11] Zaitsev, D. I., Kurdachenko, L. A., Tushev, A. V., Modules over nilpotent groups of finite rank, Algebra i Logika 24 (1985), 631–666.

Soluble Products of Groups

*Lev S. Kazarin**

Abstract. If G is a finite soluble group and π is a set of odd primes, then $l_\pi(G) \leq d_\pi(G)$. If $G = AB$ for two subgroups of coprime orders, then $h(G) \leq d(A) + d(B) + l_2(G)$ and $d(G) \leq 2d(A)d(B) + d(A) + d(B)$. The last statement holds also for periodic soluble groups $G = AB$ satisfying $\pi(A) \cap \pi(B) = \emptyset$.

1991 Mathematics Subject Classification: 20D10, 20D40

1. Introduction

Let $G = AB$ be a soluble group which is a product of two of its subgroups A and B. What can be said about the structure of the group G if the structure of the factors A and B is known? This is an important question in the theory of groups.

Some characteristic invariants of a soluble group are the following:

$h(G)$ — the nilpotent length or Fitting length of G,

$d(G)$ — the derived length of G,

$l_p(G)$ — the p-length of G for some prime p,

$l_\pi(G)$ — the π-length of G for a subset π of the set of all primes.

These invariants were investigated by Hall and Higman in their classical paper [5] but were also studied when the finite group G is the product of two of its nilpotent subgroups. By the Kegel–Wielandt theorem [12] such groups are always soluble. Some further results in this direction can be found in [15], [1] and [2]. All groups in Sections 2 and 3 of this paper are finite.

The main purpose of this paper is to improve the known bounds on the derived length and the Fitting length in the case when the group $G = AB$ is a product of two subgroups A and B of coprime orders. For example, we obtain a linear bound for the derived length of G when at least one of the parameters $d(A)$ or $d(B)$ is fixed. We also obtain an improvement on the known bounds for the π-length of a group in [3]. This

* The author wishes to thank Dipl.-Math. B. Höfling, Mainz, for his helpful suggestions and aid in typesetting this paper.

Offprint from
Infinite Groups 94, Eds.: de Giovanni/Newell

bound is the same as in [13] but under substantially fewer restrictions. The main results will be stated in Section 4. Our notation is standard. In particular, recall that for each subset π of the set of all primes, $d_\pi(G) = d_\pi$ denotes the derived length of the Hall π-subgroups of G and $d_{\pi'}$ is the derived length of the Hall π'-subgroup of G, where π' is the complementary set of primes for π. If $\pi = \{p\}$, then we write $d_p = d_p(G)$ and $d_{p'} = d_{p'}(G)$. Note also that $O_\pi(G)$ is the largest normal π-subgroup of G, $F(G)$ is its Fitting subgroup; $X \leftthreetimes Y$ denotes a semidirect product of the groups X and Y, where X is normal. All groups in this paper are finite unless stated otherwise.

2. Preliminary Results

The following lemma was proved by Johnson [9] for finite soluble groups.

Lemma 1. *Let the π-soluble group G be a product of its subgroups A and B. Then $O_\pi(A) \cap O_\pi(B) \subseteq O_\pi(G)$.*

Proof. Without loss of generality assume that $O_\pi(G) = 1$. Let $x \in O_\pi(A) \cap O_\pi(B)$ and $g \in G$. Then $g = ab$ for some $a \in A$, $b \in B$. Hence

$$\langle x, x^g \rangle = \langle x^{b^{-1}}, x^a \rangle^b \subseteq \langle O_\pi(A), O_\pi(B) \rangle^b.$$

As G is a D_π-group, by Theorem VI.4.6 of [7], some Hall π-subgroup K of G is a product of a Hall π-subgroup S of A and a Hall π-subgroup T of B. Since $O_\pi(A) \subseteq S$ and $O_\pi(B) \subseteq T$, the subgroup $\langle O_\pi(A), O_\pi(B) \rangle \subseteq ST = K$ is also a π-group. Now $\langle x, x^g \rangle$ is a π-group for every $g \in G$. If $g \in O_{\pi'}(G)$ then $\langle x, x^g \rangle = \langle x \rangle$ and so $x \in C_G(O_{\pi'}(G)) \subseteq O_{\pi'}(G)$ by [5], Lemma 1.2.3. Since g is a π-element, we conclude that $x = 1$, as required. □

Lemma 2. *Let A be a Hall π-subgroup of a soluble group G, K an abelian normal p-subgroup of A and let $p > 2$. If $\{2\} \cap \pi(G) \subseteq \pi$, then $K \leq O_{p',p}(G)$.*

Proof. Suppose that G is a minimal counterexample for Lemma 2. Then $O_{p'}(G) = 1 = O_{p'}(A)$. If X is a proper Hall σ-subgroup of G containing some Sylow p-subgroup of A, where $\sigma = (\pi(G) \backslash \pi) \cup \{p\}$, then also $O_{p'}(X) \leq O_{p'}(G) = 1$. By the minimality of G we have $K \leq O_p(X)$. Clearly $G = AX$. Then $K \leq O_p(A) \cap O_p(X)$. Applying Lemma 1, we conclude that $K \leq O_p(G)$.

Now G has no proper Hall σ-subgroup and $|\pi(G)| = 2$, we have $\pi(A) = \{p\}$. In this case G is a group of odd order. By Lemma 5.1 of [8], Chapter IX, we conclude that $K \leq O_{p',p}(G)$. □

The following lemma is also well-known.

Lemma 3. *Let G be a π-soluble group, $p \in \pi$ and suppose that $O_{\pi'}(G) = 1$. Then there exist groups H, X, Y and a homomorphism $\phi: G \to H = X \times Y$ such that $\phi(G)X = \phi(G)Y = H$, $\ker\phi = 1$, $O_{\pi'}(X) = O_{\pi'}(Y) = 1 = O_p(Y)$ and $O_p(X) = F(X)$.*

Proof. Put $S = O_{p'}(G)$, $\sigma = \pi' \cup \{p\}$ and $T = O_\sigma(G)$. Clearly, $S \cap T \leq O_{\pi'}(G) = 1$. Let $X = G/S$, $Y = G/T$ and $H = X \times Y$ and let ϕ be the homomorphism $G \to H$ which is determined by the rule $\phi(g) = (gS, gT) \in H$ for every $g \in G$. Then $\ker\phi = 1$. By the choice of X and Y, we have $\phi(G)X = \phi(G)Y = H$. Now we obtain that $F(X) = O_p(X)$; also $O_{\pi'}(X) \leq O_{\pi'}(Y) = 1$ and $O_\sigma(Y) = O_p(Y)$. □

Lemma 4. *Let G be a soluble group, A a Hall π-subgroup of G for some set π of primes, K an abelian normal subgroup of A and suppose that $O_{\pi'}(G) = 1$. Then K is contained in the Fitting subgroup $F(G)$, provided at least one of the following conditions holds:*

(a) $\{2,3\} \cap \pi = \emptyset$;

(b) $\{2,3\} \cap \pi(G) \subseteq \pi$ *and* $|K| \equiv 1 \pmod 2$;

(c) *the Sylow 2-subgroups of G are abelian.*

Proof. Let $p \in \pi(K)$ and put $R = O_p(K)$. Consider a group $H = X \times Y$ and a homomorphism ϕ as in Lemma 3. We can identify the group G with its image $\phi(G)$ in H. It suffices to prove that $R \leq O_p(X) \cap G \leq O_p(G)$. Obviously, RX/X is contained in a Hall subgroup AX/X of the group $Y \simeq H/X$ and $F(Y)X/X \leq AX/X$. Since $[R, A] \leq R$, it follows that RX/X centralizes $F(Y)X/X$. By [4], Theorem 6.1.3, $F(Y)X/X$ contains its centralizer and so $RX/X = 1$ and $R \leq X$. Then R is contained in a Hall π-subgroup of X and is normal in it. Now Lemma IX.5.1 of [8] implies that $R \leq O_p(X)$ in each of the cases (a)–(c) with the exception of the case $p = 3$ and $\{2,3\} \subseteq \pi$. But application of Lemma 2 gives $R \leq O_p(X)$ in this case also. □

The next lemma is Lemma 5 of [3].

Lemma 5. *Let G be a π-soluble group and π_1, π_2 be subsets of π. If $\pi_1 \cup \pi_2 \supseteq \pi$ then $l_\pi(G) \leq l_{\pi_1}(G) + l_{\pi_2}(G)$.*

Lemma 6. *Let G be a π-soluble group. Then $l_\pi \leq 2d_\pi$. If $\{2,3\} \cap \pi = \emptyset$, then $l_\pi \leq d_\pi$.*

Proof. If $\{2,3\} \cap \pi = \emptyset$, then the statement follows from Lemma 7 of [3]. If $|\pi \cap \{2,3\}| = 1$, it follows from Lemmas 5, 2 and 4 and Proposition 1 of [3] that $l_\pi \leq 2d_\pi$.

Now suppose that $\{2,3\} \subseteq \pi$ and let $\sigma = \pi \setminus \{2\}$. We shall prove that $l_\sigma \leq d_\pi$. Assume that this is false and let G be a counterexample of minimal order. Without loss of generality we may assume that $O_{\sigma'}(G) = 1$ and that G has a unique minimal normal subgroup N, so that $N \le O_p(G) = F(G) = O_{p',p}(G)$. Obviously N is contained in a Hall π-subgroup A of G and $p \in \sigma$. Put $K = A^{(d_p-1)}$, then K is an abelian normal subgroup of G. Lemma 4 implies that $K \le F(G) = O_p(G)$ Therefore $d_\pi(G/O_p(G)) < d_\pi(G)$. By the minimality of G, we have

$$l_\sigma(G) = 1 + l_\sigma(G/O_\sigma(G)) \le 1 + (d_\pi - 1) = d_\pi,$$

and we are done. □

The following lemma is another version of Lemma IX.3.3 of [8].

Lemma 7. *Let the extraspecial* 2*-group Q be a central product $Q_1 * Q_2 * \cdots * Q_r$ of quaternion groups $Q_1, Q_2, \ldots, Q_r$ of order* 8 *and let A be a group of automorphisms of Q. Then A has a Sylow* 3*-subgroup S such that for each $\alpha \in S$ and $i \in \{1, 2, \ldots, r\}$ there exists $j \in \{1, 2, \ldots, r\}$ with $Q_i^\alpha = Q_j$.*

Our last lemma in this section is Theorem IX.5.4 of [8].

Lemma 8. *Let P be a Sylow* 3*-subgroup of the soluble group G. Then the derived length of $PO_{3',3}(G)/O_{3',3}(G)$ is less than the derived length of P.*

3. The Key Lemma

We shall prove here an analogue of Lemmas 4 and 8 of the previous section.

Lemma 9. *Let A be a Hall π-subgroup of the soluble group G and suppose that $O_{\pi'}(G) = 1$. If $2 \notin \pi$, then the smallest nontrivial term of the derived series of A is contained in the Fitting subgroup of G.*

Proof. Till the end of this section, we assume that the group G is a minimal counterexample to Lemma 9. We divide the proof into a series of small steps. The notation of each of these steps remains fixed throughout the proof of the lemma.

If $t = d_\pi(G) = d_\pi$, then $K = A^{(t-1)}$ is a nontrivial abelian normal subgroup of A. The first two of the following assertions are immediate consequences of Lemmas 3 and 4.

(1) *G has a unique minimal normal subgroup N, $N \le O_3(G)$ and $3 \in \pi$.*

(2) *$N = O_3(G)$, $\Phi(G) = 1$ and $N = C_G(N)$.*

(3) *$K \le O_3(A)$ and $O_{3'}(A) = 1$.*

Proof. This follows from (1), (2) and the fact that $C_G(N) \leq N$. □

(4) $G = AS$, *where* S *is a Sylow* 2*-subgroup of* G *and* $|\pi(A)| > 1$.

Proof. Suppose that $G \neq AS$ for some Sylow 2-subgroup S of G. If G has odd order, then Lemma 9 follows from Lemma 4. Therefore $2 \in \pi(G)$. If $|\pi(A)| = 1$, then G is not a counterexample by Lemma 8. Hence $|\pi(A)| > 1$. Suppose that $q \in \pi(G) \setminus (\pi(A) \cup \{2\})$. Then there exist two proper Hall subgroups X and Y of G containing A such that $G = XY$. Obviously $O_{3'}(X)$ is normalized by a Sylow 3-subgroup of A so that $[O_{3'}(X), O_3(G)] = 1$. Since $O_{3'}(X)$ is self-centralizing by (2), we have $O_{3'}(X) = 1$. By the minimality of G, we have $K < O_3(X)$ and $K < O_3(Y)$. Applying Lemma 1, we have $K \leq O_3(G)$ which contradicts the choice of G. □

(5) *If* $R = A^{(t-2)}$, *then* $R \leq O_3(R) \leftthreetimes L$ *where* $L \neq 1$ *is an abelian* $3'$*-group,* $R' = K = O_3(R)'[O_3(R), L]$ *and* $O_3(R)' \leq N$.

Proof. Since $R' = K$ is a 3-group by (3), $R = O_3(R) \leftthreetimes L$ where L is an abelian Hall $3'$-subgroup of R. Obviously, $O_3(R)$ is normal in A. If $P \in \mathrm{Syl}_3(A)$, then $[R, O_3(R), O_3(R), O_3(R)'] = 1$ because $[R, O_3(R)] \leq O_3(R)$ and $O_3(R)' \leq K$ which is an abelian group. By Lemma IX.5.3 of [8], it follows that $O_3(R)' \leq O_3(G) = N$ (see (2)). So we have $L \neq 1$. Since $R/[O_3(R), L]$ is a nilpotent group and $K \geq [O_3(R), L]$, we can see that $K = [O_3(R), L]O_3(R)'$. □

Thus the proof of Lemma 9 is complete if we can show that the following statement holds for a group G satisfying the hypothesis of Lemma 9:

Suppose that $R = KL$ *is a normal subgroup of* A *such that* $K = O_3(R)$ *is abelian,* $K = [K, L] = [R, R]$ *and* $(|L|, 3) = 1$. *Then* $K \leq F(G)$.

Therefore let G be a minimal counterexample to the last statement. It is easy to see that statements (1)–(4) hold also in this case.

(6) *There is a maximal subgroup* M *of* G *such that* $G = N \leftthreetimes M$. *Moreover,* $A = A_0N$ *for some Hall* π*-subgroup* A_0 *of the group* M. *Also* $O_2(M) = Q \neq 1$, $G = NO_2(M)R$ *and* $A = NR$.

Proof. Using (2), it is easy to see that there exists a maximal subgroup M of G not containing N, so that $G = NM$. Since $M \cap N \trianglelefteq G$ and N is a minimal normal subgroup of G, we have $M \cap N = 1$. Clearly N is contained in each Hall π-subgroup of G and a Hall π-subgroup of M is contained in a Hall π-subgroup of G. Since the Hall π-subgroups of G are conjugate, we may assume that a Hall π-subgroup A_0 of the group M is contained in A. By the modular law, $A = A \cap MN = (M \cap A)N = A_0N$. Clearly, $O_3(M)N$ is a normal 3-subgroup of G, hence contained in $N = O_3(G)$. It follows that $O_3(M) \leq M \cap N = 1$.

Put $F = F(M)$. Since $|F|$ is coprime to 3 and $\pi(A) = \pi(G) \setminus \{2\}$ according to (4), it follows that $[O(F), K] \subseteq N$. Using [5], Lemma 1.2.3 and the fact that K is not contained in N, we have $O(F) \neq F$. Hence $O_2(M) \neq 1$ and $[O_2(M), K]$ is not contained in N.

Set $Y = NR$. Obviously, $[N, R] \subseteq N \cap R \subseteq O_3(R) \cap N$. Because of (5), $NR = O_3(NR) \rtimes L$. Moreover $O_3(NR)' \leq O_3(R) \cap N \leq Z(O_3(NR))$. By Theorem 5.3.6 of [4], we have $[O_3(NR), L] = [O_3(NR), L, L]$. It follows that $[O_3(NR), L] \subseteq [O_3(R), L] \subseteq K$. In particular $(NR)' = [O_3(R), L]O_3(NR)'$ is an abelian normal subgroup of RN containing K. If $O_2(M)Y \neq G$, then $K \leq O_{3',3}(O_2(M)Y) = N$ contrary to the choice of G. Hence $O_2(M)Y = G$ and $A = NR$. □

(7) *The following assertions hold:*

(i) $A_0 = U \rtimes L$, *where* $U = O_3(A_0) = [U, L]$;

(ii) $\big[[N, L], U\big] = 1$;

(iii) A' *is abelian and the Sylow* 3*-subgroup* NU *of the group* A *has nilpotency class* 2.

Proof. By hypothesis, $K = [R, R] = [O_3(R), L]$ and $O_3(R)' \leq N$. Hence the Hall π-subgroup A_0 of M can be expressed in the form $A_0 = U \rtimes L$ where $U = O_3(A_0) = [U, L]$. Since $R = O_3(R)L < A$ and $R \cap N \leq Z(O_3(R)N)$, $O_3(RN)' \leq R \cap N$, the group $A' = [O_3(R), L]O_3(RN)'$ is abelian. Since

$$U \leq [O_3(R), L] \leq [O_3(R)N, L] \leq A',$$

the group U centralizes $[N, L]$. Furthermore, $O_3(O^3(R))N = NU = T$ is a Sylow 3-subgroup of A and $O_3(O^3(R))$ and N are normal in T. Therefore $T' \leq Z(T)$. □

(8) *The subgroup* $Q = O_2(M)$ *is a nonabelian special* 2*-group and* A_0 *acts irreducibly on* $Q/\Phi(Q)$. *Moreover* $[U, \Phi(Q)] = 1$.

Proof. Assume that A_0 induces a reducible group of automorphisms on $Q/\Phi(Q)$. By Maschke's theorem, there exist two proper A_0-admissible subgroups Q_1 and Q_2 of A such that $Q_1Q_2 = Q \neq Q_1$ and $Q \neq Q_2$. It follows that $X = N \rtimes Q_1 \rtimes A_0$ and $Y = N \rtimes Q_2 \rtimes A_0$ are proper subgroups of G and $G = XY$. By the choice of G we have $U \leq O_3(X) \cap O_3(Y)$. By Lemma 1 we have $U < O_3(G) < N$, a contradiction. Hence A_0 induces an irreducible group of automorphisms on $Q/\Phi(Q)$.

If Q is abelian, it follows from Lemma 4 (c) that $U \leq A' \leq N$ which contradicts the choice of G. Thus Q is nonabelian.

Since A_0 acts irreducibly on $Q/\Phi(Q)$, we obtain that $[Q, U] = Q$. Now $X = N \rtimes \Phi(Q) \rtimes A_0$ is a proper subgroup of G. By the minimality of G, we have $U \leq O_3(X)$ and $[U, \Phi(Q)] = 1$. Since $[U, \Phi(Q), Q] = 1 = [Q, \Phi(Q), U]$, we also have $[U, Q, \Phi(Q)] = 1$ by the Three-Subgroup Lemma. From $|U, Q] = Q$, it follows that $[Q, \Phi(Q)] = 1$ and $\Phi(Q) \leq Z(Q)$. Since Q is nonabelian and A_0 acts irreducibly on $Q/\Phi(Q)$, we have $Z(Q) = \Phi(Q) = Q'$ and Q is special. □

(9) *L is a cyclic p-group for some $p \in \pi$.*

Proof. Obviously, if $[a, RN] \leq N$ for each $a \in L^{\#}$ whose order is a prime power, then $U \leq N$ and G is not a counterexample to Lemma 9. Therefore $L = \langle a \rangle$ for some p-element $a \in L$. □

(10) *$H = Q \rtimes U \rtimes L$ is a primitive irreducible soluble linear group as a subgroup of $GL(N)$. Moreover $(x-1)(u-1) = 0$ for each $(x, u) \in L \times N$.*

Proof. It follows from (1), (2), (3) and (6) that H is a subgroup of a soluble irreducible subgroup of $GL(N)$ over $GF(3) = Z_3$. The equality $(x-1)(u-1) = 0$ is another form of (7). In the following, we use some results and arguments from §§ 14–15 of [16]. Let us suppose that $\Phi(Q) = Z(Q)$ is a noncyclic group. Then N, viewed as a right Z_3H-module, is a sum $V_{\lambda_1} \oplus V_{\lambda_2} \oplus \cdots \oplus V_{\lambda_k}$ of subspaces V_{λ_i} $(1 \leq i \leq k)$ such that $vz = \lambda_i(z)v$ for each $(v, z) \in V \times \Phi(Q)$, $\lambda_i(z) \in Z_3^*$. Since the elements of QU are permutable with z we have for each $y \in QU$:

$$\lambda_i(z)(vy) = (vz)y = v(yz) = (vy)z.$$

Hence $vy \in V_{\lambda_i}$ if $v \in V_{\lambda_i}$ and so the subspaces V_{λ_i} are invariant under QU for $i = 1, 2, \ldots, k$. It follows that $L = \langle a \rangle$ permutes the subspaces V_{λ_i}. In particular $V_{\lambda_i}^a \neq V_{\lambda_j}$ if $i \neq j$. Let $u \in U$ and $v \in V_{\lambda_i}$ $(i = 1, 2, \ldots, k)$. Since $(a-1)(u-1) = 0$ for each $u \in U$ that satisfies $v(a-1)(u-1) = 0$ and so $va(u-1) = v(u-1)$ for each $v \in V_{\lambda_i}$ and $u \in U$. Now $va(u-1) \in V_{\lambda_i}^a = V_{\lambda_j}$ and $v(u-1) \in V_{\lambda_i}$. Since $V_{\lambda_i} \cap V_{\lambda_j} = 0$ for $i \neq j$, we have $v(u-1) = 0$ for every $v \in V_{\lambda_i}$ and $u \in U$. But then $vu = v$ for all $v \in N$, so $u \in C_H(N) = 1$ and $U = 1$. However in this case G is not a counterexample to Lemma 9.

Hence we can assume that $\Phi(Q) = Z(Q)$ is a cyclic group and H is a primitive irreducible soluble subgroup of $GL(N)$. □

(11) *$Q = Q_1 * Q_2 * \cdots * Q_p$ where the Q_i are quaternion groups $(i = 1, 2, \ldots, p)$ and QU can be embedded into a central product of p copies of groups isomorphic with $SL(2, 3)$. In addition, the element a permutes the Q_i cyclically.*

Proof. Without loss of generality, we may suppose that $L = \langle a \rangle$ where $a^p \in C_L(U)$ and that U is an elementary abelian 3-group. Since U is not a cyclic group $(p > 3)$ and the action of UL on $Q/\Phi(Q)$ is irreducible, it follows that

$$\bar{Q} = Q/\Phi(Q) = \bar{Q}_1 \times \bar{Q}_2 \times \cdots \times \bar{Q}_s,$$

where the $\bar{Q}_i$ are minimal U-invariant and UL-isomorphic subgroup of the group $\bar{Q}$. Moreover, the $\bar{Q}_i$ are irreducible under the action of $U/C_U(Q_i)$. Since $U/C_U(\bar{Q}_i)$ is an irreducible abelian subgroup of $GL(\bar{Q}_i)$, it is clear that the dimension of $\bar{Q}_i$ over $GF(2)$ is equal to 2. So $|\bar{Q}_i| = 4$.

As $\langle a \rangle$ permutes the $\bar{Q}_i$ cyclically, it follows that $s = p$. Finally, the inverse images Q_i of the $\bar{Q}_i$ in Q are groups of order 8 which admit a cyclic group of order 3 as an automorphism group for each $i \in \{1, 2, \ldots, s\}$. Observe that Q_i is normal in QU. If Q_i is an abelian group, then $[Q_i, U]$ is a normal 4-subgroup of QU, and some element of U permutes the subspaces of N on which elements of $[Q_i, U]^{\#}$ act trivially. This leads to the conclusion that G contains a wreath product $Z_3 \wr Z_3$ and so a Sylow 3-subgroup of G has nilpotency class at least 3. This contradicts (7). It follows that the Q_i are quaternion groups of order 8. The other assertions are now clear. □

(12) $|L| = p$.

Proof. This follows from (11). We have $L = \langle a \rangle$ and $a^p \in Z(H)$. Since $C_N(a^p)$ and $[N, \langle a^p \rangle]$ are admissible under H, the irreducibility of H implies that $C_N(a^p) = 1$ (and so $[N, \langle a^p \rangle] = N \leq K$) or $C_N(a^p) = N$ (and so $a^p = 1$). The first possibility leads to a contradiction since $U \leq C_G(K) \leq N$. □

(13) *G is not a counterexample to Lemma* 9.

Proof. Since $H = QUL$ acts irreducibly on N and is a primitive soluble subgroup of $GL(N)$, Q acts irreducibly on N. In particular the linear span of Q over $GF(3)$ is a full matrix algebra (see [16]). Note that every irreducible faithful representation of $Q_i \leq Q_1 * Q_2 * \cdots * Q_p$ of N over $GF(3)$ has degree 2, since $Z(Q_i) = Z(Q)$ is represented by scalars and $[Z(Q), N] = N$. Hence the representation of A on N over $GF(3)$ is absolutely irreducible.

Let $F = GF(3)$, then, using Theorem 16.3 of [16], we obtain that $\dim_F(N) = 2^p$ and in an appropriate basis of N, the elements of Q can be expressed as Kronecker matrices $h_1 \otimes h_2 \otimes \cdots \otimes h_p$, where $h_i \in SL(2, 3)'$. The irreducibility of Q on N implies that QU is also irreducible on N. It follows from Theorem 16.3 of [16] that the elements of QU can be expressed in the same basis as $h_1 \otimes h_2 \otimes \cdots \otimes h_p$, where $h_i \in SL(2, 3)$. Thus, N is a tensor product $V \otimes V \otimes \cdots \otimes V$ of p copies of a two-dimensional subspace over F. The action of the elements of QU on $x_1 \otimes x_2 \otimes \cdots \otimes x_p \in N$ ($x_i \in V$ for $i = 1, 2, \ldots, p$) can be written as follows:

$$(x_1 \otimes x_2 \otimes \cdots \otimes x_p)(h_1 \otimes h_2 \otimes \cdots \otimes h_p) = x_1 h_1 \otimes x_2 h_2 \otimes \cdots \otimes x_p h_p.$$

Let a be a generator of L which permutes the subgroups Q_i $(i = 1, 2, \ldots, p)$ cyclically so that $Q_i^a = Q_{i+1}$ if $0 < i < p$ and $Q_p^a = Q_1$. Let θ be a permutation of the set $\{1, 2, \ldots, p\}$ according to this action (i.e. $\theta(i) = i + 1 (\mathrm{mod}\, p)$). Define an element a^* by the rule

$$(x_1 \otimes x_2 \otimes \cdots \otimes x_p)a^* = x_{\theta(1)} \otimes x_{\theta(2)} \otimes \cdots \otimes x_{\theta(p)},$$

Then we see that

$$(a^*)^{-1}(h_1 \otimes h_2 \otimes \cdots \otimes h_p)a^* = h_{\theta(1)} \otimes h_{\theta(2)} \otimes \cdots \otimes h_{\theta(p)}.$$

Using only elements in Q and the absolute irreducibility of Q, it is not hard to verify that $a^* = \lambda a$ where $\lambda \in F^*$. As $|a^*| = |a| = p$, we have $\lambda = 1$. Now we can calculate the dimension of a fixed-point subspace of an element a on N as a vector space over F. Obviously,

$$x_1 \otimes \cdots \otimes x_p = (x_1 \otimes x_2 \otimes \cdots \otimes x_p)a = x_2 \otimes x_3 \otimes \cdots \otimes x_p \otimes x_1$$

if and only if $x_1 \otimes x_2 \otimes \cdots \otimes x_p$ is proportional to $x_1 \otimes x_1 \otimes \cdots \otimes x_1$, where $x_1 \in V$. So $\dim_F C_N(a) = 2$. But $\dim_F C_N(a) = \dim_F \ker(a-1)$ and so $\dim_F(\mathrm{Im}(a-1)) = \dim_F N(a-1) = \dim_F N - 2$. As $N(a-1) \le C_N(U)$ according to (10),

$$\dim_F C_N(U) \ge \dim_F N - 2.$$

Note that if $u = h_1 \otimes h_2 \otimes \cdots \otimes h_p$ and $u_1 = h_1' \otimes h_2' \otimes \cdots \otimes h_p'$ are contained in U, then it follows from the commutativity of U and the properties of a Kronecker product of matrices that

$$\begin{aligned} uu_1 &= (h_1 \otimes h_2 \otimes \cdots \otimes h_p)(h_1' \otimes h_2' \otimes \cdots \otimes h_p') \\ &= (h_1h_1' \otimes h_2h_2' \otimes \cdots \otimes h_ph_p') \\ &= (h_1'h_1 \otimes h_2'h_2 \otimes \cdots \otimes h_p'h_p) \\ &= u_1u. \end{aligned}$$

Since $h_ih_i' = h_i'h_i$ and due to the irreducibility of h_i on V when $h_i \neq 1$, it follows that h_i' is a power of h_i if $|h_i| = |h_i'| = 3$. Now we can identify the set U with a subspace S of $R = F^p$ which is invariant under cyclic shifts. Define the weight $w = w(y)$ of an element $y = (f_1, f_2, \ldots, f_p) \in R$ as the number of nonzero f_i. If $u = h_1 \otimes h_2 \otimes \cdots \otimes h_p \in U$ has weight r (in S), then its Jordan normal form has at least r boxes of size 2. In particular if $r < 2$ then the element $h_1 \otimes h_2 \otimes \cdots \otimes h_p$ cannot belong to U since $\dim_F C_N(U) \ge \dim_F N - 2$.

We can now calculate the average weight of an element in S. Note that a subgroup U_i $(i \le p)$ of a group U consisting of those elements $u \in U$ for which the corresponding h_i operates as the identity in the expression $u = h_1 \otimes h_2 \otimes \cdots \otimes h_p$ is conjugate to U_i for each $j \in \{1, 2, \ldots, p\}$. Considering subspaces of S, we find that a subspace $S_i \le S$ consisting of all $y = (f_1, f_2, \ldots, f_p)$ with the property that $f_i = 0$ is a shift of the analogous subspace S_1. We may assume that $U \neq 1$, so that $|S| > 1$. Arrange all elements of S in an $m \cdot p$-dimensional table whose rows consist of elements of S and $m = |S|$. Since S_i is obviously a subgroup of the additive group S of index 3, $\frac{2}{3}$ of each column in this table consists of nonzero elements. Therefore the number of nonzero components in the whole table is $\frac{2}{3}mp$. On the other hand, this number is a sum of weights of all elements in S. So the average weight of an element in S

is $\frac{2}{3}p$. It follows that there exists an element $y \in S$ whose weight is at least $\frac{2}{3}p$. Since $w(y) \leq 2$, we have $p \leq 3$, contradicting the definition of p.

This is a final contradiction and we conclude that the counterexample to Lemma 9 does not exist. □

4. Main Results

Theorem 1. *Let G be a π-soluble finite group. If $2 \notin \pi$, then $l_\pi(G) \leq d_\pi(G)$.*

Proof. If $3 \notin \pi$, then this follows from Lemma 6. So let $3 \in \pi$ and let the group G be a minimal counterexample for the theorem. Then $O_{\pi'}(G) = 1$ and G has a unique minimal normal subgroup N. As G is π-soluble, N is an elementary abelian p-group and $p \in \pi$. In this case $F(G) = O_p(G)$. It follows from [5], Lemma 1.2.3 that $C_G(F(G)) \leq F(G)$. Set $d_\pi(G) = t$. If A is any Hall π-subgroup of G and $K = A^{(t-1)}$, then K is an abelian normal subgroup of A. Let r be a prime in $\pi(G) \setminus \pi$ and put $\sigma = \pi \cup \{r\}$. By a well-known theorem by P. Hall, G is a D_σ-group. In particular, G has a soluble Hall σ-subgroup S containing a subgroup A for each choice of r. As $F(G) \leq S$ and $C_G(F(G)) \leq F(G)$, we have $O_{\pi'}(S) = 1$. By Lemma 9 we see that $K \leq O_\pi(G)$ so that $d_\pi(G/O_\pi(G)) \leq d_\pi(G) - 1$. By the minimality of G, we have $l_\pi(G/O_\pi(G)) \leq d_\pi(G) - 1$ and so $l_\pi(G) \leq d_\pi(G)$. □

Theorem 2. *Let the finite soluble group G be a product of two subgroups A and B of coprime orders. Then the Fitting height $h(G)$ does not exceed $d(A) + d(B) + l_2(G)$. In particular if $2 \in \pi(A)$, then $h(G) \leq 2d(A) + d(B)$.*

Proof. Let the group G be a minimal counterexample for Theorem 2. Since the functions d and l do not increase when we pass to subgroups and homomorphic images, it is not hard to show that G has a unique minimal normal subgroup $N \leq O_p(G) = F(G) = O_{p',p}(G)$. Let $\bar{G} = G/F(G)$, $\bar{A} = AF(G)/F(G)$ and $\bar{B} = BF(G)/F(G)$. Obviously, $h(\bar{G}) = h(G) - 1$ and $h(\bar{G}) \leq d(\bar{A}) + d(\bar{B}) + l_2(G)$. It is enough to prove that at least one of the parameters on the right side of the above inequality is strictly less for $\bar{G}$ than for G.

If $p = 2$ then $\bar{G} = G/O_{2',2}(G)$ and $l_2(\bar{G}) < l_2(G)$. Therefore suppose that $p > 2$. Put $\pi(A) = \pi$, $t = d(A)$ and $K = A^{(t-1)}$. Obviously $p \in \pi$ and $[O_{p'}(K), F(G)] = 1$. Since $C_G(F(G)) \leq F(G)$, so K is a p-group. By Lemma 4, $K \leq F(G)$ holds in each of the following cases: $\{2, 3\} \cap \pi = \emptyset$; $\{2, 3\} \subseteq \pi$. But if $2 \in \pi$, then it follows from Lemma 2 that $K \leq O_{p',p}(G) = F(G)$. So $d(\bar{A}) \leq d(A) - 1$. As $d(A)$ and $d(B)$ are symmetric in the inequality, this proves that counterexamples to Theorem 2 do not exist and we are done. By the theorem of

Brûhanova [2], $l_2(G)$ does not exceed the derived length of its Sylow 2-subgroup. This gives the last bound of Theorem 2. □

Remark. It is not hard to show that $h(G) \leq \sum l_p(G)$, where p ranges over $\pi(G)$. In particular if the least common multiple of the orders of the elements of G is $p_1 p_2 \dots p_n$ where the p_i are primes (which are not necessarily distinct), then $h(G) \leq n+k$ where k is the number of Fermat primes in this decomposition. If G is a group of exponent $2^\alpha q^\beta$, then $h(G) \leq \alpha + 2\beta$.

Theorem 3. *Let the finite soluble group G be the product of its subgroups A and B with coprime orders. Then the derived length of G does not exceed*

$$2d(A)d(B) + d(A) + d(B).$$

If G has odd order then

$$d(G) \leq d(A)d(B) + \max\{d(A), d(B)\}.$$

Proof. Without loss of generality we may assume that $2 \in \pi = \pi(A)$ and $\pi' = \pi(G) \setminus \pi(A) = \pi(B)$. Set $l_\pi = l_\pi(G)$, $l_{\pi'} = l_{\pi'}(G)$, $d(A) = d_\pi$ and $d(B) = d_{\pi'}$. We consider the upper $\pi'\pi$-series of G:

$$1 = N_0 \leq P_1 \leq \dots \leq P_{l-1} \leq N_{l-1} \leq P_l \leq N_l \leq P_{l+1} = G,$$

where $N_i/P_i = O_{\pi'}(G/P_i)$ for $i = 1, 2, \dots, l$ and $P_i/N_{i-1} = O_\pi(G/N_{i-1})$ for $i = 1, 2, \dots, l+1$. Obviously $l \leq l_{\pi'}$ and $l \leq l_\pi$. Note that Lemma 9 implies that $d(N_i/P_i) \leq d_{\pi'} - i + 1$ for all $i \in \{1, 2, \dots, l\}$ and $d(P_i/N_{i-1}) \leq d_\pi$. Therefore we have

$$\begin{aligned} d(G) &\leq (l+1)d_\pi + \sum_{i=0}^{l-1}(d_{\pi'} - i) = (l+1)d_\pi + \frac{1}{2}(2d_{\pi'} - 1)l \\ &= d_\pi + \frac{1}{2}(2d_\pi + 2d_{\pi'} - l + 1)l = \phi(l). \end{aligned}$$

Obviously $\phi'(l) = \frac{1}{2}(2d_\pi + 2d_{\pi'} - 2l + 1)$. As $l \leq l_{\pi'} \leq d_{\pi'}$ by Theorem 1, we have $\phi'(l) \geq 0$ for each admissible l and the maximum of the $\phi(l)$ is reached when l has is maximal value. Let $d_{\pi'} \leq 2d_\pi$, then

$$d(G) \leq d_\pi + \frac{1}{2}(2d_\pi + d_{\pi'} + 1)d_{\pi'} \leq 2d_\pi d_{\pi'} + d_\pi + d_{\pi'}$$

and we are done. If $d_{\pi'} < 2d_\pi$, then $l \leq 2d_\pi$ and

$$d(G) \leq d_\pi + \frac{1}{2}(2d_{\pi'} + 1)2d_\pi = 2d_\pi d_{\pi'} + 2d_{\pi'}.$$

This proves the first part of Theorem 3.

Now suppose that G has odd order and is a counterexample of minimal order to the second statement of Theorem 3. Then G has a unique minimal normal subgroup N and $N \leq O_p(G) = F(G)$. Suppose that $p \in \pi = \pi(A)$. If $t = d_\pi$, then by Lemma 9 we have $A^{(t-1)} = K \leq F(G)$. Thus $d_\pi(G/O_\pi(G)) \leq d_\pi - 1$. Similarly $d_{\pi'}(G/O_{\pi,\pi'}(G)) \leq d_{\pi'} - 1$. By the minimality of G, we have $d(G/O_{\pi,\pi'}(G)) \leq (d_\pi - 1)(d_{\pi'} - 1) + \max\{d_\pi - 1, d_{\pi'} - 1\}$. Note that $d(O_{\pi,\pi'}(G)) \leq d_\pi + d_{\pi'}$ so that

$$\begin{aligned} d(G) &\leq d(G/O_{\pi,\pi'}(G)) + d(O_{\pi,\pi'}(G)) \\ &= d_\pi d_{\pi'} - d_\pi - d_{\pi'} + 1 + \max\{d_\pi, d_{\pi'}\} - 1 + d_\pi + d_{\pi'} \\ &= d_\pi d_{\pi'} + \max\{d_\pi, d_{\pi'}\}. \end{aligned}$$

The symmetry between A and B gives the proof in the general case. □

Theorem 4. *Let the periodic soluble group G be the product of two subgroups A and B with $\pi(A) \cap \pi(B) = \emptyset$. Then the derived length $d(G)$ does not exceed $2d(A)d(B) + d(A) + d(B)$.*

Proof. In fact it is not hard to see that $\pi(G) = \pi(A) \cup \pi(B)$. Now we can prove by induction that each finite $\pi(A)$-subgroup of G is contained in a conjugate of A and each finite $\pi(B)$-subgroup of G is contained in a conjugate of B. As G is soluble, each finite subgroup H of G is factorized by two subgroups one of which is a Hall $\pi(A)$-subgroup and the other is a Hall $\pi(B)$-subgroup. Then we may apply Theorem 3 to complete the proof. □

The bound of Theorem 3 can be improved if one of the subgroups A and B is abelian and the other is nilpotent. A group is called 2-finite if every subgroup that is generated by two elements is finite.

Corollary. *Let $G = AB$ be a 2-finite group which is a product of an abelian group A and a nilpotent group B of nilpotency class c. If $\pi(A) \cap \pi(B) = \emptyset$, then G is soluble and the derived length of G does not exceed $3 + 2\log_2(c)$. If $2 \notin \pi$, then $d(G) \leq 2 + 2\log_2(c)$.*

Proof. The group G is soluble by [10]; see also [11], Theorem 3.4.13. Since the π-length of a finite soluble group with abelian Hall π-subgroup is at most 1 by [5], Lemma 1.2.3, we have $d(G) \leq 2d(B) + 1$ and $d(G) \leq 2d(B)$ if $2 \notin \pi$. So the result follows from the fact that $d(B) \leq \log_2(c) + 1$. □

References

[1] Berger, T. R., Gross, F., 2-Length and the Derived Length of a Sylow 2-subgroup. Proc. London Math. Soc. (3) 34 (1977), 520–534.

[2] Brûhanova, E. G., Connection between the 2-length and the derived length of a Sylow 2-subgroup of a finite solvable group. Math. Notes 29 (1981), 85–90.

[3] Chernikov, N. S., Petravchuk, A. P., Characterization of a periodic locally soluble group by its soluble Sylow π-subgroups of finite exponent. Ukrain. Math. J. 39 (1987), 619–624.

[4] Gorenstein, D., Finite Groups. Harper Row, New York 1968.

[5] Hall, P., Higman, G., On the p-length of p-soluble groups and reduction theorems for Burnside's Problem. Proc. London Math. Soc. (3) 6 (1956), 1–42.

[6] Heineken, H., Products of finite nilpotent groups. Math. Ann. 287 (1990), 643–652.

[7] Huppert, B., Endliche Gruppen I. Springer, Berlin 1967.

[8] —, Blackburn, N., Finite Groups II. Springer, Berlin 1982.

[9] Johnson, P. M., A property of factorizable groups. Arch. Math. (Basel) 60 (1993), 414–419.

[10] Kazarin, L. S., On products of an abelian group and a group with non-trivial centre. Deposited at VINITI No. 3565-81, Novosibirsk 1981.

[11] —, Kurdachenko, L. A., Finiteness conditions and factorizations in infinite groups. Russian Math. Surveys 47 (1992), 81–126.

[12] Kegel, O. H., Produkte nilpotenter Gruppen. Arch. Math. (Basel) 12 (1961), 90–93.

[13] Kramer, O. U., On the theory of soluble factorizable groups. Bull. Austral. Math. Soc. 16 (1976), 97–110.

[14] Mazurov, D., On the p-length of a soluble group. In: All-Union Symposium on Group Theory, Collection of Scientific Works, Kiev, Naukova Dumka 1980, 51–60.

[15] Pennington, E., On Products of Finite Nilpotent Groups. Math. Z. 134 (1973), 81–83.

[16] Suprunenko, D. A., Matrix Groups. M. Nauka, 1972.

Torsionfree Varieties of Metabelian Groups

L. G. Kovács and M. F. Newman

Abstract. It is proved here that the free groups of the variety $\mathfrak{N}_c\mathfrak{A}_s \wedge \mathfrak{A}^2$ are torsionfree. As usual, $\mathfrak{A}$ denotes the variety of abelian groups, $\mathfrak{A}_s$ the variety of abelian groups of exponent dividing s, and $\mathfrak{N}_c$ the variety of nilpotent groups of class at most c.

1991 Mathematics Subject Classification: 20E10

We call a variety of groups ***torsionfree*** if its free groups are torsionfree. Apart from this, we follow the notation and terminology of Hanna Neumann's book [4]. Recent work by Samuel M. Vovsi and the first author [3] on the growth of varieties of groups depends (among other things) on results of J. R. J. Groves [2]. In turn, these make use of the classification of the torsionfree varieties of metabelian groups, Theorem 6.1.2 in R. A. Bryce [1]. All but one step of the proof of this classification was given in Appendix I of [1], but our 'forthcoming' paper which would have contained the missing step was never written. In view of the renewed interest, it seems desirable to place the missing step on record.

Theorem (6.1.2 in [1]). *The varieties of groups* $\mathfrak{N}_c\mathfrak{A}_s \wedge \mathfrak{A}^2$ $(c, s \geq 1)$ *are torsionfree and join-irreducible. Every torsionfree proper subvariety of* $\mathfrak{A}^2$ *can be uniquely expressed as an irredundant join of some of these torsionfree join-irreducibles.*

All joins mentioned here are the joins of finitely many join-irreducibles. The uniqueness claim is a particularly important part of the Theorem. It implies (by very simple and general lattice-theoretic considerations) that one join of join-irreducibles, $\bigvee_i \mathfrak{U}_i$, is contained in another, $\bigvee_j \mathfrak{V}_j$, if and only if to each i there is a j such that $\mathfrak{U}_i \subseteq \mathfrak{V}_j$. In particular, a join of join-irreducibles is irredundant if and only if its components are pairwise incomparable. It is also well-known that comparability is easy to settle here: $\mathfrak{N}_c\mathfrak{A}_s \wedge \mathfrak{A}^2 \subseteq \mathfrak{N}_{c'}\mathfrak{A}_{s'} \wedge \mathfrak{A}^2$ if and only if s divides s' and $c \leq c'$. It is with these points in mind that one speaks of a 'classification'. We conclude the paper with a short argument which justifies these points independently: instead of appealing to the uniqueness claim, it actually implies that claim.

Offprint from
Infinite Groups 94, Eds.: de Giovanni/Newell

The outstanding step in the proof of the theorem itself is to show the first of its claims, namely that each $\mathfrak{N}_c\mathfrak{A}_s \wedge \mathfrak{A}^2$ is torsionfree. The key point of the proof is the following.

Lemma. *Let G be a free group of the variety $\mathfrak{N}_c \wedge \mathfrak{A}^2$ freely generated by $g_1, \dots, g_n$, let m be an integer with $1 \le m < n$, and denote by H the subgroup of G generated by the commutator subgroup G' and the elements $g_{m+1}, \dots, g_n$. Then the commutator factor group H/H' is torsionfree.*

Proof. All we really need to show is that G'/H' is torsionfree. We shall be working with basic commutators with reference to the given free generating set of G. As is well known (see 36.32 in [4]), the left-normed basic commutators of weight at least 2 and at most c form a basis of G' as free abelian group. Let X denote the union of this set with $\{ g_j \mid m < j \le n \}$. Clearly, H is just the subgroup generated by X. Let Y be the set of those left normed basic commutators of weight at least 2 whose last entry has subscript larger than m: thus a typical element y of Y has the form $[g_{i_1}, \dots, g_{i_t}]$ with $t \ge 2$ and $i_t > m$. Here $[g_{i_1}, \dots, g_{i_{t-1}}]$ lies in X: by definition when $t > 2$, and when $t = 2$ then because y is basic. Thus $Y \subset H'$.

We claim that the subgroup $\langle Y \rangle$ generated by Y is normal in H. Of course Y is centralized by G', so what we have to show to prove this claim is that if $y \in Y$ and $j > m$ then $[y, g_j] \in \langle Y \rangle$. If $i_t \le j$ then $[y, g_j]$ is basic as written and lies in Y. If $i_2 \le j < i_t$ then $[y, g_j]$ is not basic as written but is still equal to an element of Y, for in a metabelian group the order of the entries of a left-normed commutator is irrelevant beyond the first two places (see 34.51 in [4]). Finally, if $j < i_2$ then $i_1 > i_2 > j > m$ and $m < j < i_2 \le \cdots \le i_t$ because y is basic, so all $t+1$ entries of $[y, g_j]$ belong to $\{ g_j \mid m < j \le n \}$. In this case of course $[y, g_j]$ can be written as a product of basic commutators of weight at least $t+1$, all entries of all the basic commutators involved coming from $\{ g_j \mid m < j \le n \}$. Then all these basic commutators lie in Y and so $[y, g_j]$ lies in $\langle Y \rangle$.

Next we claim that if $u, v \in X$ then $[u, v] \in \langle Y \rangle$. If $u, v \in G'$ then $[u, v] = 1$ while if $u, v \notin G'$ then either $[u, v]$ or $[u, v]^{-1}$ lies in Y, so it suffices to deal with the case of $u = [g_{i_1}, \dots, g_{i_t}]$ with $t \ge 2$, $v = g_j$ with $j > m$. If now $i_t \le j$ then $[u, v]$ is basic as written and so lies in Y; otherwise $i_t > j > m$ so $u \in Y$ and then $[u, v] \in \langle Y \rangle$ follows by the previous paragraph.

We have proved that the subset Y of H' generates a normal subgroup in H which contains the commutator of each pair of elements from the generating set X of H: thus $\langle Y \rangle = H'$. Since Y is a subset of a basis of the free abelian group G', it follows that G'/H' is also free abelian. □

Proof of the first claim of the Theorem. It suffices to deal with noncyclic relatively free groups of finite rank. Let F be a noncyclic absolutely free group of rank m, and let A be the verbal subgroup $\mathfrak{A}_s(G)$. By Schreier's Theorem, A is also absolutely free of finite rank and its rank is greater than m: denote this rank by n. Set $N = A''\mathfrak{N}_c(A)$ and

$G = A/N$: then G is an $(\mathfrak{N}_c \wedge \mathfrak{A}^2)$-free group of rank n. The quotient A/F' is free abelian of rank m (because it is a subgroup of finite index in the free abelian group F/F' of rank m), so A/A' splits over F'/A', and F'/A' is free abelian of rank $n-m$. Choose a basis for A/F', and choose a preimage $\{g_1, \ldots, g_m\}$ for that basis in A/N. Similarly, choose $g_{m+1}, \ldots, g_n$ in F'/N so that the image of $\{g_{m+1}, \ldots, g_n\}$ modulo A'/N is a basis for F'/A'. Then $\{g_1, \ldots, g_n\}$ generates G modulo G'. As is well known (see 31.25 in [4]), this implies that $\{g_1, \ldots, g_n\}$ generates G itself, and (see 32.1, 41.4, and 41.33 in [4]) generates it freely. We may therefore set $H = F'/N$, note that $H' = F''N/N$, and apply the Lemma to conclude that $F'/F''N$ is torsionfree. Of course then $F/F''N$ is also torsionfree. Since $F''N = F''\mathfrak{N}_c(A)$ by the definition of N and since $\mathfrak{N}_c(A) = \mathfrak{N}_c\mathfrak{A}_s(F)$, we have that $F''N = (\mathfrak{N}_c\mathfrak{A}_s \wedge \mathfrak{A}^2)(F)$. This completes the proof of the Theorem. □

As promised in the introduction, we close with a simple proof of the fact that

$$\mathfrak{N}_c\mathfrak{A}_s \wedge \mathfrak{A}^2 \subseteq \bigvee_i (\mathfrak{N}_{c(i)}\mathfrak{A}_{s(i)} \wedge \mathfrak{A}^2) \tag{1}$$

cannot hold unless for some i we have $s|s(i)$ and $c \leq c(i)$.

Suppose that (1) holds.

By Dirichlet's Theorem, there are infinitely many primes p such that $s|(p-1)$. For such a p, in the holomorph of a group of order p one can find an element g of order p and an element h of order s. The subgroup $\langle g, h\rangle$ is in $\mathfrak{A}_p\mathfrak{A}_s$ and hence also in $\mathfrak{N}_c\mathfrak{A}_s \wedge \mathfrak{A}^2$. Consider the left-normed commutator

$$[x, y^{t(1)}, \ldots, y^{t(n)}]$$

where $n = \sum(c(i)+1)$ and for each i the sequence $t(1), \ldots, t(n)$ has $c(i)+1$ terms equal to $s(i)$. This commutator is a law in each of the $\mathfrak{N}_{c(i)}\mathfrak{A}_{s(i)}$, but setting $x = g$, $y = h$ shows that it is not a law in $\langle g, h\rangle$ unless some $s(i)$ is divisible by s. This proves that there is at least one i with $s|s(i)$.

Suppose now that $c(i) < c$ whenever $s|s(i)$: we shall show that this leads to a contradiction. Let s' denote the least common multiple of the $s(i)$ that are divisible by s. The inclusion (1) remains valid if we replace all the corresponding join components on the right hand side by $\mathfrak{N}_{c-1}\mathfrak{A}_{s'}$. Instead of changing notation, we assume without loss of generality that s divides $s(1)$ but does not divide any of the other $s(i)$, and that $c(1) < c$.

Choose p large — say, so that also $p > c + \sum(c(i)+s(i))$. Let m be an integer such that, in the holomorph considered above, the conjugate g^h is the same as the power g^m. The wreath product W of two groups of order p is nilpotent of class p (which is larger than c), and it has an automorphism of order s which acts trivially on the top group and mth poweringly on the base group. Let $P = W/\mathfrak{N}_c(W)$; let g be the image in P of a generator of one of the coordinate subgroups of W, and k the image of a generator of the top group. It follows that P has an automorphism which sends g to g^m and fixes k. Let G be the semidirect product of P by $\langle h\rangle$, with h acting on P as the automorphism just described. The normal closure of g is abelian (of order p^c), and the factor group over that is cyclic (of order ps), so $G \in \mathfrak{N}_c\mathfrak{A}_s \wedge \mathfrak{A}^2$. It is easy to see that a nontrivial element of the normal closure of g and a nontrivial element of $\langle h \rangle$ can never commute. Consider the left-normed commutator

$$[x_1^{s(1)}, \ldots, x_c^{s(1)}, y^{t(1)}, \ldots, y^{t(n)}]$$

where this time $n = \sum_{i \neq 1}(c(i) + 1)$ and for each $i \neq 1$ the sequence $t(1), \ldots, t(n)$ has $c(i)+1$ terms equal to $s(i)$. This commutator is also a law in each of the $\mathfrak{N}_{c(i)}\mathfrak{A}_{s(i)}$, but setting

$$x_1 = g, \ x_2 = \cdots = x_c = k, \ y = h$$

shows that it is not a law in G. The desired contradiction has been reached and the proof is complete. □

References

[1] Bryce, R. A., Metabelian groups and varieties. Philos. Trans. Roy. Soc. London Ser. A 266 (1970), 281–355.

[2] Groves, J. R. J., Varieties of soluble groups and a dichotomy of P. Hall. Bull. Austral. Math. Soc. 5 (1971), 391–410.

[3] Kovács, L. G., and Vovsi, S. M., Growth of varieties of groups and group representations, and the Gel'fand–Kirillov dimension. J. Algebra. To appear.

[4] Neumann, H., Varieties of groups. Springer-Verlag, Berlin Heidelberg New York 1967.

Rings with Noetherian Groups of Units

*Jan Krempa**

Abstract. Let R represent an associative, but nonnecessarily commutative ring with $1 \neq 0$, G a nontrivial group, and RG the group ring of G over R. We are going to survey and to extend some results about the following problem: *Find the necessary and sufficient conditions under which the unit group of RG is Noetherian.* The same problem will also be considered here for crossed products.

1991 Mathematics Subject Classification: 20C07

1. Preliminaries

In this paper we assume that, if not stated otherwise, rings are associative with $1 \neq 0$. Subrings with the same unities will be called unital. If A is a ring then $\mathrm{U}(A)$ will always denote the unit group of the ring A, A^+ the additive group of A, and $\mathrm{J}(A)$ the (Jacobson) radical of A. If $B \subset A$ is a radical subring then the set $1 + B$ will be considered as a subgroup of $\mathrm{U}(A)$. $\inf\{a \times b\}$ Further we will say that a ring A is ***semisimple*** if $\mathrm{J}(A) = 0$, ***reduced*** if A has no nontrivial nilpotents and ***indecomposable*** if it has no proper central idempotents. In the sequel we will use the following notation:

$\mathbb{N}$ the set of natural numbers;
$\mathbb{Z}$ the ring of rational integers;
$\mathbb{Q}$ the field of rational numbers;
$\mathbb{F}_q$ the finite field with q elements;
C_n The cyclic group of order n.

Let us recall that a group G is Noetherian if it satisfies the maximum condition for subgroups or, equivalently, if any subgroup of G is finitely generated. It is well known, (see [14]), that Noetherian groups form quite a large class which is closed under many natural operations.

In this note we are going to investigate rings with Noetherian unit groups. In this section we will show that such rings are rather close to reduced ones. In the next two sections we will discuss necessary and sufficient conditions for group rings and

*Supported by Polish research grant of KBN

Offprint from
Infinite Groups 94, Eds.: de Giovanni/Newell

crossed products to have Noetherian groups of units. In this way we continue research of Bovdi [2], Karpilovsky [6], and others.

Andruszkiewicz in [1, Corollary 3] proved the following extension of results obtained by Watters in [18]:

Lemma 1.1. *Let S be a radical subring of the ring R. Then the following conditions are equivalent:*

1. *$1+S$ is a finitely generated nilpotent group;*
2. *$1+S$ is a Noetherian group;*
3. *S^+ is finitely generated.*

Under any of these conditions the ring S is nilpotent.

Corollary 1.2. *Let R be a ring and $I \subset \mathrm{J}(R)$ be its ideal. Then $\mathrm{U}(R)$ is Noetherian if and only if I^+ is finitely generated and $\mathrm{U}(R/I)$ is Noetherian.*

The above corollary will allow us often to consider only semisimple or semiprime rings. For orders (over $\mathbb{Z}$) we have the following result mainly due to Hartley [4].

Theorem 1.3. *Let R be an order, and let A denote its ring of quotients. Then $\mathrm{U}(R)$ is Noetherian if and only if A is a direct sum of fields and totally definite quaternion algebras.*

Proof. After a reduction as in [5] we can assume that R is an order in a finite dimensional division $\mathbb{Q}$-algebra A. If A is a field then $\mathrm{U}(R)$ is Noetherian by the Dirichlet Unit Theorem. If A is a totally definite quaternion algebra then, as in [3], $\mathrm{U}(R)$ is also Noetherian. In any other case by [4] $\mathrm{U}(R)$ contains a free nonabelian subgroup, hence it is not Noetherian. □

Remark. If R is an arbitrary ring with R^+ finitely generated then by Corollary 1.2 we can assume that R is semiprime. If T denotes the torsion part of R^+ then T as a ring is finite semiprime, hence with unity element. It means that $R = T \oplus S$ where S is an order. Moreover $\mathrm{U}(R)$ is Noetherian if and only if $\mathrm{U}(S)$ is Noetherian. In this way rings R with $\mathrm{U}(R)$ Noetherian and R^+ finitely generated can be completely described.

As an elementary observation about units of rings without any restrictions on additive groups we have:

Proposition 1.4. *Let A be a ring such that $\mathrm{U}(A)$ is Noetherian. Then:*

1. *If $B \subset A$ is a subring with unity then $\mathrm{U}(B)$ is Noetherian;*
2. *If $e_1, e_2, \dots$ is a sequence of nonzero orthogonal idempotents in A then $\mathrm{U}(e_kAe_k) = \{e_k\}$ for almost all k.*

Proof. Let $B \subset A$ be a subring with unity e. Then we have an embedding ϕ of $\mathrm{U}(B)$ into $\mathrm{U}(A)$ given by $\phi(u) = u + 1 - e$ for $u \in \mathrm{U}(B)$.

If $e_1, e_2, \ldots$ is a sequence of nonzero orthogonal idempotents in A then we have an ascending chain of subgroups $H_k \subset \mathrm{U}(A)$ given by:

$$H_k = \Big\{u_1 + \cdots + u_k + (1 - \sum_{l=1}^{k} e_l);\ u_i \in \mathrm{U}(e_i A e_i) \text{ for } i = 1, \ldots, k\Big\}.$$

This by assumption easily completes the proof. □

For proofs of further results we will use a lemma which is also of independent interest.

Lemma 1.5. *Let A be a ring in which every subring with zero multiplication is finitely generated. If $a \in A$ is such that $a^2 = 0$ then the group $(AaA)^+$ is finitely generated.*

Proof. Let $B = \{b \in A;\ aba = 0\}$ and let $C = \{c \in A;\ ac = 0\}$. Clearly $C \subset B$ and $Ba \subset C$.

By assumption $(aAa)^2 = 0$ hence the group $(aAa)^+ \simeq A^+/B^+$ is finitely generated. Moreover $(aB)^2 = 0$ hence by the same argument $(aB)^+ \simeq B^+/C^+$ is also finitely generated. It means that $(aA)^+ \simeq A^+/C^+$ is finitely generated too. Analogously one can show that $(Aa)^+$ is finitely generated.

Now it is evident that if $x_1, \ldots, x_m$ are such that $ax_1, \ldots, ax_m$ generate $(aA)^+$ and $y_1, \ldots, y_n$ are such that $y_1 a, \ldots, y_n a$ generate $(Aa)^+$ then the set $\{y_k a x_l\}$ generates $(AaA)^+$. □

Theorem 1.6. *Let A be a semiprime ring with $\mathrm{U}(A)$ Noetherian. Then for some $0 \leq m < \infty$ we have a decomposition:*

$$A = I \oplus S_1 \oplus \cdots \oplus S_m$$

where the ideal I is reduced and all ideals S_k are simple finite noncommutative.

Proof. If a finite noncommutative ring $S \subset A$ is an ideal then it is semiprime, which implies that it has the unity and $\mathrm{U}(S)$ is nontrivial. Hence by Proposition 1.4 there exists only a finite number of such ideals. So we can write

$$A = I \oplus S_1 \oplus \cdots \oplus S_m$$

for some $0 \leq m < \infty$ where S_l is a simple finite noncommutative ring for every $1 \leq l \leq m$ and I is an ideal of A which has the unity and does not contain ideals of type S_l.

Now we will show that I is reduced. Let $a \in I$ be an element of finite additive order such that $a \neq 0$ but $a^2 = 0$. By the above lemma the group $(AaA)^+$ is finite. Semiprimeness of A implies that as a ring AaA is semisimple, hence noncommutative, because it contains the nonzero nilpotent a. By the Wedderburn Theorem AaA contains an ideal $S \subset I$ which is simple, finite and noncommutative, which is impossible by the choice of I.

Now let $a \in I$ be an element of infinite additive order such that $a \neq 0$ but $a^2 = 0$. Again, as above, the group $(AaA)^+$ is finitely generated and as a ring AaA is semiprime and noncommutative. Let B be the unital subring of A generated by AaA. Then $B = AaA + \mathbb{Z} \cdot 1$, hence B^+ is finitely generated. Let T be the torsion part of B. Then due to the case considered above we can assume that T is a finite reduced ring, hence $B = T \oplus B'$ for some ideal $B' \subset B$. Clearly $a \in B'$ and it can be verified that B' is an order with $\mathrm{U}(B')$ Noetherian. This contradicts Theorem 1.3. □

Corollary 1.7. *Let R be a semiprime ring without proper finite ideals. If $\mathrm{U}(R)$ is Noetherian then R is reduced.*

2. Group Rings

In this section R will represent a ring of coefficients, G a nontrivial group, and RG the group ring of G over R. We will apply some standard notation, terminology, and results, on group rings used for example in [6, 12, 16].

In this section we are going to discuss necessary and sufficient conditions for $\mathrm{U}(RG)$ to be Noetherian. Commutative aspect of this question is described for example in [6, 10]. Noncommutative case of it was discussed for example in [2, 9]. We extend here some results from these two papers.

Using some obvious units of RG connected with nilpotents and idempotents of R and Proposition 1.4 we have the following observation from [9], (see [6], Lemma 3.4.4):

Proposition 2.1. *Let $\mathrm{U}(RG)$ be Noetherian and let $A = A(RG)$ be the subring of R generated by all coefficients of all elements from $\mathrm{U}(RG)$. Then:*

1. *$\mathrm{U}(R)$ and G are Noetherian;*
2. *A is a finitely generated unital subring of R and $\mathrm{U}(RG) = \mathrm{U}(AG)$;*
3. *A contains all radical subrings and all idempotents of R;*
4. *R has no infinite sequences of nonzero orthogonal idempotents; in particular it is a finite direct sum of indecomposable rings.*

From Lemma 1.5 and Theorem 1.6 we can obtain more essential facts.

Theorem 2.2. *Let $\mathrm{U}(RG)$ be Noetherian. Then RG is reduced in any of the following cases:*

1. *G is infinite;*
2. *R is semiprime and has no finite nonzero ideals.*

Proof. Let RG be not reduced, $0 \neq a \in RG$ be such that $a^2 = 0$ and let I be the ideal of RG generated by a. By Lemma 1.5 this ideal has finitely generated additive group. Looking at supports of elements from I it can be seen that G is finite. This proves the first case.

Now let G be finite and R semiprime with no finite nonzero ideals, and let J be the ideal of R generated by all coefficients of all elements of I. Then J^+ is finitely generated. If T is the torsion part of J then T is a finite ideal of R hence by assumption $T = 0$. Let A denote the unital subring of R generated by J. Then $a \in I \subset JG \subset AG$ and AG is an order with Noetherian group of units. It is impossible by Theorem 1.3. Hence in this case RG is reduced too. □

Our next two results were in another way proved by Bovdi in [2].

Theorem 2.3. *Let G have a finite subgroup which is not normal. Then* $\mathrm{U}(RG)$ *is Noetherian if and only if R and G are finite.*

Proof. By assumption there exists $g \in G$ of order $n < \infty$ such that the subgroup $\langle g \rangle$ is not normal in G. Then there exists $h \in G$ which does not normalize this subgroup. If we put $a = (1 - g)h(1 + g + \cdots + g^{n-1})$ then $a \neq 0$ but $(Ra)^2 = 0$, hence by the above theorem G is finite. Moreover $R^+ \simeq (Ra)^+$ hence by Lemma 1.1 R^+ is finitely generated. By Theorem 2.2 or Theorem 1.3 the inclusion $\mathbb{Z} \subset R$ is impossible, hence R^+ has to be periodic. This means that R is finite. □

Corollary 2.4. *Let RG be an infinite ring with* $\mathrm{U}(RG)$ *Noetherian and $H \subset G$ the subgroup generated by all elements of finite order. Then any subgroup of H is finite and normal in G.*

Remark. From the above corollary one can see that if G is an infinite group with finite subgroups, then crossed products with torsion free groups should be involved in studying Noetherian $\mathrm{U}(RG)$. We will do it in the next section.

Now let us return to ordinary group rings. For convenience of further notation let us agree that if $n \in \mathbb{N}$ then $R_{[n]} = \{r \in R \mid nr = 0\}$.

Proposition 2.5. *Let R be semiprime without proper finite ideals and let G be a finite group such that* $\mathrm{U}(RG)$ *is Noetherian. Then:*

1. *If m is the order of G then $R_{[m]} = 0$;*
2. *If G is nonabelian then R^+ is torsion free.*

Proof. Let m be the order of G and $a = \sum_{g \in G} g$. Then $a^2 = ma$ hence $(R_{[m]}a)^2 = 0$. By Lemma 1.1 it gives that $R_{[m]}$ is a finite ideal of R. Hence by assumptions on R we have $R_{[m]} = 0$.

Let p be a prime number such that $R_{[p]} \neq 0$. By the above part of the proof p does not divide the order of G. Hence the group algebra $\mathbb{F}_p G$ is semisimple and by assumption on G it is noncommutative. Then there exists a nilpotent $0 \neq a \in \mathbb{F}_p G$.

Now we obtain that $R_{[p]}^+$ is finitely generated, i.e. it is finite, because $(R_{[p]}a)^2 = 0$. It means by assumption that $R_{[p]}$ has to be 0. □

Now let us indicate that finite abelian groups can be replaced by cyclic groups in our considerations. Below we remind the key result in this direction, a proof of which can be found in §4.4 of [6] or in [10].

Lemma 2.6 ([9]). *Let G be a finite abelian group of exponent n. If R is a ring in which n is regular then there exists an R-embedding of RG into a finite direct product of copies of $R[C_n]$.*

In some special cases we can eliminate G from our considerations

Proposition 2.7. *Let G be a finite elementary 2-group and R a semiprime ring. Then $\mathrm{U}(RG)$ is Noetherian if and only if $\mathrm{U}(R)$ is Noetherian and $R_{[2]}^+$ is finite.*

Proof. Follow carefully the proof of analogous result presented in [6, 10] for finite generation of $\mathrm{U}(RG)$. □

Going back to nonabelian groups we have

Theorem 2.8. *Let G be a finite nonabelian group and let R be infinite. Then $\mathrm{U}(RG)$ is Noetherian if and only if G is a Hamiltonian 2-group and $\mathrm{U}(RQ_8)$ is Noetherian, where Q_8 denotes the quaternion group of order 8.*

Proof. Let $J(R) = J$. By Lemma 1.1 J is nilpotent and J^+ is finitely generated. This by assumption implies that R/J is infinite too. So we can assume that R is semiprime. Now from Proposition 2.1 $R = S \oplus T$ where T is finite and S has no finite ideals. By Proposition 2.5 we know that S^+ is torsion free. So we can simply take R semiprime with R^+ torsion free.

Let $\mathrm{U}(RG)$ be Noetherian. By the above agreement $\mathbb{Z} \subset R$ hence $\mathrm{U}(\mathbb{Z}G)$ is Noetherian. By [5] or Theorem 1.3 G has to be a Hamiltonian 2-group. Clearly $Q_8 \subset G$, hence $\mathrm{U}(R[Q_8])$ is Noetherian by the same assumption.

Now let $\mathrm{U}(R[Q_8])$ be Noetherian and let G be Hamiltonian 2-group. From the structure of Hamiltonian 2-groups and Proposition 2.7 we obtain that $\mathrm{U}(RG)$ is Noetherian. □

In the case of some torsion free groups we can apply the structure theorem for units proved in [8] (see [6, 11]). Together with Proposition 2.1 this result immediately gives

Theorem 2.9 ([9]). *Let G be a u.p.-group. Then $\mathrm{U}(RG)$ is Noetherian if and only if the following conditions are satisfied:*

1. *$\mathrm{U}(R)$ and G are Noetherian;*
2. *R is reduced and has only a finite number of idempotents.*

3. Crossed Products

In this section we will discuss crossed products with Noetherian groups of units. We will follow notation and terminology from [13].

From the previous section we know that if $U(RG)$ is Noetherian then in many important cases RG is reduced. Using the same arguments as in the proof of Theorem 2.2 one can prove a more general result:

Proposition 3.1. *Let a crossed product $R * G$ be given such that $U(R * G)$ is Noetherian. Then $R * G$ is reduced in any of the following cases:*

1. *G is infinite;*
2. *R is semiprime and has no finite nonzero ideals.*

Further we will use some results about reduced rings. First of them is well known (see [15, §2.6]) and [7].

Theorem 3.2 (Andrunakievič and Riabuhin). *Let R be reduced and P its minimal prime ideal. Then the factor ring R/P is a domain. In particular every reduced ring is a subdirect product of domains.*

Lemma 3.3. *Let R be a reduced ring, P a minimal prime ideal of R and $S = R \setminus P$. Then the following statements are true:*

1. *If $x \in R$ then $x \in P$ if and only if there exists $s \in S$ such that $sx = 0$;*
2. *If $B \subset P$ is a nonempty finite subset then there exists $s \in S$ such that $sB = 0$.*

Proof. From the above theorem we can easily deduce that for every $n \geq 2$ and any permutation σ we have a quasi-identity:

$$x_1 x_2 \dots x_n = 0 \Rightarrow x_{\sigma(1)} x_{\sigma(2)} \dots x_{\sigma(n)} = 0. \tag{3.1}$$

Now let $x \in R$. If for some $s \in S$ we have $sx = 0$ then from Theorem 3.2 it follows that $x \in P$.

If $x \in P$ let $T \subset R$ be the multiplicative subsemigroup of R generated by S and x. From minimality of P as a prime ideal we then have $0 \in T$. It means that $0 = t_1 \dots t_n$ where either $t_i \in S$ or $t_i = x$ for all $1 \leq i \leq n$. By Theorem 3.2 at least one t_i is equal to x hence by the quasi-identity 3.1 we can write $0 = sx^k$ for some $s \in S$ and $k \geq 1$. It consequently gives $(sx)^k = 0$ hence $sx = 0$ because R is reduced. In this way the first claim of the lemma follows.

Now let $B = \{b_1, \dots, b_n\} \subset P$. Then we have elements $s_1, \dots, s_n \in S$ such that $s_i b_i = 0$ for all $i = 1, \dots n$. Using quasi-identity 3.1 it can be seen that $sB = 0$ where $s = s_1 \dots s_n$. By Theorem 3.2 $s \in S$ which completes the proof. □

Now let us return to crossed products.

Lemma 3.4. *Let $R * G$ be given. Then $R * G$ is reduced if and only if R is reduced, all minimal prime ideals of R are G-invariant and the factor ring $(R * G)/(P * G)$ is reduced for every minimal prime ideal P of R.*

Proof. Let $R * G$ be reduced. Clearly R is reduced. Let P be a minimal prime ideal of R and $a \in P$. Then by the above lemma there exists $s \in R \setminus P$ such that $sa = as = 0$. If $g \in G$ then for some $c \in \mathrm{U}(R)$ we have

$$(sa^g g)^2 = sa^g s^g a^{g^2} cg^2 = s(as)^g a^{g^2} cg^2 = 0$$

hence by assumption $sa^g = 0$. By the same lemma we then get $a^g \in P$. It means that P is G-invariant and hence $P * G$ is an ideal of $R * G$.

Now let $f \in R * G$ be such that $f^2 \in P * G$. Because f^2 involves only finitely many coefficients from R, and they belong in fact to P, then by Lemma 3.3 there exists $s \in R \setminus P$ such that $sf^2 = 0$. Because $R * G$ is reduced then from the quasi-identity 3.1 it follows that $(sf)^2 = 0$ and consequently $sf = 0$. Hence by Lemma 3.3 $f \in P * G$ which means that the factor ring $R * G/P * G$ is reduced.

The converse implication is obvious. □

Lemma 3.5. *Let R be a reduced ring and let a group G act on R by automorphisms. Then all minimal prime ideals of R are G-invariant if and only if every annihilator in R is G-invariant.*

Proof. Let all minimal prime ideals of R be G-invariant and let I be an annihilator of an ideal J in R. Let us also choose $x \in R \setminus I$. Then there exists $y \in J$ such that $xy \neq 0$. By Theorem 3.2 there exists a minimal prime ideal $P \subset R$ such that $xy \notin P$ hence $x \notin P$ and $y \notin P$. On the other hand $IJ \subset P$ which gives $I \subset P$ because $J \not\subset P$. In this way we showed that I is an intersection of a family of minimal prime ideals of R, which by assumption are G-invariant. This shows that I is G-invariant.

Now let us assume that all annihilators in R are G-invariant and let P be a minimal prime ideal of R. Let us put $S = R \setminus P$ and for every $s \in S$ let $I_s = \{x \in R : sx = 0\}$. Because R is reduced then we have $sRI_s = 0$ which gives $I_s \subset P$. From Lemma 3.3 we obtain that P is the sum, and in fact the union of ideals I_s which by assumption are G-invariant, hence P is G-invariant. □

For investigations of $\mathrm{U}(R*G)$ we introduce some further notation and terminology. As in the case of group rings and semigroup rings, (see [11]), any unit of the form $rg \in R * G$ where $r \in R$ and $g \in G$ will be named a ***monomial unit***.

Let $M(R * G)$ denote the set of all monomial units from $R * G$. It is clear that $\mathrm{M}(R * G)$ is a subgroup of $\mathrm{U}(R * G)$ such that $M(R * G) = \mathrm{U}(R) \rtimes G$.

It is still not known if for torsion free groups and domains, or even fields we have $\mathrm{U}(R * G) = M(R * G)$. Probably the best result in this direction is proved by Strojnowski. He showed in [17] that any u.p.-group is a t.u.p.-group. In this way he in fact obtained the following result (see [11]):

Theorem 3.6. *If R is a domain and G is a u.p.-group then* $\mathrm{U}(R * G) = M(R * G)$ *for any crossed product $R * G$.*

We will apply the above result to our considerations. For this we restrict our attention to the case when G is a u.p.-group.

Let us note here that, under this restriction, if R is reduced then ordinary group ring RG is reduced too. If R is a domain then $R * G$ is a domain hence it is reduced, but $R * G$ need not be reduced for arbitrary reduced R and arbitrary action of G on R (see Lemma 3.4).

Lemma 3.7. *Let G be a u.p.-group and R a ring such that $R * G$ is reduced. If $u = \sum_{g \in G} a_g g \in \mathrm{U}(R * G)$ and $u^{-1} = \sum_{g \in G} b_g g$ then $a_g b_h = 0$ whenever $gh \neq 1$ and $a_g a_h = 0$ for every $g \neq h$.*

Proof. Let X be the support of u and Y be the support of u^{-1}. We will proceed by induction on $n = |X| + |Y|$. If $n = 2$ then both formulas are certainly true.

Let $n > 2$. Because G is a u.p.-group then there exist $x \in X$ and $y \in Y$ such that the element xy is uniquely presented in the product XY, hence $a_x (b_y)^x = 0$. Let $I = \{r \in R : r(b_y)^x = 0\}$ and let $J = \{r \in R : Ir = 0\}$. Because R is reduced then I, J are ideals in R and they are annihilators. Because $R * G$ is reduced too, hence by Lemmas 3.4 and 3.5 these annihilators are G-invariant. It means that $a_x \in I$ and $b_y \in J$. Moreover both factor rings $R * G/I * G$ and $R * G/J * G$ are reduced. By the induction hypothesis we then have for every $g, h \in G$, $gh \neq 1$ that $a_g b_h \in I \cap J$, but this intersection is 0, hence $a_g b_h = 0$. The proof of the second claim is similar. □

Theorem 3.8. *Let G be a u.p.-group. Then* $\mathrm{U}(R * G) = M(R * G)$ *if and only if $R * G$ is reduced and R is indecomposable.*

Proof. Let $\mathrm{U}(R * G) = M(R * G)$ and let us assume that there exists $a \in R * G$ such that $a \neq 0$ but $a^2 = 0$. Then $1 + a$ is a unit, hence by assumption $a \in R$. Now let $1 \neq g \in G$. If $aa^g = 0$ then $(ag)^2 = 0$. If $aa^g \neq 0$ then $(aa^g g)^2 = 0$. This in both cases gives a nonzero nilpotent outside of R from which one can immediately produce an unit which is not monomial, a contradiction. Hence $R * G$ is reduced. Now if $e \in R$ is a nontrivial idempotent then by Lemma 3.5 e is G-invariant as the unity element of the annihilator of $1 - e$, so $1 - e + eg$ is a unit for every $g \in G$. This contradicts the assumption.

Conversely, let $R * G$ be reduced and let R be indecomposable. Moreover let $u = \sum_{g \in G} a_g g \in \mathrm{U}(R * G)$ and let $u^{-1} = \sum_{h \in G} b_h h$. Let us assume that $|X| \geq 2$ where X denotes the support of u. By Lemma 3.7 we have

$$1 = uu^{-1} = \sum_{g,h \in G} a_g g\, b_h h = \sum_{g \in G} a_g g\, b_{g^{-1}} g^{-1} = \sum_{g \in G} a_g (c_g)$$

for some $c_g \in R$. It means by Lemma 3.7 that R can be decomposed into a direct sum of ideals generated by all a_x where $x \in X$. This contradicts the assumption on R, hence we must have $|X| = 1$ and $u \in M(R * G)$. □

Theorem 3.9. *Let R be a ring and G a u.p.-group. Then* $\mathrm{U}(R * G)$ *is Noetherian if and only if the following conditions are satisfied:*

1. $\mathrm{U}(R)$ *and G are Noetherian;*
2. *R is reduced and has only a finite number of idempotents;*
3. *Every minimal prime ideal of R is G-invariant.*

Proof. Let $\mathrm{U}(R*G)$ be Noetherian. Because $\mathrm{U}(R) \rtimes G = M(R*G) \subset \mathrm{U}(R*G)$ then by assumption $\mathrm{U}(R)$ and G are Noetherian. Because G is infinite then by Proposition 3.1 $R * G$ is reduced and hence R is reduced too. Now, as in the above proof, Lemmas 3.4 and 3.5 give that all idempotents of R are G-invariant. This means that they are central. Let $0 \neq e \in R * G$ be an idempotent. Then we have

$$G \subset \mathrm{U}(eRe * G) = \mathrm{U}(e(R * G)e)$$

Hence by Proposition 1.4 R has only finitely many idempotents.

Conversely, let us assume that R and G satisfy the conditions listed in the formulation of the theorem. By assumption and Lemmas 3.3, 3.4 and 3.5 we can restrict our considerations to the case when R is reduced, indecomposable and all its minimal prime ideals are G-invariant. Because G is a u.p.-group then from Theorem 3.8 we have that $\mathrm{U}(R * G) = M(R * G) = \mathrm{U}(R) \rtimes G$ and the last group is Noetherian by assumption. □

References

[1] Andruszkiewicz, R. R., Finiteness conditions induced by properties of maximal commutative subalgebras. Arch. Math. 48 (1987), 303–307.

[2] Bovdi, A. A., On the structure of the multiplicative group of a group algebra with finiteness conditions. In: Algebraic Structures, Mat. Issled. 56 (1980), 14–27.

[3] Gonçalves, J. Z., Free subgroups in the group of units of group rings II. J. Number Theory 21 (1985), 121–127.

[4] Hartley, B., Free groups in normal subgroups of unit groups and arithmetic groups. Contemporary Math. 93 (1989), 173–177.

[5] —, Pickel, P. F., Free subgroups in the unit groups of integral group rings. Canad. J. Math. 32 (1980), 1342–1352.

[6] Karpilovsky, G., Unit Groups of Group Rings. Longman, Essex 1989.

[7] Klein, A. A., A simple proof of a theorem on reduced rings. Canad. Math. Bull. 23 (1980), 495–496.

[8] Krempa, J., Homomorphisms of group rings. Banach Center Publication vol. 9, PWN Warsaw (1982), 233–255.

[9] —, Finitely generated groups of units in group rings. Preprint of the Institute of Mathematics, Warsaw University, Warsaw 1985.

[10] —, Unit groups and commutative ring extensions. Comm. Algebra 16 (1988), 2349–2361.

[11] —, On finite generation of unit groups for group rings. In: Groups '93 - Galway/St. Andrews, London Math. Soc. Lecture Note Ser. 212, Cambridge University Press, 352–367.

[12] Passman, D. S., The Algebraic Structure of Group Rings. Wiley-Interscience Publications, New York 1977.

[13] —, Infinite Crossed Products. Academic Press, New York 1989.

[14] Robinson, D. J. S., Finiteness Conditions and Generalized Soluble Groups. Springer, Berlin 1972.

[15] Rowen, L. H., Ring Theory vol. I. Academic Press, New York 1988.

[16] Sehgal, S. K., Units in Integral Group Rings. Longman, Essex 1993.

[17] Strojnowski, A., A note on u.p.-groups. Comm. Algebra 8 (1980), 231–234.

[18] Watters, J. F., On the adjoint group of a radical ring. J. London Math. Soc. 43 (1968), 725–729.

On Normal Closures of Elements in Generalized FC-Groups

Leonid A. Kurdachenko

1991 Mathematics Subject Classification: 20F24

Let G be a group, x an element of G and $x^G = \{x^g = g^{-1}xg \mid g \in G\}$. Then $C_G(x^G)$ is a normal subgroup of G, and the factor group $G/C_G(x^G)$ is called the ***cocentralizer*** of x in G. If $\mathfrak{X}$ is a class of groups, the group G is said to be an ***$\mathfrak{X}C$-group*** (or to have ***$\mathfrak{X}$-conjugacy classes***) if the cocentralizer of each element x of G belongs to $\mathfrak{X}$, i.e. $G/C_G(x^G) \in \mathfrak{X}$ for each x in G. Let $\operatorname{Coc}\mathfrak{X}$ be the class of all $\mathfrak{X}C$-groups. Then $\operatorname{Coc}\mathfrak{F}$ is the class of all FC-groups and so, if $\mathfrak{F} \subseteq \mathfrak{X}$, then $\operatorname{Coc}\mathfrak{X}$ can be considered as a generalization of the class of FC-groups. The structure of FC-groups has been studied in many papers (see [14]). Properties of $\mathfrak{X}C$-groups, for some other group classes $\mathfrak{X}$, have also been considered: CC-groups or groups in the class $\operatorname{Coc}\big((\check{\mathfrak{M}}\cap\mathfrak{A})\mathfrak{F}\big)$ have been studied for example in [4], [5], [6], [9], [10] (here $(\check{\mathfrak{M}}\cap\mathfrak{A})\mathfrak{F}$ is the class of Chernikov groups); PC-groups or groups in the class $\operatorname{Coc}(\mathfrak{P}F)$ have been studied in [3] (here $\mathfrak{P}F$ is the class of polycyclic-by-finite groups); groups in the class $\operatorname{Coc}(\mathfrak{S}_2\mathfrak{F})$ have been considered in [8] (here $\mathfrak{S}_2\mathfrak{F}$ is the class of soluble-by-finite minimax groups). In [2] the class $\operatorname{Coc}\mathfrak{X}$ is considered, where $\mathfrak{X} = \mathbf{QS}G$ (i.e. $\mathfrak{X}$ consists of all sections of G) and G is a group in $\mathfrak{S}_0\mathfrak{F}$ or in $\hat{\mathfrak{S}}\mathfrak{F}$, where $\mathfrak{S}_0$ is the class of soluble groups with finite sectional rank and $\hat{\mathfrak{S}}$ is the class of soluble groups of finite rank (in the sense of Mal'cev–Prüfer). Therefore it is interesting to consider the classes $\operatorname{Coc}(\mathfrak{S}_0\mathfrak{F})$ and $\operatorname{Coc}(\hat{\mathfrak{S}}\mathfrak{F})$. The first question here is the following. Let $G \in \operatorname{Coc}(\mathfrak{S}_0\mathfrak{F})$ (respectively $G \in \operatorname{Coc}(\hat{\mathfrak{S}}\mathfrak{F})$) and $g \in G$. Does $\langle g\rangle^G$ lie in $\mathfrak{S}_0\mathfrak{F}$ (respectively in $\hat{\mathfrak{S}}\mathfrak{F}$)? For the subclass $\operatorname{Coc}(\mathfrak{S}_2\mathfrak{F})$ this is true. But in the general case it is not true, as it is shown by the following example.

Let $p, q_1, \dots, q_n, \dots$ be infinitely many distinct primes, let $G_n = \langle a_n\rangle \wr \langle b_n\rangle$, where $|a_n| = p$, $|b_n| = q_n$ ($n \in \mathbb{N}$) and let $H = \operatorname{Cr}_{n\in\mathbb{N}} G_n$. Put $A = \operatorname{Cr}_{n\in\mathbb{N}} A_n$, where A_n is the base group of G_n, $B = \operatorname{Dr}_{n\in\mathbb{N}} \langle b_n\rangle$, and consider the semidirect product $G = A \rtimes B$. If

Offprint from
Infinite Groups 94, Eds.: de Giovanni/Newell

$g \in A$, then $A \leq C_G(g^G)$, in particular $G/C_G(g^G)$ has finite rank. If $g \in B$, then there exists a non-negative integer l such that $\mathrm{Cr}_{n \geq l} A_n \leq C_G(g^G)$ and $\mathrm{Dr}_{n \geq l} \langle b_n \rangle \leq C_G(g^G)$, in particular $G/C_G(g^G)$ is finite. It follows that the cocentralizer of every element $g \in G$ has finite rank. Hence $G \in \mathrm{Coc}\,\hat{\mathfrak{S}}$. If $a = (a_n)_{n \in \mathbb{N}}$, then $\langle a \rangle^G$ is an infinite elementary abelian p-group, and so $\langle a \rangle^G \notin \hat{\mathfrak{S}}$.

On the other hand, there exist subclasses of $\mathrm{Coc}(\mathfrak{S}_0\mathfrak{F})$ for which the answer is positive. In this article we shall consider generalized nilpotent groups in the class $\mathrm{Coc}(\mathfrak{S}_0\mathfrak{F})$.

Most of our notation is standard. In particular we refer to [13].

Let R be a ring, G a group and A an RG-module. If $x \in G$, then $C_A(x^G)$ is an RG-submodule of A. In fact, if $a \in C_A(x^G)$ and $y, g \in G$, then

$$(ay)^{x^g} = (ay)(g^{-1}xg) = a(yg^{-1}xgy^{-1})y = ay$$

and so $ay \in C_A(x^G)$. In the following $\omega(RG)$ will denote the augmentation ideal of RG, that is the ideal of RG which is generated by all elements of the form $g - 1$, $g \in G$.

Let $\mathfrak{M}$ be a class of R-modules. We say that the RG-module A belongs to the class $\mathrm{Coc}_{RG}\,\mathfrak{M}$ if $A/C_A(x^G) \in \mathfrak{M}$ for every $x \in G$. Recall also that an abelian group G is an ***$\mathfrak{A}_0$-group*** if it has finite sectional rank.

Lemma 1. *Let G be an $\mathfrak{A}_0$-group, A a $\mathbb{Z}$-torsion $\mathbb{Z}G$-module and $A \rtimes G$ a locally nilpotent group. If $A \in \mathrm{Coc}_{\mathbb{Z}G}\,\mathfrak{A}_0$, then $A(\omega\mathbb{Z}G) \in \mathfrak{A}_0$.*

Proof. Let $t(G)$ be the torsion subgroup of G. Since $G/t(G)$ has finite rank, then $G/t(G)$ contains a finite maximal $\mathbb{Z}$-independent subset $\{g_1 t(G), \dots, g_n t(G)\}$. Put $H = \langle g_1, \dots, g_n \rangle$. Since G is abelian, then

$$A(\omega\mathbb{Z}H) \leq A(g_1 - 1) + \cdots + A(g_n - 1).$$

But $A(g_i - 1)$ is isomorphic with $A/C_A(g_i)$ for every $i \leq n$, and so each $A(g_i - 1)$ belongs to $\mathfrak{A}_0$, so that also $A_1 = A(\omega\mathbb{Z}H) \in \mathfrak{A}_0$. Since $H \leq C_G(A/A_1)$, the factor group $G/C_G(A/A_1)$ is a periodic $\mathfrak{A}_0$-group. Therefore it can be assumed that G is a periodic $\mathfrak{A}_0$-group. Since A is $\mathbb{Z}$-torsion, then $A = \bigoplus_{p \in \pi(A)} A_p$, where A_p is the p-component of A, and so $A(\omega\mathbb{Z}G) = \bigoplus_{p \in \pi(A)} A_p(\omega\mathbb{Z}G)$. Thus we can suppose that A is a p-group for some prime p. Moreover $G = G_p \times G_{p'}$, where G_p is the Sylow p-subgroup and $G_{p'}$ is the Sylow p'-subgroup of G. Since $A \rtimes G$ is locally nilpotent, then $G_{p'} \leq C_G(A)$, and we can also suppose that G is a p-group, so that in particular G is a Chernikov group. Let K be the largest divisible subgroup of G. Then the index $|G : K|$ is finite. For every $x \in G$ consider the map

$$\varphi_x : A \to A$$

defined by $\varphi_x(a) = a(x-1)$. It is easy to see that φ_x is a $\mathbb{Z}G$-endomorphism of A, $\ker\varphi_x = C_A(x) = \mathrm{Ann}_A(x-1)$ and $\mathrm{Im}\,\varphi_x = A(x-1)$. Therefore $A/C_A(x) \simeq A(x-1)$ and $A_x = A/C_A(x) \in \mathfrak{A}_0$. Since the additive group of A is a p-group, then A_x is a Chernikov p-group. Then $K \leq C_G(A_x)$ for every $x \in G$ (see for instance [11], Corollary of Theorem 3.29.2). Thus for every $a \in A$ and $y \in K$ we have $a(y-1) \in C_A(x)$, and so

$$a(y-1) \in \bigcap_{x\in G} C_A(x) = C_A(G).$$

Let $a \in A$ and put $A_1 = a\mathbb{Z}G$ and $A_2 = A_1 \cap C_A(G)$. Then K is contained in $C_G(A_2) \cap C_G(A_1/A_2)$. From Proposition 1.C.3 of [7] it follows that $K/C_K(A_1)$ is isomorphic with a subgroup of $\mathrm{Hom}(A_1/A_2, A_2)$. Since a has finite order, the additive group of A_1 has finite exponent, and so also $\mathrm{Hom}(A_1/A_2, A_2)$ has finite exponent. Since $K \neq C_K(A_1)$ is divisible, it follows that $K \leq C_G(a\mathbb{Z}G)$. Therefore $K \leq C_G(A)$, so that the factor group $G/C_G(A)$ is finite. Then $G = EC_G(A)$, where $E = \{y_1, \ldots, y_t\}$ is a finite subgroup. Since G is abelian, then

$$A(\omega\mathbb{Z}H) \leq A(y_1 - 1) + \cdots + A(y_t - 1).$$

It follows that $A(\omega\mathbb{Z}H)$ is a Chernikov group, and since $A(\omega\mathbb{Z}H) = A(\omega\mathbb{Z}G)$, also $A(\omega\mathbb{Z}G)$ is a Chernikov group. □

Corollary. *Let G be a $\mathfrak{S}_0$-group, A a $\mathbb{Z}$-torsion $\mathbb{Z}G$-module and $A \rtimes G$ a locally nilpotent group. If $A \in \mathrm{Coc}_{\mathbb{Z}G}\,\mathfrak{A}_0$, then $A(\omega\mathbb{Z}G) \in \mathfrak{A}_0$.*

Proof. Let

$$1 = G_0 \leq G_1 \leq \cdots \leq G_n = G$$

be a normal series of G with abelian factors. We use induction on n. If $n = 1$, then the corollary follows from Lemma 1. Let now $n > 1$ and put $A_1 = A(\omega\mathbb{Z}G_1)$. Since $G_1 \lhd G$, then A_1 is a $\mathbb{Z}G$-submodule of A, and $G_1 \leq C_G(A/A_1)$. By induction we obtain that $A_2/A_1 = A/A_1(\omega\mathbb{Z}G) \in \mathfrak{A}_0$, in particular $A_2 \in \mathfrak{A}_0$. Since $G = C_G(A/A_2)$, then $A(\omega\mathbb{Z}G) \leq A_2$ and so $A(\omega\mathbb{Z}G) \in \mathfrak{A}_0$. □

Lemma 2. *Let G be a torsion-free finitely generated nilpotent group and A a $\mathbb{Z}G$-module. If $A \in \mathrm{Coc}_{\mathbb{Z}G}\,\mathfrak{A}_0$, then $A(\omega\mathbb{Z}G) \in \mathfrak{A}_0$.*

Proof. Let

$$1 = C_0 \leq C_1 \leq \cdots \leq C_n = G$$

be the upper central series of G. We use induction on n. If $n = 1$, G is a finitely generated torsion-free abelian group, and so $G = \langle g_1\rangle \times \cdots \times \langle g_n\rangle$. Then we have

$$A(\omega\mathbb{Z}G) \leq A(g_1 - 1) + \cdots + A(g_n - 1).$$

Since $A(g_i - 1) \simeq A/C_A(g_i) \in \mathfrak{A}_0$ for every $i \leq n$, then $A(\omega\mathbb{Z}G) \in \mathfrak{A}_0$. Let now $n > 1$ and put $A_1 = A(\omega\mathbb{Z}C_1)$. Then $A_1 \in \mathfrak{A}_0$, and A_1 is a $\mathbb{Z}G$-submodule of A. Moreover $C_1 \leq C_G(A/A_1)$ and so A/A_1 can be considered as a $\mathbb{Z}(G/C_1)$-module. Since G/C_1 is torsion-free, by induction we have that $A_2/A_1 = (A/A_1)(\omega\mathbb{Z}G)$ belongs to $\mathfrak{A}_0$, and so $A_2 \in \mathfrak{A}_0$. Clearly $A(\omega\mathbb{Z}G) \leq A_2$, so that $A(\omega\mathbb{Z}G) \in \mathfrak{A}_0$. □

Lemma 3. *Let G be a nilpotent $\mathfrak{S}_0$-group, A a $\mathbb{Z}$-torsion-free $\mathbb{Z}G$-module and $A \rtimes G$ a locally nilpotent group. If $A \in \mathrm{Coc}_{\mathbb{Z}G}\,\mathfrak{A}_0$, then $A(\omega\mathbb{Z}G) \in \mathfrak{A}_0$.*

Proof. Let

$$1 = C_0 \leq C_1 \leq \cdots \leq C_n = G$$

be the upper central series of G. Since $C_{i+1}/C_i \in \mathfrak{A}_0$ for every $i \leq n-1$, there exists a subgroup D_{i+1} such that

$$D_{i+1} = \langle g_{i+1,1}, \ldots, g_{i+1,k_i}, C_i \rangle \leq C_{i+1}$$

and C_{i+1}/D_{i+1} is a torsion group. Let $H = \langle g_{1,1}, \ldots, g_{n,k_{n-1}} \rangle$. Then H is finitely generated and therefore contains a torsion-free subgroup F such that $|H : F|$ is finite. Since F is subnormal in G, there exists a series of finite length

$$F = F_0 \lhd F_1 \lhd \cdots \lhd F_k = G,$$

where the factor groups F_{i+1}/F_i are periodic.

Let $A_1 = A(\omega\mathbb{Z}F_0)$. It follows from Lemma 2 that $A_1 \in \mathfrak{A}_0$, and so A_1 has finite rank, since A is $\mathbb{Z}$-torsion-free. Let B_1/A_1 be the torsion subgroup of A/A_1; then B_1 has finite rank. Since $F_0 \lhd F_1$, then A_1 is a $\mathbb{Z}F_1$-submodule of A, and B_1 is also a $\mathbb{Z}F_1$-submodule. But $F_0 \leq C_{F_1}(A/B_1)$, and so A/B_1 is a $\mathbb{Z}(F_1/F_0)$-module. Since $A/B_1 \rtimes F_1/F_0$ is a locally nilpotent group, the set of all elements of finite order of $A/B_1 \rtimes F_1/F_0$ is a normal subgroup, and hence $F_1 = C_{F_1}(A/B_1)$; in particular $A_2 = A(\omega\mathbb{Z}F_1) \leq B_1$. Let B_2/A_2 be the torsion subgroup of A/A_2; then B_2 has finite rank and B_2 is a $\mathbb{Z}F_2$-submodule. As before we obtain that $A_3 = A(\omega\mathbb{Z}F_2)$ has finite rank, and after finitely many steps $A(\omega\mathbb{Z}G)$ has finite rank. □

Lemma 4. *Let G be a locally nilpotent $\mathfrak{S}_0$-group, and A a $\mathbb{Z}G$-module satisfying the following conditions:*

(a) *the $\mathbb{Z}$-torsion subgroup T of A is contained in $C_A(G)$;*

(b) *A/T has rank 1;*

(c) *$A/T = C_{A/T}(G)$.*

If $A \in \mathrm{Coc}_{\mathbb{Z}G}\,\mathfrak{A}_0$, then $A(\omega\mathbb{Z}G) \in \mathfrak{A}_0$.

Proof. Since A/T has rank 1, then $A/T = \bigcup_{n\in\mathbb{N}} \langle \bar{a}_n \rangle$, where

$$\langle \bar{a}_1 \rangle \leq \langle \bar{a}_2 \rangle \leq \cdots \leq \langle \bar{a}_n \rangle \leq \cdots$$

and $\bar{a}_n = a_n + T$, and so for every $n \in \mathbb{N}$ there exist $t_n \in T$ and $k_{n+1} \in \mathbb{N}$ such that $k_{n+1}a_{n+1} = a_n + t_n$. Put $L_n = a_n(\omega\mathbb{Z}G)$ for every $n \in \mathbb{N}$. For each g in G we have

$$k_{n+1}a_{n+1}(g-1) = a_n(g-1) + t_n(g-1) = a_n(g-1),$$

and so $k_{n+1}L_{n+1} = L_n$. Then $L_n \leq L_{n+1}$ for every $n \in \mathbb{N}$. Moreover, it follows from condition (a) that $A(\omega\mathbb{Z}G) = \bigcup_{n\in\mathbb{N}} L_n$. Consider the map

$$\psi_n : G \to A$$

defined by $\psi_n(g) = a_n(g-1)$. For every g, y in G we have

$$gy - 1 = (g-1)(y-1) + (g-1) + (y-1)$$

and so

$$a_n(gy-1) = a_n(g-1)(y-1) + a_n(g-1) + a_n(y-1) = a_n(g-1) + a_n(y-1).$$

This means that ψ_n is a homomorphism of G into the additive group of A, and $\operatorname{Im}\psi_n = a_n(\omega\mathbb{Z}G) = L_n$. It follows that $L_n \in \mathfrak{A}_0$ for every $n \in \mathbb{N}$, so that also $A(\omega\mathbb{Z}G) = \bigcup_{n\in\mathbb{N}} L_n \in \mathfrak{A}_0$. □

Corollary. *Let G be a locally nilpotent $\mathfrak{S}_0$-group and A a $\mathbb{Z}G$-module containing a series of finite length of submodules*

$$0 = A_0 \leq A_1 \leq \cdots \leq A_n = A$$

satisfying the following conditions:

(a) *A_1 is the $\mathbb{Z}$-torsion subgroup of A;*

(b) *A_{i+1}/A_i is a torsion-free group of rank 1 for every i such that $1 \leq i \leq n-1$;*

(c) *$A_i/A_{i-1} \leq C_{A/A_{i-1}}(G)$ for every $i \leq n$.*

If $A \in \operatorname{Coc}_{\mathbb{Z}G}\mathfrak{A}_0$, then $A(\omega\mathbb{Z}G) \in \mathfrak{A}_0$.

Lemma 5. *Let G be a $\mathfrak{S}_0$-group, A a $\mathbb{Z}G$-module such that $A = a\mathbb{Z}G$ for some $a \in A$, and let $A \rtimes G$ be a locally nilpotent group. If $A \in \operatorname{Coc}_{\mathbb{Z}G}\mathfrak{A}_0$, the $A \in \mathfrak{A}_0$.*

Proof. Let P be the torsion subgroup of G and T the torsion subgroup of A. It follows from Corollary to Lemma 1 that $T(\omega\mathbb{Z}G) \in \mathfrak{A}_0$. Therefore we can suppose that $T \leq C_A(G)$. Since the semidirect product $(A/T) \rtimes P$ is locally nilpotent, then $P \leq C_{A/T}(G)$. Hence we can consider A/T as a $\mathbb{Z}(G/P)$-module. Since G/P is torsion-free, it is nilpotent (see for instance [12], Corollary 1 to Theorem 6.36), and so it follows from Lemma 3 that A/T has finite rank. Moreover $A(\omega\mathbb{Z}P) \in \mathfrak{A}_0$ by Corollary to Lemma 4, and so we can suppose that $A = C_A(P)$. The semidirect product $(A/T) \rtimes (G/P)$ is a torsion-free locally nilpotent group of finite rank, so that it is nilpotent, and hence A has a series of finite length of submodules

$$0 = A_0 \leq A_1 \leq \cdots \leq A_n = A$$

such that

(a) $A_1 = T$;

(b) $A_i/A_{i-1} \leq C_{A/A_{i-1}}(G)$ for every $i \leq n$;

(c) A_{i+1}/A_i is a torsion-free group of rank 1 for every i such that $1 \leq i \leq n-1$.

Application of the Corollary to Lemma 4 yields that $A(\omega\mathbb{Z}G) \in \mathfrak{A}_0$. Since $A = a\mathbb{Z}G = a\mathbb{Z} + A(\omega\mathbb{Z}G)$, it follows that $A \in \mathfrak{A}_0$. □

Corollary. *Let G be a $\mathfrak{S}_0$-group and A a finitely generated $\mathbb{Z}G$-module such that $A \rtimes G$ is a locally nilpotent group. If $A \in \mathrm{Coc}_{\mathbb{Z}G}\, \mathfrak{A}_0$, then $A \in \mathfrak{A}_0$.*

Theorem 1. *Let G be a locally nilpotent $\mathfrak{S}_0C$-group, a an element of G and $A = a^G$. Then $A \in \mathfrak{S}_0$.*

Proof. Put $C = C_G(A) = C_G(a^G)$; then $G/C \in \mathfrak{S}_0$. Since $A \cap C \leq Z(A)$ and $A/A \cap C \simeq AC/C$ is a $\mathfrak{S}_0$-group, it follows from Theorem 2.16 of [2] that also $[A, A]$ belongs to $\mathfrak{S}_0$. Therefore we can assume that A is an abelian group, and we can consider A as a $\mathbb{Z}\bar{G}$-module, where $\bar{G} = G/C$. Moreover $A \in \mathrm{Coc}_{\mathbb{Z}\bar{G}}\, \mathfrak{A}_0$, since G is a $\mathfrak{S}_0C$-group. It follows from Lemma 5 that $A \in \mathfrak{A}_0$. □

Let R be a ring, G a group and A an RG-module. If $B \leq C$ are RG-submodules of A, the factor C/B is said to be ***RG-central*** (or ***RG-trivial***) if $C(g-1) \leq B$ for every g in G, i.e. $C(\omega RG) \leq B$. The RG-module A is a ***$\bar{Z}$-module*** (over RG) if every RG-composition factor of A is RG-central, while A is said to be an ***RG-hypercentral module*** if A has an ascending series of RG-submodules

$$0 = A_0 \leq A_1 \leq \cdots \leq A_\alpha \leq A_{\alpha+1} \leq \cdots \leq A_\gamma = A,$$

where $A_{\alpha+1}/A_\alpha$ is RG-central for every $\alpha < \gamma$.

Lemma 6. *Let G be a $\mathfrak{S}_0$-group and A a $\bar{Z}$-module over $\mathbb{Z}G$. If B is a $\mathbb{Z}G$-submodule of A such that $B \in \mathfrak{A}_0$, then $B \cap C_A(G) \neq 0$.*

Proof. Let T be the $\mathbb{Z}$-torsion subgroup of B. If $T \neq 0$, then T contains a finite non-zero $\mathbb{Z}G$-submodule, and so also a non-zero finite simple $\mathbb{Z}G$-submodule C. Since A is a $\bar{Z}$-module, then $C \leq C_A(G)$ and the result follows. Let now B be $\mathbb{Z}$-torsion-free. Since B has finite rank, then B contains a $\mathbb{Z}G$-submodule $D \neq 0$ such that $D \otimes_{\mathbb{Z}} \mathbb{Q}$ is a simple $\mathbb{Q}G$-submodule. Let $0 \neq d \in D$ and put $D_1 = d\mathbb{Z}G$. It follows from Theorem 2 of [1] that $G/C_G(D_1)$ is (free abelian)-by-finite. Since $G/C_G(D_1) \in \mathfrak{S}_0$, then $G/C_G(D_1)$ is a finitely generated abelian-by-finite group. Then D_1 contains a subgroup C_1 such that D_1/C_1 is a torsion group with $\pi(D_1/C_1)$ finite and C_1 is free abelian (see for instance [12], Corollary 1 to Lemma 9.53). Put $\pi = \mathbb{P} \setminus \pi(D_1/C_1)$. Then π is infinite and so $\bigcap_{p\in\pi} pC_1 = 0$. For each $p \in \pi$ we have $pD_1 \cap C_1 = pC_1$, and so

$$\Big(\bigcap_{p\in\pi} pD_1\Big) \cap C_1 = \bigcap_{p\in\pi}(pD_1 \cap C_1) = \bigcap_{p\in\pi} pC_1 = 0.$$

Since D_1 is $\mathbb{Z}$-torsion-free, it follows that $\bigcap_{p\in\pi} pD_1 = 0$. As D_1/pD_1 is finite, there exists a $\mathbb{Z}G$-composition series

$$pD_1 = E_0 < E_1 < \cdots < E_n = D_1$$

where $n \leq r = r_0(D_1)$. Since A is a $\bar{Z}$-module over $\mathbb{Z}G$, then E_i/E_{i-1} is contained in $C_{A/E_{i-1}}(G)$ for every $i \leq n$. Hence for every element $d_1 \in D_1$ and any elements $g_1, \ldots, g_r \in G$ we have

$$d_1(g_1 - 1)(g_2 - 1)\ldots(g_r - 1) \in pD_1,$$

that is $D_1(\omega\mathbb{Z}G)^r \leq pD_1$. Since this inclusion holds for each $p \in \pi$, we have

$$D_1(\omega\mathbb{Z}G)^r \leq \bigcap_{p\in\pi} pD_1 = 0.$$

Let m be the smallest positive integer such that $D_1(\omega\mathbb{Z}G)^m = 0$. Then $D_2 = D_1(\omega\mathbb{Z}G)^{m-1} \neq 0$ and $D_2(\omega\mathbb{Z}G) = 0$. This means that $D_2 \leq C_A(G)$ and so $B \cap C_A(G) \neq 0$. □

Corollary 1. *Let G be a $\mathfrak{S}_0$-group and A a $\bar{Z}$-module over $\mathbb{Z}G$. If A has an ascending series of submodules*

$$0 = A_0 \leq A_1 \leq \cdots \leq A_\alpha \leq A_{\alpha+1} \leq \cdots \leq A_\gamma = A,$$

where $A_{\alpha+1}/A_\alpha \in \mathfrak{A}_0$ for every $\alpha < \gamma$, then A is a $\mathbb{Z}G$-hypercentral module.

Corollary 2. *Let G be a $\bar{Z}$-group in $\mathfrak{S}_0$. Then G is a hypercentral group.*

Corollary 3. *Let G be a $\bar{Z}$-group in $\mathfrak{S}_0$ and A a $\bar{Z}$-module over $\mathbb{Z}G$. If A is in* $\mathrm{Coc}_{\mathbb{Z}G}\,\mathfrak{A}_0$*, then A is a $\mathbb{Z}G$-hypercentral module.*

Proof. The group G is hypercentral by Corollary 2. Let $1 \neq \bar{z} \in Z\big(G/C_G(A)\big)$. The map $\varphi_{\bar{z}} : A \to A$, defined by

$$\varphi_{\bar{z}}(a) = a(\bar{z} - 1),$$

is a $\mathbb{Z}G$-endomorphism of A, and so $A_1 = \operatorname{Im}\varphi_{\bar{z}}$ and $\ker\varphi_{\bar{z}} = \operatorname{Ann}(\bar{z} - 1)$ are $\mathbb{Z}G$-submodules of A. Since $A_1 \simeq A/\operatorname{Ann}(\bar{z} - 1)$, then $A_1 \in \mathfrak{A}_0$. In this way we can construct an ascending series of submodules of A, all factors of which belong to $\mathfrak{A}_0$. Therefore A is a $\mathbb{Z}G$-hypercentral module by Corollary 1. □

Lemma 7. *Let G be a $\mathfrak{S}_0$-group. The following statements are equivalent:*

(a) *G is a Baer-nilpotent group;*

(b) *G is an Engel group;*

(c) *G is a locally nilpotent group;*

(d) *G is a $\bar{Z}$-group;*

(e) *G is a hypercentral group;*

(f) *G is a N-group;*

(g) *G is a $\tilde{N}$-group.*

Proof. (a)$\Rightarrow$(c) Let F be a finitely generated subgroup of G. Since every finite quotient of F is nilpotent, also F is nilpotent (see for instance [12], Theorem 10.51). Therefore G is locally nilpotent. Thus (a), (b) and (c) are equivalent. It follows from Corollary 2 that (d)$\Rightarrow$(e). The other implications are clear. □

Theorem 2. *Let G be a $\mathfrak{S}_0 C$-group, a an element of G and $A = \langle a \rangle^G$.*

(a) *If G is a Baer-nilpotent group, then A is a locally nilpotent $\mathfrak{S}_0$-group;*

(b) *if G is an Engel group, then A is a locally nilpotent $\mathfrak{S}_0$-group;*

(c) *if G is a locally nilpotent group, then A is a locally nilpotent $\mathfrak{S}_0$-group;*

(d) *if G is a $\bar{Z}$-group, then A is a locally nilpotent $\mathfrak{S}_0$-group;*

(e) *if G is a hypercentral group, then A is a locally nilpotent $\mathfrak{S}_0$-group;*

(f) *if G is an N-group, then A is a locally nilpotent $\mathfrak{S}_0$-group;*

(g) *if G is an $\tilde{N}$-group, then A is a locally nilpotent $\mathfrak{S}_0$-group.*

Proof. Put $C = C_G(A) = C_G(a^G)$. Then $\bar{G} = G/C \in \mathfrak{S}_0$, so that also $A/A \cap C \simeq AC/C$ is a $\mathfrak{S}_0$-group. Since $A \cap C \leq Z(A)$, it follows from Theorem 2.16 of [2] that $[A, A] \in \mathfrak{A}_0$. Thus we can suppose that A is abelian, so that we can consider A as a $\mathbb{Z}\bar{G}$-module, and moreover $A \in \mathrm{Coc}_{\mathbb{Z}\bar{G}}\, \mathfrak{A}_0$. We will show that in each case the group $A \rtimes \bar{G}$ is locally nilpotent. Then it will follow from Theorem 1 and Lemma 7 that A is a locally nilpotent $\mathfrak{S}_0$-group. Let G be a Baer-nilpotent group, let $a_1, \ldots, a_k$ be elements of A and $g_1, \ldots, g_t$ elements of G, and put $H = \langle a_1, \ldots, a_k, g_1, \ldots, g_t \rangle$. Every finite quotient of H is nilpotent, and so H is also nilpotent (see [12], Theorem 10.51). Thus $A \rtimes \bar{G}$ is locally nilpotent. If G is a $\bar{Z}$-group, then A is a $\bar{Z}$-module over $\mathbb{Z}\bar{G}$. Then A is a $\mathbb{Z}\bar{G}$-hypercentral module by Corollary 3, and so $A \rtimes \bar{G}$ is locally nilpotent. The other statements are now clear. □

Theorem 3. *Let G be a $\mathfrak{S}_0 C$-group. The following statements are equivalent:*

(a) *G is a Baer-nilpotent group;*

(b) *G is an Engel group;*

(c) *G is a locally nilpotent group;*

(d) *G is a $\bar{Z}$-group;*

(e) *G is a hypercentral group;*

(f) *G has an ascending central series*

$$1 = Z_0 \leq Z_1 \leq \cdots \leq Z_\alpha \leq Z_{\alpha+1} \leq \cdots \leq Z_\gamma = G$$

where $\gamma \leq 3\omega$;

(g) *G is an N-group;*

(h) *G is an $\tilde{N}$-group.*

Proof. (a)⇒(c) Let $g_1, \ldots, g_n$ be elements of G, and put $H = \langle g_1, \ldots, g_n \rangle$ and $L = H^G = \langle g_1 \rangle^G \ldots \langle g_n \rangle^G$. It follows from Theorem 2 that L is a locally nilpotent $\mathfrak{S}_0$-group, and in particular H is nilpotent. Therefore G is locally nilpotent. Thus (a), (b) and (c) are equivalent.

(d)⇒(e) Clearly it is enough to show that $Z(G) \neq 1$. Let a be a non-trivial element of G and put $A = \langle a \rangle^G$. It follows from Theorem 2 that A is a locally nilpotent $\mathfrak{S}_0$-group. Then A contains a G-invariant abelian non-trivial subgroup A_0, and Lemma 6 yields that $A_0 \cap Z(G) \neq 1$.

(e)⇒(f) Let a be a non-trivial element of G, and put $A = \langle a \rangle^G$. If T is the torsion subgroup of A, then $T = \underset{p \in \pi(A)}{\mathrm{Dr}} T_p$, where T_p is the Sylow p-subgroup of A. Since $A \in \mathfrak{S}_0$, then T_p is a Chernikov p-group and so it contains an ascending series of G-invariant subgroups

$$1 = L_0 \leq L_1 \leq \cdots \leq L_n \leq \cdots \leq L_\omega \leq L_{\omega+1} \leq \cdots \leq L_{\omega+n} = T_p,$$

where the factors are G-composition factors. Thus $L_\alpha \leq Z_\alpha(G)$ and $T \leq Z_{2\omega}(G)$. The factor group A/T has a series of finite length of G-invariant subgroups, whose factors are torsion-free abelian groups of finite rank. It follows from a result of Charin (see for instance [12], Lemma 6.37) that $A \leq Z_{2\omega+k}(G)$ for some $k \in \mathbb{N}$. Hence $G = Z_{3\omega}(G)$. The other implications are obvious. □

Corollary. *Let G be a CC-group. The following statements are equivalent:*

(a) *G is a Z-group;*

(b) *G is a residually central group;*

(c) *G is a hypercentral group;*

(d) *G has an ascending central series*

$$1 = Z_0 \leq Z_1 \leq \cdots \leq Z_\gamma = G$$

where $\gamma \leq 2\omega$;

(e) *G is a Baer-nilpotent group;*

(f) *G is an Engel group;*

(g) *G is a locally nilpotent group;*

(h) *G is a $\bar{Z}$-group;*

(i) *G is an N-group;*

(j) *G is an $\tilde{N}$-group.*

Proof. It is enough to prove that (b)⇒(c). Since G is a CC-group, for every $x \in G$ we have that $[G, x]$ is a Chernikov group. If $x \notin Z(G)$, there exists a minimal normal subgroup N of G contained in $[G, x]$. But $N \leq Z(G)$ and so $Z(G) \neq 1$. It follows that G is hypercentral. □

We note that the analogous of Theorem 2 for $\hat{\mathfrak{S}}C$-groups is not true, as the following counterexample shows.

Example. Let $\{p_n \mid n \in \mathbb{N}\}$ be the set of all primes in their natural order, $A_n = \langle a_{n1} \rangle \times \cdots \times \langle a_{nn} \rangle$ an elementary abelian p_n-group, $\langle b_n \rangle$ an infinite cyclic group. The group $A_n \times \langle b_n \rangle$ admits an automorphism φ_n such that

$$b_n\varphi_n = b_n a_{nn},\ a_{nn}\varphi_n = a_{nn}a_{nn-1},\ \ldots,\ a_{n2}\varphi_n = a_{n2}a_{n1},\ a_{n1}\varphi_n = a_{n1}.$$

Put $C_n = \big(A_n \times \langle b_n \rangle\big) \rtimes \langle \varphi_n \rangle$ and $H = \operatorname*{Cr}_{n \in \mathbb{N}} C_n$. Let $T = \operatorname*{Dr}_{n \in \mathbb{N}} A_n$, $\Phi = \operatorname*{Dr}_{n \in \mathbb{N}} \langle \varphi_n \rangle$ and $b = (b_n)_{n \in \mathbb{N}}$. Consider now in H the subgroup

$$G = \big(T \times \langle b \rangle\big) \rtimes \Phi.$$

If $g \in T \times \langle b \rangle$, then $T \times \langle b \rangle \leq C_G(g)$ and so $G/C_G(g)$ is a group of rank 1. If $g \in \Phi$, then $g = \varphi_1^{s_1} \ldots \varphi_l^{s_l}$, where $s_1, \ldots, s_l$ are non-negative integers, and so $\operatorname*{Cr}_{n \geq l} C_n \leq C_H(g)$. Hence $G/C_G(g^G)$ is finitely generated and nilpotent. It follows that G is a locally nilpotent $\hat{\mathfrak{S}}C$-group. But $\langle b \rangle^G = T \times \langle b \rangle$ has infinite rank.

References

[1] Charin, V. S., On automorphism groups of nilpotent groups. Ukrain. Mat. Zh. 6 (1954), 295–304.

[2] Franciosi, S., de Giovanni, F., Kurdachenko, L. A., The Schur property and groups with uniform conjugacy classes. J. Algebra. To appear.

[3] Franciosi, S., de Giovanni, F., Tomkinson, M. J., Groups with polycyclic-by-finite conjugacy classes. Boll. Un. Mat. Ital. (7) 4B (1990), 35–55.

[4] —, —, —, Groups with Chernikov conjugacy classes. J. Austral. Math. Soc. Ser. A 50 (1991), 1–14.

[5] Gonzales, M., Otal, J., P. Hall's covering group and embedding of countable CC-subgroups. Comm. Algebra 18 (1990), 3405–3412.

[6] —, —, Peña, J. M., CC-groups with periodic central factor. Manuscripta Math. 69 (1990), 93–105.

[7] Kegel, O. H., Wehrfritz, B. A. F., Locally Finite Groups. North Holland, Amsterdam, 1973.

[8] Kurdachenko, L. A., On groups with minimax conjugacy classes. In: Infinite Groups and the Adjoining Algebraic Structures, Kiev (1990), 160–177.

[9] Otal, J., Peña, J. M., Tomkinson, M. J., Locally inner automorphisms of CC-groups. J. Algebra 141 (1991), 382–398.

[10] Polovickii, Ya. D., Groups with extremal conjugacy classes, Siberian Mat. Zh. 5 (1964), 891–895.

[11] Robinson, D. J. S., Finiteness Conditions and Generalized Soluble Groups, Part 1. Springer, Berlin 1972.

[12] —, Finiteness Conditions and Generalized Soluble Groups, Part 2. Springer, Berlin 1972.

[13] —, A Course in the Theory of Groups. Springer, Berlin 1982.

[14] Tomkinson, M. J., FC-groups. Pitman, Boston, 1984.

Centrality in Soluble Groups

John C. Lennox

1991 Mathematics Subject Classification: 20E15, 20F14

P. Hall's Questions

The aspects of centrality which we would like to discuss in this survey spring essentially from P. Hall's seminal work in 1954 on certain classes of finitely generated (f.g.) soluble groups.

Theorem 1 [4]. *Every f.g. abelian by polycyclic group satisfies* Max-*n, the maximal condition for normal subgroups.*

It is an immediate consequence of Theorem 1 that the upper central series of a f.g. abelian by polycyclic group G terminates after a finite number of steps, that is, G has ***finite upper central height.*** Indeed, the same holds for any normal subgroup H of G since each term of the upper central series of H is characteristic in H and therefore normal in G.

In view of this in 1968 Hall asked whether something much stronger were true:

Question 1. Do subgroups of f.g. abelian by polycyclic groups have finite upper central heights?

He also asked:

Question 2. Do f.g. abelian by polycyclic groups satisfy Max-c, the maximal condition on centralizers?

These questions, in their full generality, are still open. However, significant partial answers have been obtained. In 1970 Lennox and Roseblade proved that Hall's question had a positive answer for the class of f.g. abelian by nilpotent groups.

Offprint from
Infinite Groups 94, Eds.: de Giovanni/Newell

Theorem 2 [11]. *Suppose that G is a f.g. abelian by nilpotent group. Then:*

(i) *There exists $n = n(G)$ such that $\zeta_\omega(H) = \zeta_n(H)$, for all subgroups H of G, that is, G is* stunted.

(ii) *There exists $e = e(G)$ such that $c_G(x^n) \leq c_G(x^e)$ for all $x \in G$, that is, G is* eremitic.

(iii) *There exists $m = m(G)$ such that any centralizer chain in G, that is, any chain of the form $c_G(X_1) \leq c_G(X_2) \leq \cdots \leq c_G(X_n) \leq \cdots$, where the X_n are subsets (or subgroups) of G, has at most m strict inequalities or gaps, that is, G has* finite central gap number.

We note that if $e = 1$ in (ii) then centralizers in G are isolated, so that (ii) tells us that centralizers in G are "nearly" isolated.

It was also shown in [11] that metanilpotent by finite groups satisfying Max-n are both stunted and eremitic.

We recall from [4] that Theorem 1 fails for f.g. centre by metabelian groups. In the case of Theorem 2 we have the following

Examples.

(i) There exists a f.g. (nilpotent of class 2) by abelian group with central height ω;

(ii) There exists a f.g. centre by metabelian group which is not eremitic;

(iii) There exists a f.g. centre by metabelian group which has Max-n but not Max-c.

The most that could be said in [11] for f.g. abelian by polycyclic groups was to show that in such a group the upper central heights of *subnormal* subgroups are boundedly finite, that there exists $e = e(G)$ such that $c_G(H^n) \leq c_G(H^e)$ for all *subnormal* subgroups H of G and that the maximal condition for centralizers of *subnormal* subgroups is satisfied. It is not known whether there is a bound on the number of gaps in a chain of centralizers of subnormal subgroups in a f.g. abelian by polycyclic group.

Progress has been made since [11] appeared and in order to chronicle it we now recall that all the results so far mentioned were obtained using parallel theorems for modules. Suppose that G is a f.g. abelian by polycyclic group. Then there is a normal abelian subgroup of G such that $\Gamma = G/A$ is polycyclic. The conjugation action of G on A induces a natural action of Γ on A since A is abelian so that A becomes a Γ (indeed a $\mathbb{Z}\Gamma$)-module. Hall proved that A is a Noetherian $\mathbb{Z}\Gamma$-module and deduced Theorem 1 from it.

Suppose now that Γ is any group and that A is a Γ-module and form the natural split extension $A \rtimes \Gamma$. For $H \leq \Gamma$ we write $A_n(H) = A \cap \zeta_n(AH) = \{a \in A : [a, x_1, \ldots, x_n] = 1$ for all n-tuples $x_1, \ldots, x_n$ of elements of $H\}$.

We say that the pair (A, Γ) is stunted if there exists a non-negative integer h such that for all $H \leq \Gamma$ we have $A_\omega(H) = A_h(H)$. We say that the pair (A, Γ) is eremitic if there exists a positive integer e such that $c_A(x^n) \leq c_A(x^e)$ for all $x \in \Gamma$.

Theorem 2 is deduced from

Theorem 3. *Suppose that Γ is f.g. nilpotent and that A is a Noetherian Γ-module. Then (A, Γ) is both stunted and eremitic.*

This result is proved by induction on the Hirsch length $h(\Gamma)$ of Γ and it ultimately depends on the properties of the centralizer $c_A(X)$, where X is chosen to have the following properties: $c_A(X) > 1$; $n_\Gamma(X)$ is isolated; and, if $X < H \leq \Gamma$ and $c_A(H) > 1$, then $|H : X|$ is finite and $n_\Gamma(H)$ is not isolated. The fact that such an X exists depends on P. Hall's theory of isolators for f.g. nilpotent groups [4]. In this situation, $c_A(X)$ is a Noetherian $n_\Gamma(X)$-module and it is easy to establish that the pair $(c_A(X), n_\Gamma(X))$ has the desired properties. The deduction of these properties for (A, Γ) depends on the very special behaviour of the Γ-module $c_A(X)^\Gamma$, generated by $c_A(X)$. They are summed up in the

Fan Out Lemma A. *Suppose that Γ and A are as in the statement of Theorem 3 and X is as described above. Suppose further that B is a Γ-submodule of A, H is a subgroup of Γ and $N = n_\Gamma(X)$. Then:*

1. $c_A(X)^\Gamma = \operatorname*{Dr}_{t \in T} c_A(X)^t$, *where T is a transversal to the cosets of N in Γ.*
2. $B \cap c_A(X)^\Gamma > 1 \iff c_B(X) > 1$.
3. $A_n(H) \cap \left(c_A(X)\right)^\Gamma = \operatorname*{Dr}_{t \in T_H} \left(c_A(X)^t \cap A_n(H)\right)$, *where $t \in T_H \iff H \leq N^t$.*

In 1981 K. A. Brown, in his work [1] on the primitivity of group rings of soluble groups, developed another version of the Fan Out Lemma. Suppose that Γ is a f.g. nilpotent group and that A is a (not necessarily f.g.) non-zero $\mathbb{Z}\Gamma$-module written additively. For any subgroup H of Γ we define $A_0(H) = \{a \in A : \dim_{\mathbb{Q}}(a\mathbb{Z}H \otimes \mathbb{Q}) < \infty\}$. If X is chosen such that $A_0(X)$ is non-zero and $A_0(H)$ is zero for all $X < H \leq \Gamma$, then, setting $N = n_\Gamma(X)$, T a right transversal to N in Γ and $V = A_0(X)\mathbb{Z}\Gamma$, we have:

Fan Out Lemma B.

1. $V = \operatorname*{Dr}_{t \in T} A_0(H) \otimes_{\mathbb{Z}N} t$.
2. *If B is a $\mathbb{Z}\Gamma$-submodule of A and $B \cap V \neq 0$ then $B_0(H) \neq 0$.*
3. *If $H \leq \Gamma$ then $V_0(H) = \operatorname*{Dr}_{t \in T_H} A_0(X) \otimes t$, where $t \in T_H \iff H \leq N^t$.*

This version of the Fan Out Lemma is easier to prove than the previous version and Theorem 2 can be deduced from it. However, once more, the result depends very much on the nilpotency of the ambient group

More recently in 1991 Nazzal and Rhemtulla [12] have extended Theorem 2 to

Theorem 4. *Suppose that G is a f.g. abelian by polycyclic group whose polycyclic quotient is either abelian by cyclic or plinth by abelian. Then G is stunted, eremitic and has finite central gap number.*

Here, following Roseblade [15] a ***plinth*** for a group Γ is a free abelian subgroup M of finite non-zero rank such that no non-trivial subgroup of M of lower rank is normal in any subgroup of finite index in Γ, that is, Γ and all of its subgroups of finite index act rationally irreducibly on M.

In [14] Robinson and Wilson proved that just non-polycyclic groups, that is, soluble groups which are not polycyclic, but have all of their proper quotients polycyclic, are either f.g. metabelian by finite or are finite extensions of f.g. abelian by polycyclic groups with plinth by abelian polycyclic quotient. Hence

Corollary 5 [12]. *All just non-polycyclic groups are stunted, eremitic and have finite central gap number.*

Nazzal and Rhemtulla's work depends firstly on the work of Rhemtulla and Wehrfritz [13] which shows that polycyclic groups possess subgroups of finite index with similar isolator properties to those established by P. Hall for f.g. nilpotent groups, and secondly on two further versions of the Fan Out Lemma. In order to describe them we need some notation derived from Roseblade [15].

Suppose that S is any ring with a 1 and that M is any S-module. If $X \subseteq S$ then we we set $^*X = \{m \in M : mx = 0 \text{ for all } x \in X\}$. Roseblade proved that the set $\prod_S(M)$ of ideals of S maximal with respect to $^*P = 0$, consists of prime ideals and is non-empty whenever M is non-zero and S is Noetherian. Nazzal and Rhemtulla's first version of the Fan Out Lemma is an extension of a result of Roseblade

Fan Out Lemma C. *Suppose that Γ is any group, A is a normal abelian subgroup of Γ, R is a commutative ring with a 1 and M is any $R\Gamma$-module. Suppose further that $P \in \prod_{RA}(M)$, $N = n_\Gamma(P)$ and that T is a set of coset representatives of N in Γ. Set $U = {}^*P$ and $Y = U(R\Gamma)$. Then:*

1. $Y = \operatorname{Dr}_{t \in T} Ut$.
2. *If M is Γ-Noetherian, then U is N-Noetherian.*
3. *If N is isolated and H is any subgroup of Γ, then $Y_n(H) = \operatorname{Dr}_{t \in T_H} U_n(H)t$ for all $n \geq 0$.*
4. *If the pair (U, N) is stunted and eremitic then so is (Y, Γ).*

If we try to prove Theorem 3 by induction on $h(\Gamma)$ all is well unless each $P \in \prod_{RA}(M)$ is orbital, that is, $|\Gamma : n_\Gamma(P)|$ is finite. It is for this reason that a further version of the Fan Out Lemma is needed. It deals in fact with the case where the ideals P are not only orbital but also faithful, that is, $(P + 1) \cap A = 1$.

Fan Out Lemma D. *Suppose that Γ is a torsion free metabelian polycyclic group with the isolator property, A is a normal abelian subgroup of Γ containing Γ' and M is a K-module, where K is a field. Suppose that there exists $x \notin c_\Gamma(A)$ such that $U = M,(x)$ is non-trivial and that $\Gamma = AN$, where $N = c_\Gamma(x)$. Set $Y = U(K\Gamma)$ and let T be a transversal to the cosets of N in Γ such that $T \subseteq A$. Then:*

1. *If every ideal $P \in \prod_{KA}(M)$ is faithful and orbital then $Y = \mathop{\mathrm{Dr}}_{t \in T} Ut$.*
2. *If M is $K\Gamma$-Noetherian, then U is KN-Noetherian.*
3. *If H is any subgroup of Γ, then $Y_n(H) = \mathop{\mathrm{Dr}}_{t \in T_H} U_n(H)t$, for all $n \geq 0$.*
4. *If the pair (U, N) is stunted and eremitic, then so is (Y, Γ).*

We record the main open question in its module form:

Question 3. Suppose that A is a f.g. $\mathbb{Z}\Gamma$-module where Γ is a polycyclic group. Is there a non-zero submodule Y such that (Y, Γ) is stunted and eremitic?

In 1977 Segal in [17] uses a very different module theoretic approach, involving Krull dimension and critical modules, to prove, amongst other results, that nilpotent by nilpotent by finite groups with Max-n are stunted and eremitic. Furthermore, his methods yield an estimate for the bound on the central heights of subgroups of these groups.

Bryant's Verbal Topology

In [2] Bryant has given an ingenious and very short alternative argument which proves that f.g. abelian by nilpotent groups satisfy Min-c, the minimal condition for centralizers (which is, as is well known and easy to see, the same as Max-c). However Bryant's method does not yield the fact that there is a bound for the number of gaps, which we know from Theorem 2.

We now give the main features of Bryant's method. Suppose that G is a group and form the free product $G * \langle x \rangle$ of G with an infinite cyclic group $\langle x \rangle$. If $w \in G * \langle x \rangle$ and $g \in G$, we define $w(g)$ to be the element of G obtained by substituting g for x in w. Given $w \in G * \langle x \rangle$, the set $\{g \in G : w(g) = 1\}$ is called a ***primitive solution set***. The ***verbal topology*** on G is defined by taking all primitive solution sets as a subbasis for the closed sets of the topology. Thus a closed set is an intersection of finite unions of primitive solution sets. A ***solution set*** is any intersection of primitive solution sets (and therefore closed). In particular, if A is any subset of G, then $c_G(A)$ is closed, since it is the solution set found by taking the intersection of the primitive solution sets corresponding to the words $[x, a]$, $a \in A$. Bryant proves

Theorem 6. *A f.g. abelian by nilpotent group satisfies the minimal condition on closed sets (and hence* Min-c*).*

We describe the two main ingredients in the proof. First of all Bryant shows that a group satisfies the minimal condition on closed sets if and only if it satisfies the minimal condition on solution sets, and then he goes on to prove

Theorem 7. *Suppose that G is a f.g. group such that every f.g. group in the variety generated by G satisfies* Max-n. *Then G satisfies the minimal condition on solution sets.*

The proof of this is so short that we reproduce it here. For any $g \in G$ the map $\varphi_g : w \mapsto w(g)$ is a homomorphism from $G * \langle x \rangle$ to G. Put $N = \bigcap_{g \in G} \operatorname{Ker} \varphi_g$. If S is a solution set, then $S = \{g \in G \mid w(g) = 1 \; \forall w \in W\}$ where $W = \bigcap_{g \in S} \operatorname{Ker} \varphi_g$. Then $W/N \lhd (G * \langle x \rangle)/N \in$ Max-n. Hence G has the minimal condition for solution sets.

In the application, f.g. abelian by nilpotent groups satisfy the hypotheses of Theorem 7. However, f.g. abelian by polycyclic groups do not and so we are left with the following open question which, if it had a positive answer, would lead at once to a proof that such groups satisfied Min-c, and hence Max-c.

Question 4. Do f.g. abelian by polycyclic groups satisfy the minimal condition on closed sets?

Other Related Results

Various other classes of groups have been considered in connection with the centrality properties mentioned in Theorem 2. For instance, in [6] Houghton, Lennox and Wiegold find necessary and sufficient conditions for wreath products to be stunted, eremitic or have finite central gap number. Linear groups have also been studied: Gruenberg [3] proved that torsion free linear groups are stunted and Wehrfritz [18] proved a result analogous to Theorem 2 for f.g. linear groups:

Theorem 8. (i) *F.g. linear groups are stunted.*
(ii) *F.g. linear groups over fields (of characteristic* 0*) are eremitic (with $e = 1$).*
(iii) *A chain of centralizers of subgroups without repetitions in a linear group of degree n has length bounded by $n^2 - 1$.*

Notes on (ii). a. This result may be compared with that of Lennox [7] who proved that a f.g. abelian by nilpotent group which is torsion free has a subgroup of finite index with $e = 1$.

b. (ii) is deduced from the following lemma whose hypotheses are satisfied by f.g. linear groups.

Lemma 9. *Suppose that G is any group which contains a subgroup T of finite index such that T is simultaneously a residually finite π-group and a residually finite π'-group for some set π of primes. Then G is eremitic.*

In 1974 Segal [16] showed that the hypotheses of Lemma 9 are also satisfied by all torsion free f.g. abelian by polycyclic groups so that we have

Theorem 10. *Torsion free f.g. abelian by polycyclic groups are eremitic.*

Maximal Nilpotent Subgroups

One immediate consequence of the fact that f.g. abelian by nilpotent groups are stunted is that every nilpotent subgroup of such a group is contained in a maximal nilpotent subgroup. This observation led to the question whether there exist f.g. soluble groups without maximal nilpotent subgroups. An affirmative answer follows without difficulty from the following result of Lennox, P. M. Neumann and Wiegold in 1990 [10]

Theorem 11. *There exists a f.g. soluble group G of derived length 4 whose hypercentre $\zeta_\omega(G)$ is not nilpotent.*

The following question is left open in [10]

Question 5. Are there soluble groups with Max-n without maximal nilpotent subgroups?

The stimulus for [10] was an earlier result [8] of Lennox in 1984:

Theorem 12. *Suppose that G is a f.g. abelian by nilpotent group such that each proper f.g. nilpotent subgroup is contained in a larger one. Then G is nilpotent.*

This theorem is deduced from a module theoretic result which in turn depends on the Fan Out Lemma.

Theorem 13 [8]. *Suppose that Γ is a f.g. nilpotent group and that A is a f.g. $\mathbb{Z}\Gamma$-module such that $c_A(x) > 1$ for all x in Γ. Then there exists a subgroup Δ of finite index in Γ such that $c_A(\Delta) > 1$.*

This result does not hold for the case where Γ is polycyclic, even if the module is f.g. as a $\mathbb{Z}$-module. Notwithstanding that, the following question is still open:

Question 6. Does Theorem 12 extend to f.g. abelian by polycyclic groups?

Groups in Which Centralizers of F.G. Subgroups Are F.G.

It is easy to see that a group G satisfies the maximal condition for centralizers if and only if for every subset X of G we have $c_G(X) = c_G(X_0)$ for some finite subset X_0 of X. Hence, if G satisfies Max-c and if every centralizer of a f.g. subgroup of G is itself f.g., then every centralizer is f.g. Thus, by Theorem 2, if G is a f.g. abelian by nilpotent group with this property, then all of its centralizers are f.g. and it readily follows that G is polycyclic. For if A is an abelian normal subgroup of G maximal with respect to having G/A nilpotent, then $A = c_G(A)$ and so A is f.g. and the result is immediate. It is therefore natural to ask

Question 7. Suppose that G is a (f.g.) soluble group such that centralizers of f.g. subgroups are f.g. Is G polycyclic?

In this direction we have

Theorem 14 [9]. *A soluble group of finite rank in which centralizers of f.g. subgroups are f.g. is polycyclic.*

Now f.g. soluble groups of finite rank are nilpotent by abelian by finite and so Theorem 14 follows from the somewhat stronger

Theorem 15 [9]. *Suppose that G is a f.g. nilpotent by polycyclic group in which the centralizer of each polycyclic subgroup is f.g. Then G is polycyclic.*

The proof, which goes by induction on the class of a nilpotent normal subgroup with polycyclic factor group, uses Malcev's theorem that a soluble group of automorphisms of a polycyclic group is polycyclic to deduce that the hypothesis is inherited by a polycyclic normal subgroup. The proof also depends on the easily established fact that the hypothesis is also inherited by centralizers of polycyclic subgroups.

Of course, if G is any f.g. group, one extreme situation in which centralizers of f.g. subgroups are f.g. is where the centralizer of any element of G has finite index in G, that is, where G is an FC-group. Such a group, being f.g., is centre by finite and hence certainly polycyclic in the case where G is soluble. This observation suggests the following modification to Question 7:

Question 8. Is a soluble group polycyclic if the centralizer of each of its elements is f.g.?

In partial answer to this question we have the following result:

Theorem 16 [9]. *A f.g. abelian by nilpotent group is polycyclic if the centralizer of each of its elements is f.g.*

This result appears as a corollary to

Theorem 17 [9]. *Suppose that G is a group with a normal subgroup A such that G/A is nilpotent and $c_G(a)$ is f.g. for all $a \in A$. Then G is polycyclic.*

References

[1] Brown, K. A., Primitive group rings of soluble groups. Arch. Math. (Basel) 36 (1981), 404–413.

[2] Bryant, R. M., The verbal topology of a group. J. Algebra 48 (1977), 340–366.

[3] Gruenberg, K. W., The hypercentre of linear groups. J. Algebra 8 (1968), 34–40.

[4] Hall, P., Finiteness conditions for soluble groups. Proc. London Math. Soc., (3) 4 (1954), 419–436.

[5] —, The Edmonton Notes on Nilpotent Groups. Queen Mary College, London 1969.

[6] Houghton, C. H., Lennox, J. C, Wiegold, J., Centrality in wreath products. J. Algebra 35 (1975), 356–366.

[7] Lennox, J. C., On a centrality property of finitely-generated torsion-free soluble groups. J. Algebra 18 (1971), 541–548.

[8] —, A fixed point theorem for modules over finitely generated nilpotent groups. Bull. London Math. Soc. 16 (1984), 289–291.

[9] —, On a property of centralizers in finitely generated soluble groups. To appear.

[10] —, Neumann, P. M., Wiegold, J., Nilpotent subgroups and the hypercentre of infinite soluble groups. Arch. Math. (Basel) 54 (1990), 417–421.

[11] —, Roseblade, J. E., Centrality in finitely generated soluble groups. J. Algebra 16 (1970), 399–435.

[12] Nazzal, S. H., Rhemtulla, A. H., Centrality in abelian by polycyclic groups. Arch. Math. (Basel) 56 (1991), 333–342.

[13] Rhemtulla, A. H., Wehrfritz, B. A. F., Isolators in soluble groups of finite rank. Rocky Mountain J. Math. 14 (1984), 415–421.

[14] Robinson, D. J. S., Wilson, J. S., Soluble Groups with many Polycyclic Quotients. Proc. London Math. Soc., (3) 48 (1984), 193–229.

[15] Roseblade, J. E., Group rings of polycyclic groups. J. Pure Appl. Algebra (1973), 307–328.

[16] Segal, D., On abelian-by-polycyclic groups. J. London Math. Soc. (2), 11 (1975), 445–452.

[17] —, On the residual simplicity of certain modules. Proc. London Math. Soc. (3), 34 (1977), 327–353.

[18] Wehrfritz, B. A. F., Remarks on centrality and cyclicity in linear groups. J. Algebra 18 (1971), 229–236.

Finite p-Groups in Which Subgroups Have Large Cores

John C. Lennox, Howard Smith and James Wiegold

1991 Mathematics Subject Classification: 20D15 20D60

1. Introduction and Preliminaries

In the notation of [1], a group G is said to be a CF-group if $|H : \mathrm{Core}_G(H)| < \infty$ for all subgroups H of G. One of the principal results of [1] is that locally finite CF-groups are abelian-by-finite, and there are other results of this type to be found in [4]. The following question arises very naturally in this context. For a finite group G, suppose that there is a uniform bound n for the indices $|H : \mathrm{Core}_G(H)|$ (we say that G is core-n in such a case); is there a function f such that G has an abelian normal subgroup of index at most $f(n)$? In this generality the problem seems to be rather difficult. The present article provides a positive answer in the simplest of situations:

Theorem. *For every odd prime p, every finite core-p p-group G is nilpotent of class at most 3, the derived group G' is of exponent at most p, and G has an abelian normal subgroup of index at most p^5.*

The case of core-2 2-groups is very different. Firstly, there is no bound on the nilpotency class, as dihedral 2-groups show. We do not know if there is always an abelian subgroup of bounded index, though methods like those used in the proof of the theorem can be used to establish analogous results for 2-groups of class 2. The bound provided by the theorem is probably too large. The worst example we know is the following (proofs being a matter of routine).

Example 1.1. Let p be an odd prime and set

$$G = \langle a, b, c \mid [a,b] = c^p,\ [b,c] = a^p,\ [c,a] = b^p,\\ [x,y,z] = 1 \text{ for } x,y,z \in \{a,b,c\} \rangle.$$

Offprint from
Infinite Groups 94, Eds.: de Giovanni/Newell

Then G is core-p and has no abelian subgroup of index p, but G does have an abelian subgroup of index p^2.

Our second example is of a core-p p-group of class exactly 3. Again we omit most of the details.

Example 1.2. Let p be a prime greater than 3 and congruent to 3 mod 4. Define G to be the group with generators a, b and defining relations as follows: every commutator of weight 4 involving a and b is trivial (so G is nil-3); $a^{p^2} = b^{p^2} = 1$; $[a, b]^p = 1$; $a^p = [b, a, b]$, $b^p = [a, b, a]$.

It is easy to check that G has order p^5, that a and b have order (exactly) p^2 and that $\langle a \rangle \cap \langle b \rangle = 1$. Further, G' is the set of elements of order at most p, and p-th powers are central. In order to check the core-p property, it is clear that one need check only those subgroups H of order p^2 or p^3. Subgroups of order p^2 are easily dealt with, so assume that $|H| = p^3$. Now either $H = G'$ or (without loss of generality) $H = \langle ab^m c, u \rangle$ for some integer m and elements c, u of G'. The conditions on the prime p may now be used to show that $\langle ab^m c \rangle$ is not normalized by $[a, b]$. It follows that $u \in [G', G]$ and hence that H has a central subgroup of order p^2.

All our methods are elementary. The only slightly unusual notation we use is for the ***breadth*** $\mathrm{br}(a)$ of an element a of a finite p-group G, defined by $p^{\mathrm{br}(a)} = |G : C_G(a)|$. The following result is immediate.

Lemma 1.3. *Let G be a finite p-group of class 2 in which the derived group G' is of exponent p. For all a in G, $\mathrm{br}(a)$ is the rank of $[a, G]$.*

We are grateful to Avinoam Mann for some helpful suggestions and for bringing to our attention the appropriate result from [2].

2. Proofs

We shall split up the rather long proof of the theorem into that of several lemmas. Throughout, p is an odd prime, and all groups considered are finite. The first result and its corollary are crucial.

Lemma 2.1. *Let G be a nonabelian core-p group of exponent p. Then $G = A \times B$, where A is abelian and B is of order p^3.*

Proof. First, we establish that G is nilpotent of class at most 2. Since the core-p property is preserved under images, we may assume that the centre Z of G is cyclic, and that G is of class at most 3 and therefore metabelian. Let X be a maximal abelian subgroup containing G'. Then $X \geq Z$, and $X = Z \times T$ for some subgroup T, which must be of order p else it would contain a nontrivrial normal subgroup and therefore intersect Z nontrivially. Thus $X \cong \mathbb{Z}_p \times \mathbb{Z}_p$ and of course $C_G(X) = X$. Moreover, X is normal in G, and G/X is embedded in Aut X. Consequently, $|G/X| = p$, so $|G| = p^3$ and G is of class 2, as required.

Next, we show that $|G'| = p$ in all cases. For, again by an obvious induction, we may assume that $|G'| = p^2$. Then, (see [5]), some element t of G must have p^2 conjugates, and thus $G' = [t, G] = \langle [t, u], [t, v] \rangle$ for suitable u, v. Now $[u, v] = [t, u]^m [t, v]^n$ for some integers m and n; replacing u by ut^{-n} and v by vt^m enables us to assume that $[u, v] = 1$. But then $\langle u, v \rangle$ is of order p^2 and so, by the core-p property, it contains a nontrivial element of the centre, and $\big| [t, G] \big| \leq p$. This contradiction proves that $|G'| = p$.

The next step is to show that G has an abelian subgroup of index p. Let N be a normal subgroup of G maximal with respect to intersecting G' trivially; if we can show that G/N has an abelian subgroup X/N of index p then we are done, since $X' \leq N \cap G' = 1$. So from now on in this part of the proof we shall assume that every nontrivial normal subgroup of G intersects G' nontrivially. Once more, let X be a maximal abelian subgroup of G containing G'; as before, X is of order p^2 and G is of order p^3, and the claim follows.

Now let Y be an abelian subgroup of index p in the arbitrary nonabelian core-p group G of exponent p. Then $Y = Z \times T$ for some subgroup T of order p, say $T = \langle t \rangle$; suppose $G = \langle Y, g \rangle$. Set $B = \langle t, g \rangle$, so that B is of order p^3 since it is not abelian. The final step of the proof is to write Z as $A \times \langle [t, g] \rangle$; then $G = A \times B$, as required. □

Corollary 2.2. *For every core-p p-group G, G' centralizes G^p and G is metabelian.*

Proof. By Lemma 2.1, G/G^p is nilpotent of class at most 2, and $G'/(G' \cap G^p)$ is cyclic. However, G' centralizes G^p since, for all x in G, $\langle x^p \rangle$ is normal in G so that $G/C_G(x^p)$ is abelian and G' centralizes every p-th power. Thus G' is abelian since $G'/Z(G')$ is cyclic, and the corollary follows. □

The proof of the next lemma is long, but essentially elementary.

Lemma 2.3. *Let G be a nonabelian core-p p-group. Then G' is of exponent p.*

Proof. By Corollary 2.2, all we need is that commutators are of order dividing p. This enables us to concentrate on 2-generator groups: set $G = \langle a, b \rangle$. We may assume that G' is of exponent dividing p^2 and that the centre of G is cyclic. This means that abelian subgroups of G are 2-generator, for any subgroup H isomorphic to $\mathbb{Z}_p \times \mathbb{Z}_p \times \mathbb{Z}_p$ in a core-p p-group must contain a normal subgroup of order at least p^2, so that we can write $H = A \times B$, where A is central of order p. Similarly $B = C \times D$, where again C is central of order p. Indeed this argument extends to show that abelian subgroups of G are cyclic or of the form $\mathbb{Z}_p \times \mathbb{Z}_{p^n}$. This applies to G', and we see that $G' \cong \mathbb{Z}_{p^2}$ or $\mathbb{Z}_p \times \mathbb{Z}_p$ in any awkward case.

Suppose that $G' \cong \mathbb{Z}_{p^2}$. Then $[a, b]$ must be of order p^2, and of course G is of class at most 3. Moreover, $\gamma_3(G)$ is of order at most p. Thus

$$[a, b^{p^2}] = [a, b]^{p^2}[a, b, b]^{p^2(p^2-1)/2} = 1$$

and b^{p^2} is central. So is a^{p^2}, and therefore without loss of generality we have $b^{p^2} = a^{p^2\lambda}$ for suitable λ (remember that $Z(G)$ is cyclic). But then we have

$$\begin{aligned}(ba^{-\lambda})^{p^2} \\ &= b^{p^2}a^{-\lambda p^2}[a^{-\lambda}, b]^{\frac{p^2(p^2-1)}{2}}[a^{-\lambda}, b, b]^{\frac{p^2(p^2-1)(p^2-2)}{6}}[a^{-\lambda}, b, a^{-\lambda}]^{\frac{p^2(p^2-1)(2p^2-1)}{6}},\end{aligned}$$

which is 1, since p is odd. The upshot is that there is no loss of generality in assuming that $b^{p^2} = 1$, so that b^p is central since it is normal and of order at most p. Thus $1 = [a, b^p] = [a, b]^p[a, b, b]^{p(p-1)/2}$, so that $[a, b]^p = 1$, a contradiction.

Next, suppose that $G' \cong \mathbb{Z}_{p^2} \times \mathbb{Z}_p$, so that G is of class at most 4. We claim that $\gamma_3(G)$ is not of exponent p^2. Otherwise, some triple commutator, $[a, b, a]$, say, has order p^2, and it generates a direct factor of G'. Thus $[a, b]^p = [a, b, a]^{p\lambda}$ for suitable λ. Commutating twice with a, we get $[a, b, a]^p = [a, b, a, a]^{p\lambda}$, $[a, b, a, a]^p = 1$, so that $[a, b, a]^p = 1$ and then $[a, b]^p = 1$, another contradiction.

Next, we show that $G^{p^2} \leq Z(G)$ and $G^p \leq Z_2(G)$. For all x, y in G,

$$[x, y^{p^2}] = [x, y]^{p^2}[x, y, y]^{\frac{p^2(p^2-1)}{2}}[x, y, y, y]^{\frac{p^2(p^2-1)(p^2-2)}{6}} = 1$$

since $\gamma_3(G)^p = 1$ and p is odd. Thus $G^{p^2} \leq Z(G)$. Similarly,

$$[x, y^p] = [x, y]^p[x, y, y]^{\frac{p(p-1)}{2}}[x, y, y, y]^{\frac{p(p-1)(p-2)}{6}}$$

and thus is central, because $(G')^p$ is of order dividing p, $[x, y, y]^p = 1$ and G is of class 4.

We claim that $\Phi(G)$ is abelian. As we saw in Corollary 2.2, G' is abelian and centralizes G^p, so that we need to show that p-th powers commute. For all x, y in G,

$$\begin{aligned}[x^p, y^p] &= [x^p, y]^p[x^p, y, y]^{\frac{p(p-1)}{2}}[x^p, y, y, y]^{\frac{p(p-1)(p-2)}{6}} \\ &= [x^p, y]^p = [x, y]^{p^2}[x, y, x]^{\frac{p^2(p-1)}{2}}[x, y, x, x]^{\frac{p^2(p-1)(p-2)}{6}} = 1.\end{aligned}$$

Thus $\Phi(G)$ is abelian, so that $\Phi(G)^p = (G'G^p)^p = (G')^p(G^p)^p$ is central. If every element of $\gamma_3(G)$ has a p-th root in $\Phi(G)$, then $\gamma_3(G)$ is central and G is of class at most 3, and we can deal with that case by an argument almost identical with that where $G' \cong \mathbb{Z}_{p^2}$. So we may assume that G is of class exactly 4, and that some element x of $\gamma_3(G)$ has no p-th root in $\Phi(G)$, and thus generates a direct factor of $\Phi(G)$. Since abelian subgroups are 2-generator and $\Phi(G)$ is not cyclic, we may write $\Phi(G) = \langle u \rangle \times \langle x \rangle$. If u has order more than p^2 then $[a, b] = u^{p\lambda}x^{\mu}$ for some λ, μ;

but this means that $G' \leq \Phi(G)^p \gamma_3(G)$, so that $G' \leq Z_2(G)$ and G is of class 3. Thus u has order exactly p^2, so that G is a group of maximal class of order p^5. It has an elementary abelian normal subgroup of order p^2 (namely $\gamma_3(G)$), so it cannot have cyclic normal subgroups of that order. By the core-p property, this means that G must be of exponent precisely p^2. However, in that case all p-th powers are central, so G^p is central; since G/G^p is of class 2 (Lemma 2.1), G is of class at most 3, and this is the final contradiction that establishes the lemma. □

Corollary 2.4. *Every core-p p-group is of class at most* 3.

Proof. We remind the reader that p is odd. Once again, we may assume that $Z(G)$ is cyclic, so that abelian subgroups are 2-generator. In this case, G' is of order p^2 at most by Lemma 2.3, so G is certainly of class at most 3. □

We are now ready to begin the proof of the existence of abelian subgroups of bounded index. The first step is to consider the class-2 case, and it turns out that breadths of elements are either fairly small or fairly large:

Lemma 2.5. *Let G be a core-p p-group of class* 2, *and suppose that G contains an element a of breadth $k \geq 4$. Then*

(i) *for every x in G there exists an integer α which is zero or prime to p such that* $\operatorname{br}(a^\alpha x) \leq 3$, *and* $\operatorname{br}(x) \leq k+2$;

(ii) *at least* $(1/p)|G|$ *elements of G are of breadth at most* 3.

Proof. For each x in G, the subgroup $H := \langle a, x\rangle$ is of rank at most 3, since it is of class 2. Thus, no subgroup of H containing our element a can be normal in G, since $[a, G]$ is of rank k by Lemmas 1.3 and 2.3. Thus, we have to conclude that $\langle a^p, x^p, a^\alpha x, [a, x]\rangle$ is normal in G for some α that is zero or prime to p. Thus $[a^\alpha x, G] \leq \operatorname{soc} H$, so that $\operatorname{br}(a^\alpha x) \leq 3$, as required. Since $a^p \in Z(G)$, we have $\operatorname{br}(a^p g) = \operatorname{br}(g)$ for all g in G, and this proves part (ii). Finally, we have $[x, G] \leq [a, G] \operatorname{soc} H$ since $[a^\alpha x, G] \leq \operatorname{soc} H$, so that $\operatorname{br}(x) \leq \operatorname{br}(a) + 3$. However, $[a, x]$ lies in $[a, G] \cap \operatorname{soc} H$, so $[a, G] \cap \operatorname{soc} H \neq 1$ if $[a, x] \neq 1$ and then $\operatorname{br}(x) \leq \operatorname{br}(a) + 2$. If $[a, x] = 1$, then H is of rank 2 and so again we have $\operatorname{br}(x) \leq \operatorname{br}(a) + 2$. □

Lemma 2.5 shows up an oddity of core-p p-groups of class 2. Suppose that there happens to be an element of breadth $k \geq 6$. Then the possible breadths of elements are $0, 1, 2, 3, k-2, k-1, k$; $0, 1, 2, 3, k-1, k, k+1$; or $0, 1, 2, 3, k, k+1, k+2$. We do not know how far such oddities spread, but the following archetypal example shows this feature in extreme form.

Example 2.6. Let A be a homocyclic group of exponent p^2 and order p^{2n}, and G the split extension of A by the p-cycle $x \mapsto x^{1+p}$. Then G is core-p and the breadths of elements are 0, 1 and n.

Our next lemma is a variation on the theme of Lemma 2.5.

Lemma 2.7. *Let G be a core-p p-group, and let a be an element of G of breadth at least* 3. *Then* $|C_G(a)'| \leq p$.

Proof. Let x be any element of $C_G(a)$. Then $H := \langle a, x\rangle$ is of rank at most 2 and $\big|[a, G]\big| \geq p^3$. By Lemma 2.3, $[a, G]$ has rank at least 3, and it follows that no subgroup of H containing a is normal in G. Thus $\langle a^p, x^p, a^\alpha x\rangle$ is normal in G for some α that is zero or prime to p, so in particular $[a^\alpha x, C_G(a)] \leq H$, that is, $[x, C_G(a)] \leq \langle a, x\rangle$. It follows that every (cyclic) subgroup of $C_G(a)/\langle a\rangle$ is normal. But p is odd and so $C_G(a)/\langle a\rangle$ is abelian and $C_G(a)'$ is cyclic. □

We need one final preliminary lemma.

Lemma 2.8. *Let G be a core-p p-group, and suppose that $|G'| = p^l$. Then G has an abelian subgroup of index at most p^l.*

Proof. Assuming that G is not abelian, there exists an element z of order p in $G' \cap Z(G)$. By induction on l, there is a subgroup B of index at most p^{l-1} in G such that $B' \leq \langle z\rangle$. If B has an abelian subgroup of index at most p then we are done. Thus we may assume that $l = 1$. Factoring by a normal subgroup maximal with respect to missing G', we may assume that every nontrivial normal subgroup contains G': that is, G is one of Newman's JN2 groups [3]. It is proved there that such groups are central products of groups of the following type:

$$M(p^n) = \langle a, b, z \mid a^p = b^p = 1,\ z^{p^{n-1}} = [a, b],\ [a, z] = [b, z] = 1\rangle$$

$$N(p^n) = \langle a, b, z \mid a^p = b^p = z,\ z^{p^{n-1}} = [a, b],\ [a, b]^p = 1\rangle.$$

In $N(p^n)$, note that $(ab^{-1})^p = 1$. Thus, if our group G has more than one of these standard groups in its central decomposition, we could write $G = \langle U, V\rangle X$, where $[U, V] = 1$ and U and V are nonabelian and can be generated by two elements each

$$U = \langle u_1, u_2\rangle, \quad V = \langle v_1, v_2\rangle, \quad \text{with } u_1^p = v_1^p = 1.$$

Now look at $H := \langle u_1, v_1\rangle$. This group is of order p^2, so that there is an element of the form $u_1 v_1^\alpha$ in $Z(G)$, where $\alpha \not\equiv 0 \pmod p$. But then $1 = [u_1 v_1^\alpha, u_2] = [u_1, u_2]$, a contradiction that completes the proof. □

We can now prove the theorem stated in the introduction. Let G be a core-p p-group. We recall that G is metabelian and so, applying a result of Gillam [2], we need only show that *some* abelian subgroup of G has the required index

If every element of G has breadth at most 2 then $|G'| \leq p^3$ (see [5]), and Lemma 2.8 applies. If G contains an element a of breadth 3 or 4 then $|G : C_G(a)| \leq p^4$ and, by Lemmas 2.7 and 2.8, $C_G(a)$ has an abelian subgroup of index at most p, and again we are done. So assume that G contains an element of breadth greater than four but none of breadth 3 or 4. Since G is nilpotent of class at most 3 and $(G')^p = 1$, we have that G^p is central. By Lemma 2.1, $G/G^p = X/G^p \times Y/G^p$, where X/G^p is abelian and Y/G^p has order 1 or p^3. In the latter case, let L/G^p be a subgroup of index p in Y/G^p. Then $(XL)' = X'L'[X, L] \leq G^p$ and so XL is nilpotent of class at most 2. In either case, therefore, G contains a nil-2 subgroup M of index at most p. Let H be the set of elements of breadth at most 3 in M. Each of these have breadth at most 4 and hence at most 2 in G. The product of any two such elements also has breadth at most 4 and hence at most 2 in G, and it follows that H is a subgroup of M. Certainly $H \neq M$, since G has elements of breadth greater than 4. Thus M satisfies the hypothesis of Lemma 2.5, and we deduce that $|M : H| = p$. Again from [5], $|H'| \leq p^3$. Lemma 2.8 now gives us an abelian subgroup A of index at most p^3 in H and hence of index at most p^5 in G. This concludes the proof of the theorem. □

References

[1] Buckley, J. T., Lennox, J. C., Neumann, B. H., Smith, H., Wiegold, J., Groups with all subgroups normal-by-finite. J. Austral. Math. Soc. Ser. A. To appear.

[2] Gillam, J. D., A note on finite metabelian p-groups. Proc. Am. Math. Soc. 25 (1970), 189–190.

[3] Newman, M. F., On a class of nilpotent groups. Proc. London Math. Soc. (3) 10 (1960), 365–375.

[4] Smith, H., Wiegold, J., Locally graded groups with all subgroups normal-by-finite. J. Austral. Math. Soc. Ser. A. To appear.

[5] Vaughan-Lee, M. R., Breadth and commutator subgroups of p-groups. J. Algebra 32 (1974), 278–285.

Fixed Points Induced by the Permutation of Generators in a Free Group of Rank 2

Olga Macedońska and Witold Tomaszewski

Abstract. Let F be a free group of rank two, and in a group F/N let the permutation of generators define an automorphism $\bar{\sigma}$ with the group of fixed points $S(N)$. The full preimage of $S(N)$ in F we denote by S. We describe those N for which $S(N)$ is trivial, those S which are preimages of $S(N)$ for some N, and show for which S the set of corresponding N forms a lattice and give various examples.

1991 Mathematics Subject Classification: 20E36

1. Introduction

Let F be the free group of rank 2, generated by x, y, and let σ be the automorphism permuting x and y. Obviously, σ has no fixed points in F except the identity. If N is a normal σ-invariant subgroup in F, such that $G = F/N$ is not cyclic, then σ induces an automorphism $\bar{\sigma}$ in G with a group of fixed points $S(N) \subseteq G$. The preimage of $S(N)$ in F will be denoted by S, and called the subgroup of fixed points for $\sigma \mod N$, so $S(N) = S/N$. B. H. Neumann [3] considered the question of an automorphism of order two leaving only the neutral element in F/N fixed, where N is of finite index in F. In the case of $\bar{\sigma}$ there always are nontrivial fixed points in G, if G is a torsion group or if G is abelian. However for a metabelian group G the situation can be different.

We follow [3] and consider σ as an element of the near-ring of mappings F into F. Multiplication of mappings is as usual, $g^{\alpha\beta} = (g^{\alpha})^{\beta}$, and the addition is defined by $g^{(\alpha+\beta)} = g^{\alpha}g^{\beta}$. The identity map we denote by 1. There is only one distributive law, namely $\alpha(\beta+\gamma) = \alpha\beta + \alpha\gamma$. Since for example $(a^{\sigma})^{\sigma-1} = aa^{-\sigma} = a^{1-\sigma}$, we have the following properties of multiplication (valid for any automorphism of order 2 in an arbitrary group).

Properties of multiplication

(i) $\sigma(\sigma-1) = 1-\sigma$;

(ii) $(-1)(\sigma-1) = -\sigma+1$;

(iii) $(-\sigma)(\sigma-1) = -1+\sigma$;

Offprint from
Infinite Groups 94, Eds.: de Giovanni/Newell

(iv) $(\sigma - 1)(-1) = 1 - \sigma = (\sigma - 1)\sigma$;
(v) $(\sigma - 1)(\sigma - 1) = (1 - \sigma)2$ where "2" maps each element to its square.

For an automorphism α in a group G and a set $A \subseteq G$, we introduce a set $A^{\alpha-1} = \{a^{\alpha-1}, a \in A\}$.

Lemma 1. *If α is an automorphism of order 2 in a group G and A is an α-invariant subgroup in G then*

(i) $A^{\alpha-1} \subseteq A$,

(ii) $A^{\alpha-1} = A^{1-\alpha} = A^{-\alpha+1} = A^{-1+\alpha}$,

(iii) $A^{\alpha-1} = (A^{\alpha-1})^{-1} = (A^{\alpha-1})^{\alpha}$,

(iv) *if α has no fixed points in G except the identity then $s^{\alpha-1} \in A$ implies $s \in A$.*

Proof. Since A is a α-invariant subgroup we have $A = A^{-1} = A^{\alpha}$ and (i) follows; (ii) and (iii) follow now from the properties of multiplication. For (iv), if $s^{\alpha}s^{-1} = a^{\alpha}a^{-1}$, then $a^{-1}s$ is a fixed point for α in F, and hence $s = a \in A$. □

All subgroups considered below are assumed to be σ-invariant.

Definition 1. For a normal subgroup $N \subseteq F$, an element $s \in F$ is called a ***nontrivial fixed point modulo N for σ*** if $s^{\sigma-1} \in N$ and $s \notin N$. Elements of N are called trivial fixed points modulo N for σ.

We note that $s^{\sigma-1} \in N$ implies that $s^{-1+\sigma}, s^{1-\sigma}, s^{-\sigma+1} \in N$.

Lemma 2. *For a normal, σ-invariant subgroup N of F, $S = \{s; s^{\sigma-1} \in N\}$, the set of fixed points for σ modulo N, is a σ-invariant subgroup containing N such that $N \cap F^{\sigma-1} = S^{\sigma-1}$.*

Proof. Let $s, t \in S$. To show that $s^{\sigma}, s^{-1}, st \in S$ we use the equalities $(s^{\sigma})^{\sigma-1} = s^{1-\sigma}$, $(s^{-1})^{\sigma-1} = s^{-\sigma+1}$, and $(st)^{\sigma-1} = (s^{\sigma-1})(t^{\sigma-1})^{s^{-1}}$ □

Corollary 1. *The subgroup $S(N)$ of fixed points for $\bar{\sigma}$ in F/N is trivial if and only if $N \cap F^{\sigma-1} = N^{\sigma-1}$.*

2. Trivial Subgroups of Fixed Points for $\bar{\sigma}$ in F/N

For most cases N contains a word of the form $s^{\sigma-1}$, such that $s \notin N$ and hence the subgroup of fixed points for $\bar{\sigma}$ in F/N is non-trivial.

Lemma 3. *If F/N is a torsion group then the subgroup of fixed points for $\bar{\sigma}$ in F/N is non-trivial.*

Proof. Since xy^{-1} has finite order mod N, there exists an integer k such that either $(xy^{-1})^{2k} \in N$, or $(xy^{-1})^{2k+1} \in N$. Then $(xy^{-1})^k$, or $(xy^{-1})^k x$, respectively, define non-trivial fixed points for σ mod N. □

Lemma 4. *If F/N contains an abelian normal subgroup H/N, where σ does not induce the inverse automorphism, then the subgroup of fixed points* mod N *is non-trivial.*

Proof. By assumption there exists h, such that $hh^\sigma \notin N$, and $[h, h^\sigma] \in N$. Since $[h^\sigma, h] = (hh^\sigma)^{-1+\sigma}$ we conclude that hh^σ is a nontrivial fixed point modulo N as required. □

Corollary 2. *In a free 2-generated soluble group F/N the subgroup of fixed points for the permutation of generators is non-trivial.*

We give now an example of a metabelian group F/N with a trivial group of fixed points for $\bar{\sigma}$. The example will use the following condition.

Theorem 1. *If T is a normal subgroup in F and T is freely generated by elements t, satisfying $t^\sigma = t^{-1}$, then in the group $T/[T, T]$, $\bar{\sigma}$ leaves only the neutral element fixed.*

Proof. We have to show that if $a \in T$ and $a^{1-\sigma} \in [T, T]$, then $a \in [T, T]$. Let $a = t_1^{\varepsilon_1} t_2^{\varepsilon_2} \dots t_k^{\varepsilon_k}$ then $a^{1-\sigma} = (t_1^{\varepsilon_1} \dots t_k^{\varepsilon_k})(t_k^{\varepsilon_k} \dots t_1^{\varepsilon_1}) \equiv a^2 \bmod [T, T]$. So if $a^{1-\sigma} \in [T, T]$, then $a \in [T, T]$, as required. □

We also need :

Lemma 5. *Let T be the subgroup generated by elements $t_i = x^i y^{-i}$, $i \in \mathbb{Z}\backslash 0$. Then T is freely generated by the set $\{t_i\}$ and T contains F'.*

Proof. To show that the t_i generate T freely, it is convenient to have only natural numbers as subscripts i, so T is generated by the set

$$\{x^i y^{-i}, x^{-i} y^i, \ i \in \mathbb{N}\} = \{t_i, t_{-i}, \ i \in \mathbb{N}\}.$$

We consider now the set $R = \{r_i, r_{-i}, i \in \mathbb{N}\}$, where $r_i = y^{i-1}(xy^{-1})y^{-i+1}, r_{-i} = x^{-i+1}(x^{-1}y)x^{i-1}$. By Lemma 3.1 [1], this set is Nielsen reduced and hence generates $gp(R)$ freely. We define the automorphism α in $gp(R)$ such that $r_i^\alpha = r_1 r_2 \dots r_i$, $r_{-i}^\alpha = r_{-i} \dots r_{-2} r_{-1}$. Since $r_i^\alpha = t_i, r_{-i}^\alpha = t_{-i}$, we conclude that the set $\{t_i, t_{-i}\}$ generates $gp(R) = T$ freely. The elements $[x^p, y^q]$ for $p, q \in \mathbb{Z}\backslash 0$, generate F' ([1],

4.3. p. 229), and from the equality

$$[x^p, y^q] = x^{-p}y^p \cdot y^{-p-q}x^{p+q} \cdot x^{-q}y^q = t_{-p} \cdot t_{-p-q}^{-1} \cdot t_{-q} \tag{2.1}$$

we get $F' \subseteq T$. □

Example 1. If $N = [T, T]$, then F/N is metabelian and $\bar{\sigma}$ leaves in F/N only the neutral element fixed.

Proof. We show first that if $a^{1-\sigma} \in [T, T]$ then $a \in T$. Any $a \in F$ can be written as $a = x^p y^q c$, where $c \in F' \subseteq T$, and hence $c = t_1^{\varepsilon_1} \dots t_k^{\varepsilon_k}$. So if $a^{-1+\sigma} \in [T, T]$ we get $a^{-1+\sigma} = c^{-1}y^{-q}x^{-p}y^p x^q c^\sigma \in F'$, and hence $p = q$. Moreover, since $a^{-1+\sigma} \in [T, T]$, the t-length of $a^{-1+\sigma}$ should be even and because of (2.1) we conclude that $p = q = 0$. So $a = c \in T$. Now by Theorem 1, $a^{1-\sigma} \in [T, T]$ implies $a \in [T, T]$ as required. □

To give a necessary and sufficient condition for N, such that the group of fixed points for $\bar{\sigma}$ in F/N is trivial, we say that a set A is 2-isolated if $a^2 \in A$ implies $a \in A$.

Lemma 6. *A subgroup B satisfies $B \cap F^{\sigma-1} = B^{\sigma-1}$ if and only if the set $B^{\sigma-1}$ is 2-isolated.*

Proof. Let $B^{\sigma-1}$ be 2-isolated and $b \in B \cap F^{\sigma-1}$. We have to show that $b \in B^{\sigma-1}$. Since $b \in F^{\sigma-1}$, we write $b = a^{\sigma-1}$ and by the multiplication property (v) $(a^{\sigma-1})^{-2} = (a^{\sigma-1})^{\sigma-1} = b^{\sigma-1} \in B^{\sigma-1}$. Since $B^{\sigma-1}$ is 2-isolated, then $(a^{\sigma-1})^{-1}$ belongs to $B^{\sigma-1}$, and because of (iii) of Lemma 1, we have $b = a^{\sigma-1} \in B^{\sigma-1}$. So $B \cap F^{\sigma-1} \subseteq B^{\sigma-1}$, and since the opposite inclusion is obvious, we have equality. Conversely, if $B \cap F^{\sigma-1} = B^{\sigma-1}$ holds and $d^2 \in B^{\sigma-1}$, we have to show that $d \in B^{\sigma-1}$. Since for some $b \in B$, $d^2 = b^\sigma b^{-1}$, we conclude (in a free group) that $d = b^\sigma = b^{-1} \in B$. If we write $b \in F$ as a word in generators x, y, then $d^\sigma = d^{-1}$ implies that for some $a \in F$ we have $d = a^{\sigma-1} \in F^{\sigma-1}$. So $d \in B \cap F^{\sigma-1} = B^{\sigma-1}$ as required. □

Theorem 2. *In a group F/N, $\bar{\sigma}$ leaves only the neutral element fixed if and only if the set $N^{\sigma-1}$ is 2-isolated.*

Proof. By Lemma 6, $N^{\sigma-1}$ is 2-isolated if and only if $N \cap F^{\sigma-1} = N^{\sigma-1}$, which means that the fixed points for σ modulo N are trivial as required. □

Example 2. For $N = [T, T]$ considered in Example 1, the set $N^{\sigma-1}$ is 2-isolated because in F/N, $\bar{\sigma}$ leaves only the neutral element fixed.

Example 3. For $N = F'$, the set $N^{\sigma-1}$ is not 2-isolated. Indeed, we have $[x, y]^2 = [y, x]^{-1+\sigma} \in N^{\sigma-1}$, but $[x, y] = (yx)^{-1+\sigma} \notin N^{\sigma-1}$ because $yx \notin F'$.

Example 4. For the subgroup $K = \langle xy \rangle * \langle yx \rangle \subseteq F$, the set $K^{\sigma-1}$ is 2-isolated.

Proof. To show that $K^{\sigma-1}$ is 2-isolated, we apply Lemma 6, i.e., we have to check that $K \cap F^{\sigma-1} = K^{\sigma-1}$. If $s \in K$, then for $\varepsilon = 1, -1$, or 0, $s = a(xy)^{\varepsilon}b$ or $a(yx)^{\varepsilon}b$, where $a, b \in K$ and a, b have the same length in generators xy, yx in K. If $s \in F^{\sigma-1}$, then $s^{\sigma} = s^{-1}$ implies $\varepsilon = 0$, and $s \in K^{\sigma-1}$. So $K \cap F^{\sigma-1} \subseteq K^{\sigma-1}$. The opposite inclusion is clear. □

3. The Preimage in F of a Group of Fixed Points

According to Definition 1 and Lemma 2, S is the full preimage of the group of fixed points for $\bar{\sigma}$ in F/N if and only if S is the subgroup of fixed points modulo N in F for σ, and if and only if $N \cap F^{\sigma-1} = S^{\sigma-1}$. We show that not every σ-invariant subgroup $S \subseteq F$ can play this role. If $gpn(S^{\sigma-1})$ denotes the normal closure of $S^{\sigma-1}$ in F, we give a necessary condition for S to be a subgroup of fixed points mod N for some N.

Lemma 7. *If S is a subgroup of fixed points modulo N, then $gpn(S^{\sigma-1}) \subseteq S$.*

Proof. By Lemma 2, $S^{\sigma-1} = N \cap F^{\sigma-1}$, and hence $gpn(S^{\sigma-1}) \subseteq N$. Moreover, since $N^{\sigma-1} \subseteq N \cap F^{\sigma-1} = S^{\sigma-1}$, we get by (iv) of Lemma 1 that $N \subseteq S$, and hence $gpn(S^{\sigma-1}) \subseteq S$ as required. □

Problem 1. Is the condition $gpn(S^{\sigma-1}) \subseteq S$ sufficient for S to be a preimage of a group of fixed points?

Example 5. The subgroup $K = \langle xy \rangle * \langle yx \rangle \subseteq F$ is not a subgroup of fixed points for σ modulo any N.

Proof. Since $(yx)^{-1+\sigma} = [x, y]$ we have $gpn(K^{\sigma-1}) = F'$. But K is not normal, so we conclude that $gpn(K^{\sigma-1}) \not\subseteq K$, and, by Lemma 7, K is not a subgroup of fixed points for σ. □

We shall show that in the case when $S^{\sigma-1}$ is 2-isolated, the condition $gpn(S^{\sigma-1}) \subseteq S$ is necessary and sufficient for S to be a subgroup of fixed points modulo some N.

Theorem 3. *If $S^{\sigma-1}$ is 2-isolated then S is a subgroup of fixed points modulo some N if and only if $gpn(S^{\sigma-1}) \subseteq S$.*

Proof. Because of Lemma 7 we have to prove only the sufficiency. By Lemma 6 we have the equality $S \cap F^{\sigma-1} = S^{\sigma-1}$ which, combined with the condition $gpn(S^{\sigma-1}) \subseteq S$, gives $gpn(S^{\sigma-1}) \cap F^{\sigma-1} \subseteq S \cap F^{\sigma-1} = S^{\sigma-1}$. Hence, by Lemma 2, S is a subgroup of fixed points mod $gpn(S^{\sigma-1})$ as required. □

4. The Lattice of Normal Subgroups for Given S

Normal subgroups normally generated by elements from $F^{\sigma-1}$ play a special role in our considerations.

Definition 2. *A* normal, σ-invariant subgroup N is called ***symmetrically defining*** if $N = gpn(N \cap F^{\sigma-1})$.

F/N is then defined by symmetric relations $w = w^{\sigma}$ [4], and N is normally generated by words $w^{-1+\sigma}$. For example F' is normally generated by $(yx)^{-1+\sigma}$.

Let $S \subseteq F$ be a preimage of a group of fixed points $S(N)$ f or $\bar{\sigma}$ in F/N, then by Lemma 2, $N \cap F^{\sigma-1} = S^{\sigma-1}$. For given S we denote by $[S]$ the set of all normal subgroups N satisfying the above equality. Obviously the set $[S]$ is an equivalence class for the relation saying that two normal subgroups N_1 and N_2 are equivalent if $N_1 \cap F^{\sigma-1} = N_2 \cap F^{\sigma-1}$.

Theorem 4. *Each equivalence class* $[S]$ *contains the minimal normal subgroup* $N_0 = gpn(S^{\sigma-1})$*, which is symmetrically defining and for each* $N \in [S]$*, the group of fixed points* $S(N)$ *in* F/N *is a quotient group of* $S(N_0)$.

Proof. To show $N_0 \in [S]$, we check that $N_0 \cap F^{\sigma-1} = S^{\sigma-1}$. It follows from $N \cap F^{\sigma-1} = S^{\sigma-1}$ that $gpn(S^{\sigma-1}) \subseteq N$, and hence we have $gpn(S^{\sigma-1}) \cap F^{\sigma-1} \subseteq N \cap F^{\sigma-1} = S^{\sigma-1}$. Since the opposite inclusion is clear, $gpn(S^{\sigma-1}) \cap F^{\sigma-1} = S^{\sigma-1}$, and hence $N_0 = gpn(S^{\sigma-1}) \in [S]$. The subgroup $N_0 = gpn(S^{\sigma-1})$ is symmetrically defining, because $S^{\sigma-1} \subseteq N_0 \cap F^{\sigma-1}$ implies that $N_0 = gpn(N_0 \cap F^{\sigma-1})$. If $N \in [S]$, we denote $N/N_0 = M$, then $S(N) = S/N = (S/N_0)/(N/N_0) = S(N_0)/M$ as required. □

Theorem 5. *If* $S^{\sigma-1}$ *is 2-isolated, then the set* $[S]$ *forms a modular lattice with zero, contained in* S.

Proof. If $N_1 \cap F^{\sigma-1} = S^{\sigma-1}$, and $N_2 \cap F^{\sigma-1} = S^{\sigma-1}$, then $N_1N_2 \cap F^{\sigma-1} \supseteq S^{\sigma-1}$. To get the opposite inclusion, we consider $a^{\sigma-1} \in N_1N_2$. Then $a^{\sigma-1} = n_1n_2$ and $(a^{1-\sigma})^2 = (a^{\sigma-1})^{\sigma-1} = n_1^{\sigma}n_2^{\sigma}n_2^{-1}n_1^{-1} = n_1^{\sigma}(n_2^{\sigma-1})n_1^{-\sigma} \cdot n_1^{\sigma-1} \in gpn(N_2^{\sigma-1}) \cdot N_1^{\sigma-1}$. Now by Lemma 2, $N_i \subseteq S$, thus $gpn(N_2^{\sigma-1})N_1^{\sigma-1} \subseteq gpn(S^{\sigma-1}) = N_0$. So $(a^{1-\sigma})^2 \in N_0 \cap F^{\sigma-1} = S^{\sigma-1}$. By our assumption $S^{\sigma-1}$ is 2-isolated, and hence $a^{1-\sigma}$ (and so $a^{\sigma-1}$) is also in $S^{\sigma-1}$, thus $N_1N_2 \cap F^{\sigma-1} = S^{\sigma-1}$ as required. The equality $(N_1 \cap N_2) \cap F^{\sigma-1} = S^{\sigma-1}$ is clear because $N_1 \cap N_2 \supseteq N_0$. The subgroup $N_0 = gpn(S^{\sigma-1})$ is the zero in the lattice. □

Example 6. If $N = F'$, then $N = N_0$, $S = gp(xy) \cdot F'$, and $S(N)$ is the cyclic subgroup $gp(\bar{x}\bar{y})$ of F/N.

Proof. Since $F' = gpn\{[x, y]\}$ and $[x, y] = (yx)^{-1+\sigma} \in F' \cap F^{\sigma-1}$ we have $F' = gpn(F' \cap F^{\sigma-1}) = N_0$. Each word in F can be written as $s = x^k y^l c$, $c \in F'$. If s is a fixed point modulo F' then $s^{\sigma-1} \in F'$, and hence $s = x^k y^k c = (xy)^k d,\ c, d \in F'$. So $S = gp(xy) \cdot F'$, and $S(N)$ is cyclic. □

Example 7. If $N = [[F, F], F]$, then $S = gp(xy^2x) \cdot N$, and $S(N)$ is the cyclic subgroup $gp(\bar{x}\bar{y}^2\bar{x})$ of F/N.

Proof. It is known by [4], that words in F/N satisfying $s = s^\sigma$ are of the form $s = x^{2k}y^{2k}[y, x]^{2k^2}$. Hence for some $c,\ d \in N$, $s = x^{2k}y^{2k}[y^{2k}, x^k]c = x^k y^{2k} x^k c = (xy^2x)^k d$, where the last equality follow s by induction because $x^k y^{2k} x^k \cdot x^l y^{2l} x^l = x^k y^{2k} x^l \cdot x^k y^{2l} x^l,\ \ y^{2k} x^l = x^l y^{2k}[y^{2k}, x^l]$, and $x^k y^{2l} = y^{2l} x^k [x^k, y^{2l}]$. So $S = gp(xy^2x) \cdot N$, and $S(N)$ is the cyclic subgroup $gp(\bar{x}\bar{y}^2\bar{x}) \subseteq F/N$. □

Example 8. If $N = F''$, then $S(N)$ is generated by the set $\{[x^i, y^j] \cdot [y^i, x^j],\ i, j \in \mathbb{Z}\backslash 0\}$ modulo N.

Proof. In Lemma 5 we considered the group T generated by elements $t_i = x^i y^{-i}$ for $i \in \mathbb{Z}\backslash 0$. It was shown in Lemma 5 that $F' \subseteq T$, and hence $F'' \subseteq [T, T]$. If s is a fixed point modulo F'', then also modulo $[T, T]$. By Theorem 1 all fixed points modulo $[T, T]$ are trivial, that is all fixed points modulo F'' are in $[T, T] \subseteq F'$. Elements of the form $c_{ij} = [x^i, y^j]$ give a set of free generators in F' (see [2], p. 229). So if s is a fixed point modulo F'', then $s = \Pi c_{ij}$, and since $c_{ij}^\sigma = c_{ji}^{-1}$, we have $s^{-\sigma} = \Pi c_{ji}$. Moreover, $ss^{-\sigma} \in F''$ implies that the exponent sum of each c in $ss^{-\sigma}$ is zero. So we conclude that each c_{ij} in s cancels with c_{ij}^{-1} in $s^{-\sigma}$, which implies that for each c_{ij}, s contains c_{ij}^σ, and hence $s = \Pi(cc^\sigma)$. So $S(N)$ is generated by the set $\{[x^i, y^j] \cdot [y^i, x^j]\}$ modulo N as required. □

It is shown in [2] that for $N = F'''$ the group of fixed points $S(N)$ also consists of elements of the form $s = cc^\sigma,\ c \in F''/F'''$.

Problem 2. Show that in $F/F^{(n)}$ the group of fixed points consists of elements $s = cc^\sigma,\ c \in F^{(n-1)}/F^{(n)}$.

References

[1] Magnus, W., Karrass, A., Solitar, D, Combinatorial Group Theory. Interscience Publishers, New York–London–Sydney 1966.

[2] Macedońska, O., Solitar, D., On binary σ-invariant words in a group. Contemporary Math. 169 (1994), 431–449.

[3] Neumann, B. H., Groups with automorphisms that leave only the neutral element fixed. Arch. Math. (Basel) 7 (1956), 1–5.

[4] Plonka, E., Symmetric words in nilpotent groups of class ≤ 3. Fundamenta Math. 97 (1977), 95–103.

Subgroup Growth: Some Current Developments

Avinoam Mann and Dan Segal*

1991 Mathematics Subject Classification: 20E26

Introduction

For any group G and positive integer n, we write $a_n(G)$ for *the number of subgroups of index n in G* (so $0 \leq a_n(G) \leq \infty$). If G is a profinite group, only *open* subgroups are counted; so one has in general $a_n(G) = a_n(\hat{G})$ where $\hat{G}$ denotes the profinite completion of G. Thus it is harmless — and often convenient — to switch between abstract groups and profinite groups when studying these numbers.

If G (or more generally $\hat{G}$) is finitely generated, then $a_n(G)$ is finite for every n, and we have an arithmetical function $n \mapsto a_n(G)$. For the particular case of free groups G, Marshall Hall in 1949 gave an exact (recursive) formula for this function. The investigation of this function in more general classes of finitely generated groups began about ten years ago. We shall give an outline of some of the results of that investigation, and indicate where it seems to be going at the present time.

This survey has no pretensions to be encyclopaedic. As the title suggests, we concentrate on the *growth*, and leave aside the fascinating topic of the *arithmetical* properties of the functions a_n. For more on these, see the recent surveys by Alex Lubotzky [L1], [L2] as well as [GSS], [dS1], [dS2], [dS3], [Mü1], [Mü2].

As we are interested in growth, we shall usually phrase the results in terms of the function defined by

$$s_n(G) = \sum_{i=1}^{n} a_i(G),$$

that is, the number of subgroups of index *at most n* in G. We call $n \mapsto s_n(G)$ the *growth* of G (not to be confused with the "word growth" of Wolf, Milnor and Gromov). Strictly speaking, this is the *subgroup growth* of G; we shall also have things to say about the

---

* The first author acknowledges the hospitality of All Souls College, Oxford, during the preparation of this paper.

Offprint from
Infinite Groups 94, Eds.: de Giovanni/Newell

"normal", "subnormal" and "maximal" subgroup growth, defined in the analogous manner.

We wish to thank Alex Lubotzky and Aner Shalev for keeping us up to date with their latest results; and the organisers of the Ravello conference for a delightful, and mathematically stimulating, week.

Contents

1. Limits to Growth

There are three inter-related kinds of question:

(a) What are the possible growths that can occur?

(b) What type of growth does a particular kind of group have?

(c) Which groups have a particular type of growth?

Under (a) come the questions: what is the *fastest* possible growth? What is the *slowest* possible growth? Do all intermediate possibilities actually occur? The answer to the first question is very simple:

Proposition 1.1. *Let $f : \mathbb{N} \to \mathbb{N}$ be any function. Then there exists a group G such that*

$$f(n) \leq s_n(G) < \infty \quad \textit{for all } n.$$

Proof. G is the direct, or Cartesian, product of finite elementary abelian p-groups $G(p)$, where $G(p)$ has rank at least $1 + \log_p f(p)$. □

Thus to get interesting results, it is reasonable to restrict attention to *finitely generated* groups (or finitely generated profinite groups), and this is what we shall do. We generally consider groups which are *residually finite*, since for any group G, $a_n(G) = a_n(G/R)$ where R is the intersection of all subgroups of finite index in G.

Since every d-generator group is an image of the free d-generator group F_d, the fastest growth is given by M. Hall's result [H2], which implies the following (see

[N2]): *if* $d \geq 2$ *then*

$$a_n(F_d) \sim n \cdot (n!)^{d-1} \quad \text{as } n \to \infty.$$

(Since each subgroup of index n in F_d gives rise to a homomorphism of F_d onto a transitive subgroup of S_n, and conversely, Hall's theorem can be interpreted as saying that "most" d-generator subgroups of S_n are transitive.) This implies

Theorem 1.2. *If G is a d-generator group, or a d-generator profinite group, and $\varepsilon > 0$ then*

$$s_n(G) \leq n^{(d-1+\varepsilon)n} \quad \text{for all large } n.$$

Thus there is an absolute upper limit for the subgroup growth of finitely generated groups, and it is "super-exponential", but only by a little; in the terminology explained below, this maximal growth is of type n^n.

At the other extreme, one has the trivial observation that if G is finite then $s_n(G)$ is eventually constant. More interesting is the question: what is the slowest possible growth for an *infinite, residually finite*, finitely generated group? The obvious candidate here is the function $s_n(G) = n$, achieved when G is the infinite cyclic group.

Theorem 1.3. *If G is an infinite finitely generated residually finite group then there exists $c > 0$ such that*

$$s_n(G) \geq cn \text{ for infinitely many values of } n.$$

We do not know an easy proof of this fact. It follows readily from what is perhaps the major result of the subject, so far; to state this, let us first make a

Definition The group G has ***polynomial subgroup growth***, or ***PSG***, if there exists $\alpha > 0$ such that

$$s_n(G) \leq n^{\alpha} \quad \text{for all } n.$$

Theorem 1.4 [LMS; DDMS, Chapter 6]. *Let G be a finitely generated residually finite group. Then G has PSG if and only if G is a virtually soluble minimax group.*

(Recall that the finitely generated residually finite soluble minimax groups are exactly the virtually torsion-free finitely generated soluble groups of finite rank [R, §10.3]; they are also exactly the finitely generated soluble linear groups over $\mathbb{Q}$. A group has a property ***virtually*** if some subgroup of finite index has that property.)

Theorem 1.4 answers questions (b) and (c) in an important case. The proof involves a number of subsidiary results, about pro-p groups, linear groups, and infinite soluble groups, as well as the classification of the finite simple groups. Some of these results have been further developed, and we shall discuss them below: see 1.7, 1.10, 2.1, 2.5, 2.6, 2.7. To deduce Theorem 1.3, note that any counterexample G must be a group with PSG; then Theorem 1.4 implies that either G is finite, or G has a subgroup H of finite index such that H has an infinite cyclic quotient. If $|G : H| = e$ then $s_n(G) \geq s_{[n/e]}(H) \geq [n/e]$, for every n. (This neat observation was made by P. M. Neumann, during a lecture by one of the authors.)

Having established the fastest and the slowest types of growth, we turn to the question: what intermediate types of growth are possible? For convenience of expression, we introduce the following terminology, slightly adapting one used by Shalev:

Definition Let f be a function. The group G has ***growth type f*** if there exist positive constants b and c such that

$$s_n(G) \leq f(n)^c \text{ for all large } n,$$
$$s_n(G) \geq f(n)^b \text{ for infinitely many } n.$$

Thus the fastest growth type for finitely generated groups is n^n, and the slowest non-constant growth type is n, which is equivalent to "PSG — (but not eventually constant)".

In the present context, it is usually easier to understand the situation with regard to profinite groups than with regard to abstract finitely generated groups.

Growth Types of Finitely Generated Profinite Groups

Theorem 1.5. *Let $g : \mathbb{N} \to \mathbb{R}$ be a positive, non-decreasing unbounded function such that $g(n^2) \leq kg(n)$ for all n, where k is a constant. Then there exists a 2-generator profinite group whose growth type is $n^{g(n)}$.*

This shows that, for finitely generated profinite groups, there is a continuous range of possible growth types between PSG and $n^{(\log n)^t}$ (for all positive t). We sketch the construction. For each prime p, put $r(p) = [g(p)]$, take $W_p = C_p \wr C_p$, and let

$$H_p = W_p/\gamma_{r(p)}(W_p).$$

It is easy to verify that $a_{p^i}(H_p) \leq p^{i(r(p)-1)}$, and that $s_{p^2}(H_p) \geq p^{r(p)-2}$ provided $r(p) \leq p$. The hypotheses on g ensure that this latter condition holds for all large p. They imply, moreover, that $r(p)-2 \geq b'g(p^2)$ for some positive constant b', provided p is sufficiently large. Now take G to be the Cartesian product, over all primes p, of the groups H_p. Then $a_n(G) = \prod a_{p^{e(p)}}(H_p)$ where $n = \prod p^{e(p)}$, so

$$s_n(G) = \sum_{m=1}^{n} a_m(G) \leq \sum_{m=1}^{n} m^{(g(m)-1)} \leq n^{g(n)}.$$

And if p is a large prime, taking $n = p^2$ gives

$$s_n(G) \geq s_{p^2}(H_p) \geq p^{b'g(p^2)} = (n^{g(n)})^b$$

where $b = b'/2$. Thus G has growth type $n^{g(n)}$ as required. □

A similar construction, using one-dimensional affine groups over suitably chosen finite fields in place of the p-groups H_p, yields a 3-generator metabelian profinite group, with the same growth type, which is linear over a field of characteristic zero. Using some hard analytic number theory, Shalev [Sh2] has shown that suitable products of groups PSL$(2, p)$ can also be made to have arbitrarily slow non-polynomial growth type. Unlike the groups constructed above, these have infinitely many non-abelian simple groups as upper composition factors (i.e. as composition factors of finite quotients). The same holds for the following examples (where A_n denotes the alternating group of degree n):

Theorem 1.6 [LPSh1]. *There exists a sequence S of positive integers such that the 2-generator profinite group*

$$G_S = \prod_{n \in S} A_n$$

has growth type $n^{\log n/(\log\log n)^2}$.

It seems likely that many, perhaps all, intermediate growth types between e^n and $n^{\log n/(\log\log n)^2}$ are achieved by 2-generator profinite groups of this form. It is interesting to note that growth types strictly less than $n^{\log n/(\log\log n)^2}$ *cannot* occur in such groups: by adapting the proof of [MS], Theorem 4.1, one can establish

Theorem 1.7. *Let G be a group with growth type strictly less than $n^{\log n/(\log\log n)^2}$, in the sense that*

$$\log s_n(G) = o(\log n/\log\log n)^2 \quad \textit{as } n \to \infty.$$

Then

(i) *only finitely many alternating groups occur as upper composition factors of G;*

(ii) *for each prime p, only finitely many groups of Lie type in characteristic p occur as upper composition factors of G.*

It is also possible to bound the multiplicities of the upper composition factors. More severe restrictions apply in the case of PSG groups; see [MS], Theorem 4.1 and [Ma1].

Now the upper composition factors of a group are subnormal subgroups of finite quotients of the group. If we place restrictions on *all* subgroups of finite quotients, i.e. on all *upper sections* of the group, we can deduce limitations to the possible subgroup growth types. At the top end of the scale is

Theorem 1.8 [PSh]. *Let G be a finitely generated profinite group. Suppose that some finite group is not isomorphic to any upper section of G. Then the subgroup growth type of G is at most e^n.*

Thus a finitely generated profinite group whose growth is of "strictly super-exponential" type shares with the free profinite groups the property of generating the variety of all profinite groups. In particular, it follows that *a finitely generated pro-(finite soluble) group has at most exponential growth type*, a result proved earlier in [Ma2]. It is clear that the group G_S of Theorem 1.6 involves every finite group as an upper section: *is its growth type* (namely $n^{\log n/(\log\log n)^2}$) *the minimal growth type for such groups?* Theorem 1.7 is not quite strong enough to yield a positive answer to this question, suggested by Aner Shalev.

Next, let us discuss pro-p groups. It follows from Theorem 1.8 that a finitely generated pro-p group has at most exponential growth type; in fact the exact maximal growth for such groups was determined by I. Ilani some time ago:

Theorem 1.9 [I]. *Let F be the free pro-p group on $d > 1$ generators. Put $\alpha = (d-1)/(p-1)$. Then, for all n,*

$$p^{\alpha(p^n-1)-n(n-1)/2} \leq a_{p^n}(F) \leq p^{\alpha(p^n-1)}.$$

It follows that $\log s_n(F) \sim n(d-1)\log p/(p-1)$, whence the maximal growth type for finitely generated pro-p groups is e^n.

One of the first steps in the proof of Theorem 1.4 was the characterisation of pro-p groups with PSG: these are exactly the pro-p groups of finite rank, or equivalently those which are p-adic analytic [LM1]; see also [DDMS], Chapter 3. In contrast with the general case of profinite groups, there is a gap between PSG and the next slowest type of growth for finitely generated pro-p groups:

Theorem 1.10 [Sh1]. *Let G be a finitely generated pro-p group of infinite rank. Then for every $c < 1/8$ there exist infinitely many n such that*

$$a_{p^n}(G) > p^{cn^2}.$$

It follows that for finitely generated pro-p groups, *there is no growth type strictly intermediate between PSG and type $n^{\log n}$*. However, in the same paper Shalev shows that growth type $n^{\log n}$ does occur:

Theorem 1.11 [Sh1]. *Let G be the principal congruence subgroup modulo (t) in the group $\mathrm{SL}(2, \mathbb{F}_p[[t]])$. Then*

$$a_{p^n}(G) \leq 4p^{2n^2} \quad \textit{for all } n.$$

G is a 3-generator pro-p group. As G has infinite rank, it follows from the preceding result that the growth type of G is exactly $n^{\log n}$. It is an interesting problem to characterise the pro-p groups with exactly this minimal type of non-polynomial subgroup growth.

Pro-p groups with somewhat faster growth are discussed in [SSh]:

Theorem 1.12 [SSh]. *For each positive integer d, there exists a finitely generated metabelian pro-p group with growth type $e^{n^{1/d}}$.*

It is not known what other growth types are possible for finitely generated pro-p groups.

Growth Types of Finitely Generated Groups

As some of the preceding examples show, a finitely generated profinite group can often be tailor-made to have a specific type of subgroup growth, simply by fitting together a suitable collection of finite groups. It is usually much harder to construct a finitely generated abstract group with a predetermined collection of finite quotients, and accordingly our knowledge of the possible growth types here is less complete.

The known growth types of finitely generated groups fall into three broad families:

I Super-exponential

Type: $n^n = e^{n \log n}$.

Known examples: groups with a non-abelian free quotient.

II Exponential and "fractionally exponential"

Types: e^{n^γ}, $0 < \gamma \leq 1$.

Known examples:

Theorem 1.13 [SSh]. *For each positive integer d there exists a finitely generated metabelian group whose subgroup growth is of type $e^{n^{1/d}}$.*

In fact, the groups constructed in [SSh] have the property that the growth of 2-step subnormal subgroups is of the same type as the growth of all subgroups; this is in marked contrast to the case of free groups (see Section 3 below). Theorem 1.13 provides the *only known infinite family of finitely generated groups with distinct growth types*. It seems likely that the methods of [SSh], wielded with more ingenuity, should lead to groups of growth type e^{n^γ} for any *rational* γ between 0 and 1. An interesting question, which seems harder, is: *can a finitely generated group have growth type e^{n^γ} where γ is irrational?*

III "Logarithmic"

Types: $\leq n^{(\log n)^t}$.

Known examples:

Theorem 1.14 [L3]. *S-arithmetic subgroups of suitable algebraic groups over global fields of positive characteristic have growth type between $n^{\log n}$ and $n^{(\log n)^2}$, provided they satisfy the congruence subgroup property (CSP).*

(Here, "suitable" means connected, simply-connected, split and almost-simple, with a few exceptions in characteristic 2.)

Theorem 1.15 [LPSh1]. *There exists a finitely generated group with growth type $n^{\log n}$.*

Theorem 1.16 [L3]. *S-arithmetic subgroups of suitable algebraic groups over algebraic number fields have growth type $n^{\log n/\log\log n}$, provided they satisfy CSP.*

(Here, "suitable" means connected, simply-connected and almost-simple.)

Theorem 1.17 [LPSh1]. *There exists a finitely generated group with growth type $n^{\log n/(\log\log n)^2}$.*

And finally, we know from Theorem 1.4 that every finitely generated virtually soluble minimax group has PSG, hence growth type n (unless the group is finite).

No other subgroup growth types beyond those in the above list are known to occur, for finitely generated groups. The challenge is to determine whether the gaps are really there. Particular attention has been given to the gap at the bottom of the list: Lubotzky speculates that *the growth type of Theorem* 1.17 *is the slowest possible type for any finitely generated group which does not have PSG.* In view of Theorem 1.4, what this amounts to is

Speculation 1.18. *Let G be a finitely generated residually finite group. If*

$$\log s_n(G) = o(\log n/\log\log n)^2 \quad \text{as } n \to \infty,$$

then G is a virtually soluble minimax group.

We have seen in Theorem 1.7 that there are at least some restrictions on the upper composition factors of any possible counterexample to 1.18. Further results and speculations related to 1.18 are discussed in the next section.

2. Small is Beautiful

We now discuss the following question: *how slow can the subgroup growth of a finitely generated group be, if it is faster than polynomial?* The best guess at a general answer was enunciated above, as Speculation 1.18; but there are some definitive results concerning restricted classes of groups.

To begin with *linear groups*: Lubotzky's Theorems 1.14 and 1.16 (above) actually specify the growth type of *congruence subgroups* in the respective S-arithmetic groups, whether or not these have the CSP (this condition states that the congruence subgroups account for *all* subgroups of finite index). In the same paper [L3], he shows that if an S-arithmetic group Γ *fails* to have the CSP, then the *subgroup* growth of Γ is of a type *strictly greater* than the *congruence subgroup* growth of Γ. This interesting result shows the equivalence of two different ways of saying that a group has "not too many" subgroups of finite index, and provides a group-theoretic criterion for the - essentially arithmetic - congruence subgroup property. It also leads to a proof of

Theorem 2.1 [L3]. *A finitely generated linear group which does not have PSG has subgroup growth type at least* $n^{\log n/\log\log n}$.

This is reduced to the "arithmetic" and "soluble" cases by the methods of [LM1]; the soluble case follows from a sharper result:

Theorem 2.2 [SSh]. *A finitely generated soluble linear group which does not have PSG has subgroup growth type at least* $e^{n^{\gamma}}$, *for some* $\gamma > 0$.

It is well known that every finitely generated linear group is *virtually residually nilpotent* ([W], Theorem 4.7). An early step in the characterisation of PSG groups (see [LM1]) was to show that in the presence of a suitable subgroup growth condition the converse also holds. A similar argument, using now Shalev's result Theorem 1.10, gives

Lemma 2.3. *Let G be a finitely generated group which is virtually residually nilpotent. If the subgroup growth of G is of type strictly less than $n^{\log n}$, then G is a linear group over $\mathbb{C}$.*

Together with Theorem 2.1 this yields the following generalisation:

Theorem 2.4. *A finitely generated virtually residually nilpotent group which does not have PSG has subgroup growth type at least* $n^{\log n/\log\log n}$.

(This incidentally shows that the Lubotzky–Pyber–Shalev example 1.17, which is finitely generated and residually finite, cannot be virtually residually nilpotent.)

Using some of the methods of [Se1], it is possible similarly to generalise Theorem 2.2:

Theorem 2.5. *Let G be a finitely generated soluble group which is virtually residually nilpotent. If G does not have PSG then G has subgroup growth type at least e^{n^γ}, for some $\gamma > 0$.*

In fact the proof shows that the *subnormal* subgroup growth is of type at least e^{n^γ}.

In the realm of finitely generated *soluble* (or prosoluble) groups, subgroup growth seems to be closely connected with questions about *upper p-rank*: the upper p-rank of G, for a prime p, is

$$\mathrm{ur}_p(G) = \sup\{\mathrm{rk}(P) \mid P \text{ is an elementary abelian upper } p\text{-section of } G\}.$$

Thus $\mathrm{ur}_p(G)$ is just the rank of a Sylow pro-p subgroup of $\hat{G}$.

Now if G is a finitely generated group such that $\mathrm{ur}_2(G)$ is finite, then G has a normal subgroup G_0 of finite index such that every finite quotient of G_0 is soluble [LM2]; in other words, $\hat{G}$ is *virtually prosoluble*. An important part of the proof of Theorem 1.4 was the reduction to this case. At the present time, such a reduction under hypotheses weaker than PSG is not available (the best we have in this direction is Theorem 1.7, above). However, if we *assume* this condition then more can be said:

Proposition 2.6. *Let G be a finitely generated group such that $\hat{G}$ is virtually prosoluble. Suppose that*

$$\log s_n(g) = o\big((\log n)^2/f(\log n)\big) \quad \textit{as } n \to \infty,$$

where $f(n) \to \infty$ as $n \to \infty$. Then $\mathrm{ur}_p(G)$ is finite for every prime p.

This applies, for example, to groups with growth of type strictly less than $n^{\log n/\log\log n}$.

To prove 2.6, one first shows, using the method of [MS], Theorem 3.9, that for each p there is a normal subgroup $G(p)$ of finite index in G such that every finite quotient of $G(p)$ is p-nilpotent. It follows that

$$\mathrm{ur}_p(G(p)) = \mathrm{rk}(\widehat{G(p)_p})$$

where $\widehat{G(p)_p}$ is the pro-p completion of $G(p)$. Now Theorem 1.10 shows that $\widehat{G(p)_p}$ is a pro-p group of finite rank. Thus we have

$$\begin{aligned}\mathrm{ur}_p(G) &\le \mathrm{rk}\big(G/G(p)\big) + \mathrm{ur}_p\big(G(p)\big)\\ &= \mathrm{rk}\big(G/G(p)\big) + \mathrm{rk}(\widehat{G(p)_p}) \quad < \infty.\end{aligned}$$

□

Now if G is a finitely generated residually finite group such that $\mathrm{ur}_p(G)$ is *boundedly* finite, over all primes p, then G is a virtually soluble minimax group [MS]. Thus the following question is highly relevant: *can a finitely generated group (or soluble group) have finite but unbounded upper p-ranks, as p ranges over all primes?*

For groups which are virtually residually nilpotent, the answer is "no"; we shall not discuss the proof here. If it could be shown that the answer is "no" for finitely generated soluble groups, it would follow that *the minimal non-polynomial growth type for such groups is at least $n^{\log n}$*: for we have

Proposition 2.7. *Let G be a finitely generated soluble group, and let p be a prime such that* $\mathrm{ur}_p(G) = \infty$. *Then*

$$s_n(G) \geq n^{c \log n} \quad \textit{for all large } n,$$

if c is any constant less than $\frac{1}{49 \log p}$.

Whether subgroup growth of type $n^{\log n}$ actually occurs among finitely generated soluble groups is not known. Indeed, we are not aware of any such group which is residually finite without being virtually residually nilpotent.

3. Normal and Subnormal Subgroups

The number of normal subgroups is generally much smaller than the number of all subgroups. While this does not seem surprising, the proof again requires the classification of finite simple groups. Let F be a free group of rank d (abstract or profinite). P. Hall [H1] noted that the number of normal subgroups $N \lhd F$ for which $F/N \cong G$, a given finite group, is $\varphi_d(G)/|\operatorname{Aut} G|$, where $\varphi_d(G)$ is the number of d-tuples generating G. Summing over all groups of order at most n, we obtain the number $t_n(F)$ of all normal subgroups of F of index $\leq n$. Using the obvious estimate $\varphi_d(G) \leq |G|^d$, and L. Pyber's theorem that the number $f(n)$ of groups of order n satisfies $f(n) \leq n^{\left(\frac{2}{27}+O(1)\right)(\log n)^2}$ [P], we obtain

$$t_n(F) \leq n^{c(\log n)^2}$$

for some constant c. More precisely, if $f(n, d)$ is the number of d-generator groups of order $\leq n$, we have

$$f(n, d) \leq t_n(F) \leq n^{d+1} f(n, d).$$

The asymptotic behaviour of $f(n, d)$ seems not to have been determined. (It is shown in [N1] that the number of d-generator groups of p-power order $\leq n$ is at most $n^{c \log n}$, where c depends on p and d.)

In any case, we can infer

Theorem 3.1. *The normal subgroup growth type of any finitely generated group is at most* $n^{(\log n)^2}$.

We turn now to the number $u_n(G)$ of *subnormal* subgroups of index at most n in a group G. The order of magnitude of this function turns out to lie strictly between those of s_n and t_n. Lubotzky, Pyber and Shalev [LPSh2] have shown that there exists c such that $u_n(F) \leq c^n$ for all n. On the other hand, it is clear that if n is a prime power p^m, then

$$u_n(F) = u_{p^m}(F) = s_{p^m}(\hat{F}_p),$$

and $\hat{F}_p$ is a free pro-p group. Thus Ilani's theorem 1.9 provides a lower bound for the growth of u_n. Putting these together gives

Theorem 3.2. *If F is a free group (or free profinite group) on $d > 1$ generators, then the subnormal subgroups of F have growth of type e^n.*

Even if we restrict attention to subnormal subgroups of defect 2, the growth is still of exponential type, and not smaller. Thus there are generally many more 2-subnormal subgroups than normal ones.

It may therefore be expected that restrictions on subnormal subgroups are much stronger than restrictions on normal subgroups. We mention two illustrations of this phenomenon. First, in his talk in this conference, D. J. S. Robinson emphasised that requiring automorphisms to stabilise all subnormal subgroups is a much heavier assumption than requiring them to stabilise all normal subgroups. Second, in [MS] it is shown that if G is a soluble, or pro-soluble, group, in which the number of 2-subnormal subgroups grows polynomially (at most) then G is already a PSG group. (This is generalised in the remark following Theorem 2.5 above.)

The growth of normal subgroups has received less attention than the growth of all subgroups, and apart from the case of free groups just discussed, most of the results so far are in the domain of soluble groups. The rest of this section is devoted to these.

One of the key insights in the development of our subject was the observation [LM1] that a pro-p group has PSG if and only if it has finite rank; that is, if and only if the ranks of its elementary abelian sections are bounded. This suggests the possibility of a similar characterisation of pro-p groups whose *normal* subgroups grow polynomially: let us call these ***PNSG groups***. A group is said to have ***finite width*** if its elementary abelian *central* sections of finite index have bounded ranks. It is not hard to prove

Lemma 3.3. *Let G be a pro-p group. If G has finite width then G has PNSG.*

It was believed for a time that the converse of Lemma 3.3 might also be true. However, this turns out not to be the case.

Example 3.4. Let F be free abelian of rank 2, and let G be the pro-p completion of $C_p \wr F$. Then (i) G has $< p^{5n}$ normal subgroups of index at most p^n, for all large n; (ii) $\gamma_{i-1}G/\gamma_i G$ has rank i for all $i > 2$, so G has infinite width.

Example 3.5. Let $G = \mathbb{Z} \wr \mathbb{Z}$. Then (i) G has $< n^{7/2}$ normal subgroups of index at most n, for all large n; (ii) $\gamma_i G/\gamma_{i+1}G \cong \mathbb{Z}$ for all $i \geq 2$; (iii) nonetheless, G has infinite width.

Examples such as these arise when one considers commutative rings of Krull dimension 2. In general, let G be a metabelian group with derived group G', and consider G' as a G/G'-module via conjugation; if G is finitely generated, put $R = \mathbb{Z}(G/G')$ and define

$$\kappa(G) = \operatorname{Dim}\big(R/\operatorname{ann}_R(G')\big);$$

if G is a pro-p group, put $R = \mathbb{Z}_p(G/G')$ and define $\kappa(G)$ by the same formula (here, Dim denotes the Krull dimension of a ring).

Theorem 3.6. *Let G be a metabelian group which is either finitely generated, or a finitely generated pro-p group. Then* (i) *G has PNSG if and only if $\kappa(G) \leq 2$;* (ii) *if G is a pro-p group, then G has finite width if and only if $\kappa(G) \leq 1$;* (iii) *if G is a pro-p group and $\kappa(G) \leq 2$, then there is a finite bound for the ranks of all torsion-free central factors of G.*

This is proved by analysing the structure of finitely generated modules over rings such as R. The main step is

Theorem 3.7 [Se2]. *Let R be either a Noetherian semi-local ring or a finitely generated (commutative) ring, and let M be a finitely generated R-module. Then the number of submodules of index $\leq n$ in M grows at most polynomially with n if and only if* $\operatorname{Dim}\big(R/\operatorname{ann}_R(M)\big) \leq 2$.

Theorem 3.6 gives a fairly complete characterisation of metabelian groups with PNSG.

Question. What is the correct generalisation of Theorem 3.6 (i) to pro-p groups which are not metabelian?

The ring-theoretic approach leads to a reasonable picture of all possible types of normal subgroup growth in metabelian groups:

Theorem 3.8. *Let G be either a finitely generated metabelian group, or a finitely generated metabelian pro-p group, and put $k = \kappa(G)$. Then there exist positive constants a and b such that, for all n,*

$$n^{a(\log n)^{1-2/k}} \leq t_n(G) \leq n^{b(\log n)^{1-1/k}}.$$

(Here $t_n(G)$ denotes the number of normal subgroups of index at most n in G.)

This follows from an analogous result about submodule growth (as in Theorem 3.7), proved in [Se2]. It is probable that this is not the final word: presumably the correct result will have the same exponent to $\log n$ on each side of the inequalities; but this has yet to be proved.

Question. What are the possible normal subgroup growth types for finitely generated pro-p groups?

4. Maximal Subgroups and Probabilistic Methods

Let F be a non-abelian free group, and let H be a subgroup of index n. Then the action of F on the right cosets of H determines a homomorphism of F into the symmetric group S_n. Such a homomorphism is determined, in turn, by specifying the images of the generators of F, and these can be any d randomly chosen elements of S_n. The homomorphism corresponds to some subgroup of index n exactly when the image of F, say T, is a transitive subgroup of S_n, and it was remarked in §1 above, that indeed most homomorphisms correspond to transitive subgroups. It was shown by J. D. Dixon [D] that something stronger holds: for most homomorphisms T is a *primitive* subgroup of S_n (indeed, either S_n or A_n). But then H is a maximal subgroup of F. The upshot is that *most subgroups of finite index in F are maximal* (this interesting observation was made by L. Pyber).

In view of Theorem 1.2, this gives

Theorem 4.1. *If F is a finitely-generated non-abelian free group, then the maximal subgroup growth of F is of type n^n.*

However, the situation changes drastically under seemingly mild restrictions on the group. Let G be a finitely generated (abstract or profinite) group. We say that G is ***non-universal*** if there exists some finite group which is not isomorphic to any upper section of G (that is, if G satisfies the hypothesis of Theorem 1.8).

Theorem 4.2 [BPSh]. *If G is a finitely generated (abstract or profinite) non-universal group then the maximal subgroup growth of G is at most polynomial.*

(In the profinite case, "maximal subgroup" means "maximal closed subgroup".)

Recall (Theorem 1.8) that under the same hypotheses, $s_n(G)$ grows at most exponentially; and usually no slower than that. Thus, writing $m_n(G)$ for the number of maximal subgroups of index at most n in G, we see that in a non-universal group G there may be a wide divergence between the growth type of $m_n(G)$ and that of $s_n(G)$. We do not know whether polynomial growth of $m_n(G)$ is in general equivalent to growth of at most exponential type for $s_n(G)$. Examples constructed by M.

Bhattacharjee [B] show that there even exist finitely generated universal (i.e. not non-universal) groups G such that $m_n(G)$ grows polynomially (so the converse of Theorem 4.2 does not hold).

As was the case for subgroup growth, there is a continuous range of possible types of maximal subgroup growth: to establish this, one considers suitable products of alternating groups, getting

Theorem 4.3 [Ma2]. *Let f be a non-decreasing function with $n \leq f(n) \leq n^n$ for all n. Then there exists a 2-generator profinite group whose maximal subgroup growth is of type f.*

The interest in the growth of $m_n(G)$ arose in the following context [Ma2]. Let G be a profinite group. As a compact topological group, G has a finite Haar measure μ; we normalise this so that $\mu(G) = 1$, and thus consider G as a probability space. Now write

$$P(G, k)$$

for *the probability that a random k-tuple of elements generates* G. This refers to *ordered* k-tuples: thus, writing μ^k for the product measure on $G^{(k)} = G \times \cdots \times G$, we have

$$P(g, k) = \mu^k\{(x_1, \ldots, x_k) \in G^{(k)} \mid \langle x_1, \ldots, x_k\rangle = G\}.$$

If G is an abstract group, we define

$$P(G, k) = P(\hat{G}, k).$$

If the group G is finite, we see that

$$P(G, k) = \varphi_k(G)/|G|^k,$$

where $\varphi_k(G)$ is Philip Hall's function defined in §3 above.

It is shown in §16.5 of [FJ] that $P(\mathbb{Z}, 1) = 0$. For each $d \geq 1$, Kantor and Lubotzky [KL] obtain the following (where ζ denotes the Riemann zeta function):

$$P(\mathbb{Z}^d, k) = 0 \quad \text{if } k \leq d$$

$$P(\mathbb{Z}^d, k) = \prod_{j=0}^{d-1} \zeta(k - j)^{-1} \quad \text{if } k > d.$$

On the other hand, $P(F_d, k) = 0$ for all k, where F_d is the free group on $d \geq 2$ generators.

A group G is said to be ***positively finitely generated*** if $P(G, k) > 0$ for some finite k. Among finitely generated groups, the free abelian groups do have this property and the free non- abelian groups do not. The former of these two observations can be seen as a very special case of Theorem 4.2, in view of the following result, which elucidates the connection with maximal subgroup growth:

Theorem 4.4 [MSh]. *Let G be a profinite group. Then G is positively finitely generated if and only if $m_n(G)$ has polynomial growth.*

As a special case, it follows that *every profinite group with PSG is finitely generated*, a fact first noted in [Ma1].

Combining Theorem 4.4 with Theorem 4.2, we see that, for example, all arithmetic groups with CSP are positively finitely generated. It is not known whether this property is equivalent to CSP for arithmetic groups.

One direction of Theorem 4.4 is easy. Suppose $m_n(G) \leq n^s$ for all n. The probability that a random k-tuple in G is contained in a given maximal subgroup M of G is $|G : M|^{-k}$. Now a k-tuple fails to generate G exactly when it is contained in some maximal subgroup. Hence the probability that a random k-tuple fails to generate G is at most

$$\sum_M |G : M|^{-k} = \sum_{n>1} m_n(G) n^{-k} \leq \sum_{n>1} n^{s-k} < 1$$

provided $k \geq s + 2$. Thus $P(G, k) > 0$ for such k. □

The proof of the reverse implication requires more probability theory, and not only the classification of the finite simple group but also some recent results on the maximal subgroups of simple groups.

As another illustration of the probabilistic approach, we sketch a proof of

Theorem 4.5. *Let G be a finitely generated (abstract or profinite) group. Denote by $a_{n,d}(G)$ the number of (open) d-generator subgroups of index n in G. If G is non-universal, then, for each fixed d, $a_{n,d}(G)$ grows at most polynomially with n.*

Proof. Since, clearly, $a_{n,d}(\hat{G}) \geq a_{n,d}(G)$ for each n, we may as well assume that G is a profinite group. Now consider the set $\mathcal{S}$ which consists of all the non-abelian upper composition factors of G, together with all the non-abelian finite simple groups involved in these. Since G is non-universal, there is a finite bound both to the degrees of any alternating groups in $\mathcal{S}$, and to the dimensions of any groups of Lie type in $\mathcal{S}$. Conversely, it is not hard to see (given the classification of finite simple groups) that any group whose non-abelian upper composition factors are restricted in this way is in fact non-universal.

Let K be the intersection of all open normal subgroups N of $\hat{F}_d$ with the property that every non-abelian composition factor of $\hat{F}_d/N$ belongs to $\mathcal{S}$. Put $L = \hat{F}_d/K$. Then all non-abelian upper composition factors of L are in $\mathcal{S}$. The above remarks show that L is non-universal, whence by Theorems 4.2 and 4.4 we conclude that L is positively finitely generated. Say $P(L, k) = p > 0$.

Now suppose H is any d-generator open subgroup of G. Then H is a homomorphic image of $\hat{F}_d$. The definition of the set $\mathcal{S}$ ensures that all non-abelian upper composition factors of H are in $\mathcal{S}$; this implies that K is contained in the kernel of any epimorphism

$\hat{F}_d \to H$, so in fact H is a homomorphic image of $L = \hat{F}_d/K$. It follows that $P(H,k) \geq P(L,k) = p$.

Writing q for the probability that a random k-tuple in G generates some d-generator open subgroup of G, we thus have

$$1 \geq q = \sum_H P(H,k)|G:H|^{-k} \geq p \sum_n a_{n,d}(G) n^{-k},$$

whence

$$\sum_n a_{n,d}(G) n^{-k} \leq p^{-1}.$$

It follows in particular that for each n, $a_{n,d}(G) \leq p^{-1} n^k$, and the theorem is proved. □

We mention the following problem, about a special class of positively finitely generated groups: *which profinite groups have the property that, for some k, a random k-tuple generates a subgroup of finite index almost surely* (i.e. with probability 1)? PSG groups have this property. Are there any others?

Finally, we briefly discuss *maximal normal subgroups*. The following observation is made in [LPSh2]. Let G be any finitely generated profinite group. Then for a sufficiently large k, *the probability that a random k-tuple in G has normal closure equal to G is positive*. Just as in Theorem 4.4 above, it suffices to show that the number of maximal (closed) normal subgroups of index $\leq n$ in G grows polynomially with n. But if N is such a subgroup then G/N is a simple group of order n, and it follows from the classification of finite simple groups that there are at most 2 simple groups of each finite order. Philip Hall's observation, mentioned at the beginning of §3, therefore implies that there are no more than $2n^{d-1}$ possibilities for N, if G is generated by d elements. We have established

Theorem 4.6. *If G is a d-generator group, then G has at most $2n^d$ maximal normal subgroups of index at most n.*

The probabilistic result above can be given the following formulation: let $r > d$ be positive integers; then *a random presentation (in the category of profinite groups) on d generators and r relations defines the trivial group with positive probability*. Actually this probability is quite high, and appears to be about $1/2$. Of course, if $r < d$ then any such presentation defines a non-trivial, indeed infinite, group. If $r = d$, we can obtain the trivial group: but the probability of this outcome, or even of obtaining a finite group, is zero.

References

[B] Bhattacharjee, M., The probability of generating certain profinite groups by two elements. Israel J. Math. 86 (1994), 311–320.

[BPSh] Borovik, A., Pyber, L., Shalev, A., Maximal subgroups of finite and profinite groups. In preparation.

[D] Dixon, J. D., The probability of generating the symmetric group. Math. Z. 110 (1969), 199–205.

[DDMS] —, du Sautoy, M. P. F., Mann, A., Segal, D., Analytic Pro-p Groups. London Math. Soc. Lecture Note Ser. 157, Cambridge 1991.

[dS1] du Sautoy, M. P. F., Finitely generated groups, p-adic analytic groups and Poincaré series. Ann. Math. 137 (1993), 639–670.

[dS2] —, Zeta functions of groups and rings: uniformity. Israel J. Math. 86 (1994), 1–23.

[dS3] —, Counting congruence subgroups in arithmetic groups. Bull. London Math. Soc. 26 (1994), 255–262.

[FJ] Fried, M. D., Jarden, M., Field Arithmetic, Springer, New York–Berlin–Heidelberg 1986.

[GSS] Grunewald, F. J., Segal, D., Smith, G. C., Subgroups of finite index in nilpotent groups. Invent. Math. 93 (1988), 185–223.

[H1] Hall, P., The Eulerian functions of a group. Quart. J. Math. Oxford 7 (1936), 134–151.

[H2] Hall, M., Subgroups of finite index in free groups. Canad. J. Math. 1 (1949), 187–190.

[I] Ilani, I., Counting finite index subgroups and the P. Hall enumeration principle. Israel J. Math. 68 (1989), 18–26.

[KL] Kantor, W. A., Lubotzky, A., The probability of generating a finite classical group. Geom. Dedicata 36 (1990), 67–87.

[L1] Lubotzky, A., Counting finite index subgroups. In: Groups '93 Galway/St. Andrews, London Math. Soc. Lecture Note Ser. 212, Cambridge 1995, 368–404.

[L2] —, Subgroup Growth. Proceedings of ICM 1994. To appear.

[L3] —, Subgroup growth and congruence subgroups. To appear.

[LM1] —, Mann, A., On groups of polynomial subgroup growth. Invent. Math. 104 (1991), 521–533.

[LM2] —, —, Residually finite groups of finite rank. Math. Proc. Cambridge Philos. Soc. 106 (1989), 385–388.

[LMS] —, —, Segal, D., Finitely generated groups of polynomial subgroup growth. Israel J. Math. 82 (1993), 363–371.

[LPSh1] Lubotzky, A., Pyber, L., Shalev, A., Two new types of subgroup growth. In preparation.

[LPSh2] —, —, —, Normal and subnormal subgroups in residually finite groups. In preparation.

[Ma1] Mann, A., Some properties of polynomial subgroup growth groups. Israel J. Math. 82 (1993), 373–380.

[Ma2] —, Positively finitely generated groups. Submitted.

[MS] —, Segal, D., Uniform finiteness conditions in residually finite groups. Proc. London Math. Soc. (3) 61 (1990), 529–545.

[MSh] —, Shalev, A., Maximal subgroups of finite simple groups and positively finitely generated groups. In preparation.

[Mü1] Müller, T., Combinatorial aspects of finitely generated virtually free groups. In: Groups St. Andrews 1989, London Math. Soc. Lecture Note Ser. 160, Cambridge 1990, 386–395.

[Mü2] —, Finite group actions, subgroups of finite index in free products and asymptotic expansions of $e^{P(z)}$. To appear.

[N1] Neumann, P. M., An enumeration theorem for finite groups. Quart. J. Math. Oxford 20 (1969), 395–401.

[N2] Newman, M., Asymptotic formulas related to free products of cyclic groups. Math. Comp. 30 (1976), 838–846.

[P] Pyber, L., Enumerating finite groups of given order. Ann. Math. 137 (1993), 203–220.

[PSh] —, Shalev, A., Groups with super-exponential subgroup growth. In preparation.

[R] Robinson, D. J. S., Finiteness Conditions and Generalised Soluble Groups, Part 2. Springer, New York–Berlin–Heidelberg 1972.

[Se1] Segal, D., Subgroups of finite index in soluble groups I. In: Proceedings of Groups - St. Andrews 1985, London Math. Soc. Lecture Note Ser. 121, Cambridge 1986, 307–314.

[Se2] —, On the growth of ideals and submodules. In preparation.

[SSh] —, Shalev, A., Groups with fractionally exponential subgroup growth. J. Pure Appl. Algebra 88 (1993), 205–223.

[Sh1] Shalev, A., Growth functions, p-adic analytic groups and groups of finite coclass. J. London Math. Soc. 46 (1992), 111–122.

[Sh2] —, Subgroup growth and sieve methods. Preprint.

[W] Wehrfritz, B. A. F., Infinite Linear Groups. Springer, New York–Berlin–Heidelberg 1973.

Power Subgroups of Profinite Groups

*Consuelo Martínez**

Abstract. This note contains a review of results related to some conditions about a profinite group G that assure that G^n, the subgroup generated by all nth-powers of the group G is a closed subgroup of G.

1991 Mathematics Subject Classification: 20E18

1. Introduction

A pro-$\mathfrak{C}$-group, where $\mathfrak{C}$ denotes a class of finite groups, is a topological group that is isomorphic to an inverse limit of groups in $\mathfrak{C}$. We understand class in the usual sense, that is, $\mathfrak{C}$ contains the trivial group 1 and with each group G in $\mathfrak{C}$ all isomorphic copies of G also belong to $\mathfrak{C}$. When the considered class $\mathfrak{C}$ is closed under taking homomorphic images, the above definition is equivalent to saying that G is a compact, totally disconnected Hausdorff topological group such that $G/N \in \mathfrak{C}$ for every open normal subgroup N of G.

When $\mathfrak{C}$ is just the class of all finite groups, the above definition gives us the notion of profinite group.

A profinite group G is (topologically) finitely generated by elements $g_1, \ldots, g_m$ if $H = \langle g_1, \ldots, g_m \rangle$ is a dense subgroup of G.

For definitions and basic properties of profinite groups see [2], [5] and [9].

If G is a finitely generated profinite group, it is not known in general whether or not the subgroup G^n, generated by all nth-powers of elements of G, is a closed subgroup of G. It is even unknown if every subgroup of finite index in a finitely generated profinite group is closed (or equivalently open). Clearly a positive solution to the first question would imply immediately a positive answer to the second one.

The second question was solved in an affirmative way in $\mathfrak{C} = \mathfrak{A}\mathfrak{N}$ the class of finite abelian-by-nilpotent groups in [1] by Anderson, in the class of finite supersoluble groups in [7] by Oltikar and Ribes and in the class $\mathfrak{C} = \mathfrak{N}^l$ of finite groups of nilpotent length or Fitting height l in [4] by Hartley.

* Partially supported by DGICYT, Ps. 90-0129.

Offprint from
Infinite Groups 94, Eds.: de Giovanni / Newell

$\mathfrak{N}^l$ denotes the class of all finite groups G having a series

$$1 = G_0 \lhd G_1 \lhd \cdots \lhd G_l = G$$

with all factors G_{i+1}/G_i nilpotent. The least l for which such series exists is known as the Fitting height of G (or nilpotent length).

In [6] the first question was studied for pro-(finite nilpotent) groups and a positive answer was obtained.

As Shalev noticed, to prove that if G is a finitely generated pro-$\mathfrak{C}$-group, then the subgroup G^n is closed in G (where $\mathfrak{C}$ is a class of finite groups), it is sufficient to show the following proposition:

Proposition. *For arbitrary integers $m \geq 1, n \geq 1$, there exists an integer $N = N(m,n)$ such that in an arbitrary m-generated group in $\mathfrak{C}$ every product of nth-powers of elements of G can be represented in the form: $a_1^n \dots a_N^n$, where $a_i \in G$, $1 \leq i \leq N$.*

In [6] such $N = N(m,n)$ was found for $\mathfrak{C} = \mathfrak{N}$, the class of finite nilpotent groups.

Our proof uses the solution to the restricted Burnside problem given by E. Zelmanov.

We shall recall next some general ideas about Burnside problems:

General Burnside Problem. Is a periodic group locally finite?

Ordinary Burnside Problem. Is a periodic group satisfying an identity $x^n = 1$ locally finite?

The general Burnside problem has positive solution for some important classes of groups, as soluble periodic groups or linear periodic groups. The ordinary Burnside problem has positive solution for the cases $n = 2, 3, 4, 6$ and no positive solutions for groups of other exponents have been obtained.

Both (the general and ordinary Burnside) problems have negative solution in general. Golod in 1964 constructed a series of finitely generated infinite p-groups for an arbitrary prime p. In 1968 Novikov and Adian constructed a finitely generated infinite group of exponent n for any odd $n \geq 4381$. That means that $B(m,n) = F/F^n$ $(d > 1)$, the free Burnside group of exponent n (quotient of $F = F(m)$ the free group on m generators by the normal subgroup F^n generated by the nth-powers of elements of F) is, in general, infinite.

However it was conceivable that $B(m,n)$ could have only finitely many non-isomorphic finite factor-groups. So the attention was focused on the

Restricted Burnside Problem. Is it true that for each $m, n \geq 2$ there are only finitely many finite m-generated groups of exponent n?

By a theorem of Poincaré the positive answer to the restricted Burnside problem is equivalent to the existence of a largest finite group $B_0(m, n)$ having all finite groups on m generators and exponent n as homomorphic images.

The restricted Burnside problem was solved positively by E. Zelmanov in [11] and [12].

It is also known that the Restricted Burnside Problem is equivalent to the following statement:

A finitely generated profinite group of exponent n is finite.

We will denote:

$R(m, n)$ the order of $B_0(m, n)$;

$K(m, n)$ the nilpotency class of the largest finite nilpotent m-generated group of exponent n.

2. Pro-(Finite Nilpotent) Groups

The main result proved in [6], as we have indicated above is the following:

Theorem 2.1. *If G is a finitely generated pro-(finite nilpotent) group, then the subgroup G^n generated by all nth-powers of elements of G is closed in G.*

According to Shalev's result, Theorem 2.1 is equivalent to:

Theorem 2.2. *For arbitrary integers $m \geq 1, n \geq 1$, there exists a integer $N = N(m, n)$ such that in an arbitrary m-generated nilpotent group every product of nth-powers of elements of G can be represented in the form: $a_1^n \dots a_N^n$, where $a_i \in G$, $1 \leq i \leq N$.*

The proof can be found in [6]. Here we will enumerate some of the main results used in the proof, in a short sketch of the proof.

1. Let G be a group generated by a finite set of elements $x_1, \dots, x_m$ and let H be a normal subgroup of G of index at most d. Then we can choose a system of coset representatives of the subgroup H in G among words $x_{i_1}^{\pm 1} \dots x_{i_k}^{\pm 1}$, $k \leq d$.

Once an adequate system of coset representatives of the subgroup of finite index has been chosen, the following step is the election of a good system of generators in the subgroup.

2. A system of generators of the subgroup H can be chosen among words in the form $x_{i_1}^{\pm 1} x_{i_2}^{\pm 1} \dots x_{i_k}^{\pm 1}$, with $k \leq 2d + 1$.
3. There is a function $f(m, n)$ such that the subgroup G^n is generated by elements v^n, where $v = x_{i_1}^{\pm 1} x_{i_2}^{\pm 1} \dots x_{i_k}^{\pm 1}$, $k \leq f(m, n)$.

This is a crucial step in the proof, in which an adequate system of generators of the subgroup G^n is taken. In the proof the solution to the Restricted Burnside Problem is used and some properties of the Frattini subgroup of a nilpotent group.

The following result of [2] is also used:

4. If G is a nilpotent group generated by elements $a_1, \dots, a_d$, then every element of the commutator subgroup (G, H) is a product of the form $(g_1, a_1) \dots (g_d, a_d)$ with $g_1, \dots, g_d \in G$.

3. Pro-(Finite Soluble of Bounded Fitting Height) Groups

Now we will consider the solution to the first question, mentioned above, in the same class $\mathfrak{N}^l$ considered by Hartley. However in the proof we will use a result of independent interest that is proved in the paper by Hartley and used to prove the mentioned result about subgroups of finite index. The result is the following (Th.2 in [4]):

Theorem. *Let* $G \in \mathfrak{N}^c$ *an* m*-generated group. Then there is a function* $H(m, c) = m + (2m - 1)(c - 1)$ *such that every element in the derived group* G' *may be expressed as a product of* $\leq H(m, c)$ *commutators of* G.

Let G be a finite m-generated group of nilpotent length c. We want to prove that every element of G^n can be expressed as a product of a bounded number of nth-powers of elements of G. It is enough to prove it for every element in $(G^n)^n$. In fact, let us suppose that we know the existence of a natural number $N' = N'(m, n, c)$, that is a function of m, n and c, such that every element in $(G^n)^n$ is a product of $\leq N'$ nth-powers of elements of G.

Since the quotient group $G/(G^n)^n$ is a finite m-generated group of exponent n, the solution to the Restricted Burnside Problem (RBP for short) assures that its order is $\leq R(m, n^2)$) (with the above notation).

Now we may use that every element of A^n, where A is a finite group of order d, can be expressed as a product of not more than d nth-powers of elements of A.

We conclude that every element of G^n can be expressed as a product of at most $R(m, n^2)$ nth-powers of elements of G times an element of $(G^n)^n$. So finally, every element of G^n is a product of not more than $N(m, n, c) = R(m, n^2) + N'(m, n, c)$ nth-powers of elements of G.

Definition. A group G is said to satisfy the property $\mathcal{P}(M)$ if an arbitrary element of G^n can be expressed as a product of at most M nth-powers of elements of G.

Some technical lemmas are needed to prove the main result. We will indicate some of them.

Lemma 3.1. *Let G be a finite m-generated soluble group of nilpotent length c and let H be a normal abelian subgroup. Let us suppose that G/H satisfies $\mathcal{P}(M)$. Then G satisfies $\mathcal{P}(8Mk + mR(m,n) + R(m,n^2))$, where $k = H(R(m,n)m, c)$.*

Definition. A left-normed commutator of length n is an element

$$a = ((\dots(((a_1, a_2), a_3), a_4), \dots), a_n).$$

We will denote it $a = (a_1, a_2, \dots, a_n)$. If H is a normal subgroup of G, we will denote by $(H, H, G, \dots, G)_t$ the normal subgroup generated by all left-normed commutators $(a_1, a_2, \dots, a_{t+2})$, where $a_1, a_2 \in H$ and $a_3, \dots a_{t+2} \in G$.

Lemma 3.2. *Let G be a finite soluble m-generated group of nilpotent length c and let H be a normal nilpotent subgroup with $H \subseteq G_t$. Then every element of $(H, H, G, \dots, G)_t$ is a product of commutators $(a, b_{i_1}, \dots, b_{i_t})$ where $a \in H_2$ and $b_{i_j} = a_i$ or a_i^{-1}, where each selection $(b_{i_1}, \dots, b_{i_t})$ appears at most once and $a_1, \dots, a_m$ generate the group G.*

Lemma 3.3. *If G is a finite soluble m-generated group of nilpotent length c and H is a normal nilpotent subgroup of G with $H \subseteq G$, then every element in $(H, H, G, \dots, G)_t$ is a product of not more than $2m$ elements of type $(a, g_{i_1}, \dots, g_{i_t})$, where $a \in H_2$ and $g_{i_j} = a_{i_j}$ or $a_{i_j}^{-1}$ and $a_1, \dots, a_m$ generate G.*

Lemma 3.4. *Let b, a_i, h_i arbitrary elements of a group G. Then a commutator $(b, a_1h_1, \dots, a_th_t) = (b, h_1, \dots, h_t)x$ where x is a product of $\mathcal{F}(t) = 2^{t+1} - 2$ conjugates of the elements $\dot{a}_i$.*

Main Theorem. *Let $m, n, c \geq 1$. Then there is a function $N^* = N^*(m, n, c)$ such that if G is is a finite soluble group m-generated with nilpotent length c, then every element of G^n can be expressed as a product of not more than N^* nth-powers of elements of G.*

We will indicate a short sketch of the proof that uses induction over c.

a. It is always possible to suppose that if G has nilpotent length c, then there is a nilpotent group H, $H \lhd G$, such that G/H has nilpotent length $c - 1$. Without lost of generality we can assume that $H \subseteq G_t$ (for some convenient t).

b. We can substitute G by $\widehat{G} = G^n$, H by $\widehat{H} = H \bigcap G^n$ and n by $r = mR(m,n)$. Computing modulo $(\widehat{H}, \widehat{H}, \widehat{G}, \dots, \widehat{G})_t$, (where $t = K(r,n)$), we can assume that

$$(\widehat{H}, \widehat{H}, \widehat{G}, \dots, \widehat{G})_t = 1.$$

c. By induction over c, we know that $\widehat{G}/\widehat{H}$ satisfies $\mathcal{P}(M)$, where $M = \widehat{N^*_{c-1}} = N^*(r, n, c-1)$ with $r = mR(m,n)$.

So an element of $\widehat{G^n} = (G^n)^n$ can be expressed as a product $p.h$, where p is a product of $\leq L_t$ nth-powers of elements of G and $h \in (\widehat{H}, \widehat{H}, \widehat{G}, \dots, \widehat{G})_t$.

Finally, if

$$r = mR(m,n),\ t = K(r+1,n),\ A = 8H(rR(r,n),c-1),\ B = rR(r,n)+R(r,n^2)$$

then $N^*(m,n,0) = N(m,n)$ and if we know $N^*(a,b,c-1)\ \forall a,b \geq 1$ then,

$$N^*(m,n,c) = R(m,n^2) + (2r)^t(2(2^t-1)N^*(m,n,c-1) + N(r+1,n))+$$
$$A^t(AN^*(r,n,c-1)+B) + \frac{A^t-1}{A-1}B.$$

References

[1] Anderson, M. P., Subgroups of finite index in profinite groups. Pacific J. Math. 62 (1976), 19–28.

[2] Dixon, J. C., du Satoy, M. P. F., Mann, A., Segal, D., Analytic Pro-p-Groups. London Math. Soc. Lecture Note Ser. 157, Cambridge 1991.

[3] Hall, M., The Theory of Groups. Chelsea Publishing Company, New York 1976.

[4] Hartley, B., Subgroups of finite index in profinite groups. Math. Z., 168 (1979), 71–76.

[5] Koch, H., Galoissche Theory der p-Erweiterungen. VEB Deutscher Verlag der Wissenschaften, Berlin 1970.

[6] Martínez, C., On power subgroups of profinite groups. Trans. Amer. Math. Soc. To appear.

[7] Oltikar, B. C., Ribes, L., On prosolvable groups. Pacific J. Math. 77 (1978), 183–188.

[8] Passman, D. S., The Algebraic Structure of Group Rings. Wiley, New York 1977.

[9] Serre, J. P., Cohomologie Galoissienne. Springer, Berlin 1964.

[10] Vaughan-Lee, M., The Restricted Burnside Problem. Clarendon Press, Oxford 1993.

[11] Zelmanov, E., The solution of the restricted Burnside problem for groups of odd exponent. Izv. Math. USSR 36 (1991), 41–60.

[12] —, The solution of the restricted Burnside problem for 2-groups. Mat. Sb. 182 (1991), 568–592.

On the 4-Levi Centre and the 4-Bell Centre of a Group

Martin L. Newell

1991 Mathematics Subject Classification: 20F45

1. Introduction

Let G be a group and let n be an integer. In [2] we considered the sets $L(n, G)$ and $B(n, G)$ of elements a in the group satisfying the commutator identities

$$[a, x^n] = [a, x]^n$$

for every x in G and

$$[a, x^n] = [a^n, x]$$

for every x in G, respectively.

We showed that $L(2, G) = B(2, G)$ and that this set is the group of 2-right Engel elements in G. Similarly we proved that $L(3, G) = B(3, G)$ and that this set forms a group of 3-right Engel elements. Here we show that the sets $L(4, G)$ and $B(4, G)$ do not always coincide and that in general neither set forms a subgroup. However both sets consist of right Engel elements, the elements in $L(4, G)$ are right Engel of length at most 6 and the elements in $B(4, G)$ are right Engel of length at most 7 and these estimates are best possible. Thus in a group satisfying the maximum condition on subgroups both of these sets are contained in some term of the upper central series. Since a group G of exponent 4 has the property that $L(4, G)$ and $B(4, G)$ are both equal to G, we see that the upper central property does not hold in every group. However we show that the sets $L(4, G)$ and $B(4, G)$ are always contained in the Hirsch–Plotkin radical of G. Although these sets do not enjoy any mutual containment property it is perhaps interesting to note that if a is in any one of these sets then its square is in the other.

Offprint from
Infinite Groups 94, Eds.: de Giovanni/Newell

2. The Group $B_4 = \langle B(4, G)\rangle$

We shall use the properties established in [2] that for a in $B(n, G)$ it follows that a^n belongs to $B(G, n-1)$ and that $a^{n(n-1)}$ is a 2-right Engel element. We begin by describing the group action of G on the subgroup $B_3 = B(3, G)$ of B_4.

Theorem 1.
(1) *$G/C_G(B_3)$ is a 2-Engel group.*
(2) *Let a belong to $B(3, G)$ and let A be its normal closure in G. Then A is nilpotent of class at most 2 and A' has exponent 3. In particular A is 3-abelian.*

Proof. Let a belong to $B(3, G)$. Then $[a, x^3] = [a^3, x]$ for every x in G and a^3 is a 2-right Engel element. We replace x by xy^{-1} in this identity and write $(xy^{-1})^3 = x^{1+y+y^2}y^{-3} = x^3[x, y]^3cy^{-3}$ where $c = [x, y, x]^{[x,y]^2}[x, y, y]$.

Since c commutes with every right 2-Engel element and $[a^3, x]$ commutes with $[x, y]$ by [1] Theorem 7.13, we deduce that

$$[a, (xy^{-1})^3] = ([a, y^3]^{-1}[a, c][a, [x, y]^3][a, x^3])^{y^{-3}}$$

and

$$[a^3, xy^{-1}] = ([a^3, y]^{-1}[a^3, x])^{y^{-1}}.$$

Therefore $[a, c][a^3, [x, y]][a^3, x] = [a^3, x]^{y^2}$. By [1] Theorem 7.13 it follows that $[a^3, [x, y]] = [a^3, x, y]^2 = [a^3, x, y^2]$ and thus $[a, c] = 1$. Replacing y by xy in c we conclude that $[x, y, x]^y$ commutes with a and replacing a by a^y it follows that $[x, y, x]$ centralises $B(3, G)$ and $G/C_G(B_3)$ is a 2-Engel group.

Since $[x^y, x, a] = 1$ it follows that $[a^x, a^y, a] = 1$ for all x and y and so A is nilpotent of class at most 2. Since $[a^3, x, x] = 1$ for all x it follows that $[a^3, x, a] = 1$ and so $[a^{3x}, a] = 1 = [a^x, a]^3$ and A' has exponent 3. This completes the proof. □

Theorem 2. *B_4 is a locally nilpotent group and $G/C_G(B_4)$ is an exponent 4 by 2-Engel group.*

Proof. Let $B_2 = B(2, G)$. Then by [1] Theorem 7.13 it follows that B_2^2 belongs to the third term of the upper central series of G. So we can assume that B_2 is an elementary abelian 2-group. By Theorem 1 B_3 is nilpotent of class at most 4 and B_3/B_2 has exponent 3. Therefore $B_3 = B_2 \times (B_3)^2$. Let $a \in B(4, G)$ and let $x \in G$. Then $a^4 \in B_3$ and $a^{12} \in B_2$. Therefore a has order 24 and $[a, x^4] = [a^4, x]$ belongs to B_3. Thus B_4/B_3 is central by exponent 4 and generated by elements of order 4. So it is a locally finite 2-group. Let $K = \langle a^3 \mid a \in B(4, G)\rangle$. Then $B_4 = KB_3^2$ since $a^8 \in B_3^2$. Let $t \in B_3^2$. Then $t^3 = 1$ and $[a, t^4] = [a, t] = [a^4, t]$ and so $[a^3, t] = 1$. Therefore K commutes with B_3^2 and hence B_4 is locally nilpotent since B_3 is nilpotent and B_4/B_3^2 is a locally finite 2-group. Since $C_G(B_4)$ is contained in $C_G(B_3)$ and

$[a, c^4] = [a^4, c] = 1$ for every element a in $B(4, G)$ and c in $C_G(B_3)$, it follows that the factor group has exponent 4. This completes the proof. □

We now investigate the structure of the group $G = \langle a, x\rangle$, where a belongs to $B(4, G)$. This group is nilpotent of class at most 7 and $\Gamma_3(G)$ is a finite group of order dividing $2^{18} \cdot 3^2$.

Theorem 3. *Let $G = \langle a, x\rangle$ where $a \in B(4, G)$. Then*

(1) *$\Gamma_3(G)/\Gamma_4(G)$ is a direct product of two cyclic groups of order* 6,

(2) *$\Gamma_4(G)/\Gamma_5(G)$ is an elementary abelian 2-group of dimension* 3,

(3) *$\Gamma_5(G)/\Gamma_6(G)$ is an elementary abelian 2-group of dimension* 6,

(4) *$\Gamma_6(G)/\Gamma_7(G)$ is an elementary abelian 2-group of dimension* 4,

(5) *$\Gamma_7(G)$ is an elementary abelian 2-group of dimension* 3,

(6) *$\Gamma_8(G) = 1$.*

Proof. Let $A = \langle a, a^x, a^{x^2}, a^{x^3}\rangle$ be the normal closure of $a \in G$. Then A is nilpotent by Theorem 2. We show that G/A' is nilpotent and hence G is nilpotent. Since $[a^{12}, x, x] = 1$ and $[a^4, x] = [a, x^4]$ is congruent to $[a, x]^4$ mod A', we conclude that

$$[a, x, x]^6[a, x, x, x]^4[a, x, x, x, x] = 1 \bmod (A')$$

and on squaring we deduce that $[a, x, x, x, x, x, x] = 1 \bmod (A')$. Therefore G/A' is nilpotent and G is nilpotent by Hall's Theorem. In the identity $[a^4, g] = [a, g^4]$ when we substitute

$$x, ax, x^2, [a, x], a^2x, ax^2, x^3, x[a, x], a^3x, a^2x^2, ax^3, a^2[a, x], x^2[a, x], ax^2[a, x]$$

for g we determine the structure of G. In particular expanding $[a^4, x] = [a, x^4]$ mod $\Gamma_6(G)$ we get

$$[a, x, a]^6[a, x, a, a]^4[a, x, a, a, a] = [a, x, x]^6[a, x, x, x]^4[a, x, x, x, x]$$

Since $\Gamma_3(G)$ centralises a^4 on taking the fourth power of this identity we see that the fourth power terms are congruent to 1 and $\Gamma_4(G)^4 = 1 \bmod \Gamma_6(G)$. Replacing x by ax we deduce that

$$[a, x, a]^6[a, x, x, x, a][a, x, a, a, x][a, x, a, a, a] = 1 \bmod \Gamma_6(G) \tag{1}$$

and

$$[a, x, x]^6[a, x, x, x, a][a, x, a, a, x][a, x, x, x, x] = 1 \bmod \Gamma_6(G) \tag{2}$$

Replacing x by $[a, x]$ in the first of these yields $[a, x, a, a]^2 = [a, x, a, a, a, a]$ mod $\Gamma_7(G)$. Similarly replacing x by x^2 in the same identity gives

$$[a, x, x, a]^2 = [a, x, x, a, a, a] = [a, x, a, x]^2 \bmod \Gamma_7(G).$$

Commutating the first by a gives

$$\begin{aligned}[t] &[a, x, x, x, a, a] \\ &\qquad = [a, x, a, a, x, a] = [a, x, x, a, a, a][[a, x, a, a], [a, x]] \bmod \Gamma_7(G). \end{aligned}$$

Commutating the first with x and the second with a we deduce that

$$[a, x, x, x, a, x] = [a, x, a, a, x, x] \bmod \Gamma_7(G)$$

and so

$$[a, x, x, x]^2 = [a, x, x, x, x, x] \bmod \Gamma_7(G).$$

Similarly replacing x by a^2x in (2) we obtain further congruences in $\Gamma_6(G)$. Let

$$\begin{array}{lll} w1 = [a, x, a, a, a, a], & w2 = [a, x, x, a, a, a], & w3 = [a, x, x, x, a, a], \\ w4 = [a, x, x, x, x, a], & w5 = [a, x, x, x, x, x], & w6 = [[a, x, x], [a, x, a]], \\ w7 = [[a, x, a, a], [a, x]], & w8 = [[a, x, x, a], [a, x]], & w9 = [[a, x, x, x], [a, x]] \end{array}$$

be the nine basis commutators for $\Gamma_6(G)$ mod $\Gamma_7(G)$. Then the congruences $w1 = w7 = [a, x, a, a]^2$, $w2 = w4 = [a, x, x, a]^2$, $w5 = [a, x, x, x]^2$, $w3 = w1.w2$, $w8 = w1.w9$, $w6 = w1.w2.w9$ hold mod$\Gamma_7(G)$ and these are invariant under the substitution ax for x. So this factor group has dimension 4 with basis $w1$, $w2$, $w5$, $w9$. Again replacing x by x^2 and $[a, x]$ we deduce that $\Gamma_7(G)$ has dimension 3 with basis $[a, x, a, a, x]^2 = [w1, x]$, $[a, x, x, a, x]^2 = [w2, x]$ and $[a, x, x, x, x]^2 = [w5, x] = [w9, x]$, $[a, x, a]^{12} = [w1, x]$, $[a, x, x]^{12} = [w1, x][w5, x]$. The other basic commutators of weight 5 namely $[a, x, a, a, a]$, $[a, x, x, a, a]$, $[a, x, x, x, a]$ are all involutions. Finally the increased weight substitutions now yield $\Gamma_8(G) = 1$ and the result is established. □

Remark. With my colleague R. S. Dark we have verified using GAP and the nilpotent algorithm that the substitutions listed above are sufficient and have obtained the actual relations in G instead of the congruences given above. If we add the relation $a^4 = 1$, then $G/C_G(A)$ is a two generated group of exponent 4 and $x^4 \in Z(G)$. Setting $x^4 = 1$ also we obtain a 2-group with the property that $[[a, x]^4, x]$ is of order 2 and hence $[a, x]$ does not belong to $B(4, G)$ and so this set is not a subgroup.

Corollary 1. *Let $a \in B(4, G)$. Then $a^2 \in B(4, G) \cap L(4, G)$.*

Proof. The identity $[a^2, x^4] = [a^2, x]^4 = [a^8, x]$ holds in the group of Theorem 3. □

3. The Group $L_4 = \langle L(4, G)\rangle$

We wish to show that the analogue of Theorem 3 holds when $a \in L(4, G)$. The main difficulty is to show that the group is nilpotent. Its structure is then easily found by computation or by using elementary commutator calculations. We begin with some preparatory Lemmas.

Lemma 1. *Let* $G = \langle a, y\rangle$. *Let* y^3, $(ay)^3$ *and* $(ay^{-1})^3$ *belong to the centre* $Z(G)$. *Then* $a^3 \in Z(G)$ *if and only if* $[a, y]^3 = 1$.

Proof. Since $(ay^{-1})^3 = a^3[a, y]^a[a, y]^{1+y}y^{-3}$ and $ay^{-1}a^{-1} = a(ay)^{-1}$, we get

$$y^{-3a^{-1}} = a^3[a, y]^a[a, y]^{1+ay}(ay)^{-3},$$

and we conclude by cancellation that $[a, y, a]^y \in Z(G)$ and so $[a, y, a] \in Z(G)$. Hence $[y, a^3] = [y, a]^{3a}$ and the result follows. □

Lemma 2. *Let G be a group and let* $a \in L(4, G)$. *Let* $y \in G$ *and let* $[a, y^3] = 1$. *Then* $[a^3, y] = 1$ *and* $\langle a, y\rangle$ *is nilpotent of class at most* 3.

Proof. Let i and j be integers. Using

$$[(a^i y^j)^4, a] = [a^i y^j, a]^4 = [y^j, a]^4 = [y^j, a] = [a^i y^j, a],$$

since $[y^{4j}, a] = [y^j, a]$ we see that $[y, a]^3 = 1$ and $(a^i y^j)^3 \in Z(\langle a, y\rangle)$. By Lemma 1 the element a^3 also belongs to this centre. But these relations mean that $\langle a, y\rangle$ has exponent 3 modulo its centre and hence is nilpotent of class at most 3. □

It was shown in [2] Theorem 1 that $L(4, G) = L(-3, G)$ and that $[x^{12}, a, x] = 1$. So replacing x by x^{-1} we deduce that $[[a, x]^{12}, x] = 1$ for all x and hence $[[a, x]^{12}, a] = 1$ on replacing x by ax both $[a, x^{12}]$ and $[a, x]^{12} \in Z(\langle a, x\rangle)$.

Lemma 3. *Let* $G = \langle a, x\rangle$ *and let* $a \in L(4, G)$. *Suppose that* $a^3 = 1$. *Then G is nilpotent of class at most* 4.

Proof. Let $z = x^3$. Then $[a, z^4] = [a, z]^4 \in Z(G)$. Therefore $[[a, z]^4, a] = 1 = [a, z, a]^4$. Since $a^3 = 1$ it follows that $[a^z, a^{-3}] = 1 = [a^z, a]^{-3}$ and so $[a, z, a] = 1$. But then $[z, a^3] = 1 = [z, a]^3$ and so $[a, z] = [a, x^3] \in Z(G)$. So by Lemma 2 $G/Z(G)$ is nilpotent of class at most 3 and the result follows. □

Lemma 4. *Let* $G = \langle a, x\rangle$ *and let* $a \in L(G, 4)$. *Suppose that* $a^4 = 1$. *Then G is nilpotent.*

Proof. Let g be an element of G such that $\langle a, g\rangle = G$. Then $[a, g^{12}] \in Z(G)$ and by Lemma 2 we have $[a^3, g^4] = [a, g^4] = [a, g]^4 \in Z(G)$. Working modulo the hypercentre of G we can assume that $g^4 = 1$. In particular $ax, ax^{-1}, a^2x, a^2x^{-1}, a^{-1}x, a^{-1}x^{-1}$ are all elements of order 4. Also for $c \in C_G(a)$ we have that $[(cg^2)^4, a] = [g^2, a]^4 = 1$. In particular $(ax^2)^4$ and $(a^2x^2)^4$ commute with a and x^2. Let A be the normal closure of $a \in G$. Then using Hall's Theorem we can assume that the hypercentre of A is trivial since G/A' is nilpotent and A is finitely generated. We now establish a number of further reductions.

(1) The dihedral group generated by a^2, x^2 has order 16 since $[a^2, x^2]^2 = (a^2x^2)^4$ commutes with a and x^2. Therefore $[a^2, x^2]$ has order 4.

(2) Since a^2x^{-1} has order 4 we get $[a^2, x]^{1+x^2} = 1$. So $[a^2, x]^{x^2} = [a^2, x]^{-1}$ and since $[a^2, x^4] = 1 = [a^2, x]^{1+x+x^2+x^3}$, it follows that $[a^2, x]$ commutes with $[a^2, x]^x$ and with $[a^2, x^2]$. Therefore

$$[a^2, x^2]^4 = 1 = [a^2, x]^4[a^2, x]^{4x}.$$

Thus we conclude that $[a^2, x]$ has order 8 and that $[a^2, x]^4$ commutes with x and with a, using $[[a^2, x]^4, a] = [a^2, x, a]^4 = [a^{2x}, a]^4 = 1$. Therefore $[a^2, x]$ has order at most 4. So $[a^2, x, x^2] = [a^2, x]^2 = [x, a^2, x^2] = [x^2, a^2, x]$. But then $[[x^2, a^2]^2, x] = [x^2, a^2, x]^2 = 1$. In (1) we saw that $[x^2, a^2]$ commutes with a and hence $[a^2, x^2]^2 = 1$. Since $1 = [a^2, x^2]^2 = [a^2, x, x]^2 = [[a^2, x]^2, x]$, it follows on replacing x by ax that $[a^2, x]^2 \in Z(G)$ and so $[a^2, x]^2 = 1 = [a^2, x^2, x] = [a^2, x, x^2] = [a^2, x, x, x]$.

(3) Since $(ax^2)^4$ commutes with a it commutes with x^2 also. Therefore $(ax^2)^4 = [a, x^2]^{a^2+1} = [a, x^2]^{1+a^2}$. From $[a, x^2]^4 = 1$ it follows that $[a, x^2]^2 = [a, x^2, x^2]$ commutes with x^2. Therefore $[a, x^2, a^2] = [x^2, a^2, a]^{-1}$ commutes with $[a, x^2]$ and $[a, x^2]^a$ and thus with $[a^2, x^2]$. Therefore $[a^2, x^2, a]^2 = [[a^2, x^2]^2, a] = 1$. Thus $[a^2, x^2, a]$ is an involution which commmutes with x^2 and a.

(4) Since the involution $[a^2, x]$ commutes with x^2 by (2), on replacing x by ax we see that $aa^{x^{-1}}$ commutes with $[a^2, x]$ as do $[a, x^{-1}]$ and $[a^{-1}, x^{-1}]$. So

$$[a^2, x, a]^{xa} = [a^2, x]^{(-1+a)xa} = [a^2, x]^{(-1+a)ax} = [a^2, x, a]^{-x}.$$

As $[a^2, x, x]$ commutes with x by (2), replacing x by ax shows that $[a^2, x, x][a^2, x, a]$ commutes with xa and so $[a^2, x, x]^a[a^2, x, a]^{xa} = [a^2, x, x][a^2, x, a]$ and

$$[a^2, x^2, a] = [a^2, x, x, a] = [a^2, x, a]^2[a^2, x, a, x].$$

Since $[a^2, x, x][a^2, x, a] = [a^2, x^2][a^{2x}, a]$ commutes with xa, replacing x by x^{-1} we get that $[a^2, x^2][a, a^{2x^{-1}}]^{-1}$ commutes with $x^{-1}a$ and so

$$\begin{aligned}[a^2, x^2, a] = [a, a^{2x^{-1}}, x^{-1}a] = [a, a^{2x}, xa] &= [a^2, x, a]^{-1}, a][[a^2, x, a]^{-1}, x]^a \\ &= [a^2, x, a]^2[a^2, x, a]^{-1}, x].\end{aligned}$$

So it follows that $[a^2, x, a, x] = [[a^2, x, a]^{-1}, x]$. Therefore $[a^2, x, a]^2$ commutes with $[a^2, x, a, x]$ and so the latter is an involution, since $[a^2, x^2, a]$ and $[a^2, x, a]^2$ are. But then $[[a^2, x, a]^2, x] = 1$. Replacing x by ax we conclude that $[a^2, x, a]^2 = 1$. Thus $[a^2, x^2, a] = [a^2, x, a, x]$. By (3) this element commutes with a and x^2. Again replacing x by ax we see that this element commutes with $a^{x^{-1}}$ and so belongs to $Z(A)$. Hence $[a^2, x^2, a] = 1$ and so $[a^2, x^2] = 1$ by (2). But then $[a^2, x, x] = 1$ by (2) and we deduce that $[a^2, x, a] = 1$ also. Hence $[a^2, x] = 1$ and so $a^2 = 1$.

(5) Now $(ax^{-1})^4 = 1 = [a, x]^{1+x^2}$ and $[a, x]$ commutes with $[a, x]^x$ and so with $[a, x, x]$. Further $(ax^2)^4 = [a, x^2]^2 = [a, x, x]^2$ commutes with a and x^2. Therefore

$[a, x^2]^2 \in Z(A)$ and thus $[a, x^2]^2 = 1 = [[a, x]^2, x]$. Replacing x by ax we deduce that $[a, x]^2 = 1 = [a, x, x]$ and that $[a, x, a] = 1$. So G is nilpotent. □

Theorem 4. *Let $G = \langle a, x \rangle$ where $a \in L(4, G)$. Then*

(1) *$\Gamma_3(G)/\Gamma_4(G)$ is a direct product of two cyclic groups of order* 6,

(2) *$\Gamma_4(G)/\Gamma_5(G)$ is an elementary abelian 2-group of dimension* 3,

(3) *$\Gamma_5(G)/\Gamma_6(G)$ is an elementary abelian 2-group of dimension* 6,

(4) *$\Gamma_6(G)/\Gamma_7(G)$ is an elementary abelian 2-group of dimension* 4,

(5) *$\Gamma_7(G)$ is cyclic of order* 2,

(6) $\Gamma_8(G) = 1$.

Proof. We show first that G is nilpotent. Let F be the normal closure of a^4 and T be the normal closure of a^3 in G. Then by the preceding Lemmas we see that G/F and G/T are nilpotent. Then the normal closure A of a in G is the product of F and T. Let $g \in G$ and let $y = a^{4g}$. Then y^3 commutes with a since $[a, g]^{12} \in Z(\langle a, g \rangle)$. But then y commutes with a^3 by Lemma 2. Hence $F \cap T \in Z(A)$ and hence A is nilpotent. Again using Hall's Theorem, it follows that G is nilpotent.

Now as in Theorem 3 the expansion of the identity $[a, x^4] = [a, x]^4$ yields the congruences

$$[a, x, x]^6 = [a, x, x, x, x] \bmod \Gamma_6(G)$$

and

$$[a, x, a]^6 = [a, x, x, x, a][a, x, a, a, x][a, x, a, a, a] \bmod \Gamma_6(G).$$

The structure is again determined by the substitutions

$$ax, x^2, [a, x], a^2x, ax^2, x^3, a[a, x], x[a, x], a^3x, a^2x^2, ax^3, a^2[a, x], ax^2[a, x]$$

for x in the identity $[a, x^4] = [a, x]^4$. In particular Γ_6 mod Γ_7 has the same structure as in Theorem 3. $\Gamma_5(G)$ has exponent 2 and $\Gamma_7(G)$ is generated by $[w9, x]$ with $[w1, x] = 1 = [w2, x] = [w5, x]$. The elements $[a, x, x]$ and $[a, x, a]$ have order 12. □

Corollary 2. *Let $a \in L(4, G)$, then $a^2 \in L(4, G) \cap B(4, G)$.*

Proof. The identity

$$[a^2, x^4] = [a^2, x]^4 = [a^8, x]$$

is easily verified in the group above. □

Corollary 3. *Let $G = \langle a, x \rangle$ where $a \in L(4, G)$ and $a^2 = 1$. Then $[a, x^8] = 1$ and $[a, x^4] = [a, x]^4 \in Z(G)$.*

Proof. Since $[a, x]^2 = [a, x, a]^{-1}$, it follows that $[a, x]^8 = 1 = [a, x^8]$. Since $[a, x]^{12} \in Z(G)$ the result follows. □

Lemma 5. *Let $G = \langle a, b, z\rangle$ where $a, b \in L(4, G)$ and $a^2 = b^2 = z^2 = 1$. Then G is nilpotent of class at most 6 and $[(az)^4, b] = 1$.*

Proof. We show that G is nilpotent. Let H be the normal closure of $\langle a, b\rangle$ in G. Since G/H' is nilpotent, it is enough to show that H is nilpotent. The relation $[a, z]^4 = 1$ means that $\langle a, z\rangle$ is dihedral of order 16 and az has order 8. Thus $[(az)^4, a^b] = [[a, z]^2, [a, b]] = ([a, b]^2[z, a^b])^4$ and this element commutes with az, a^b, a, b and so belongs to $Z(G)$. We can assume that the hypercentre of H is trivial and hence $[a, z]^2$ commutes with $[a, b]$. Now $[(ab)^4, a^z] = [[a, b]^2, [a, z]] = ([a, z]^2[b, a^z])^4$. By the previous deduction we can assume that the final pair of elements commute and thus $[a, b]^2$ belongs to $Z(H)$ and is trivial. Now $[(az)^4, b] = [[a, z]^2, b] = ([a, b][z, b]^{-1})^{4z}$ and this element commutes with az, b, a^z and so belongs to $Z(G)$. Therefore we can assume that $[a, z]^2 = 1 = [b, z]^2$. For $g = [a, z]b$ we have $g^4 = [a, z, b]^2$ commutes with $[a, z]$ and b. Clearly $[g^4, a] = [g, a]^4 = 1$ and so it commutes with a also. It also commutes with b^z and so $[a, z, b]^2 = 1 = [b, z, a]^2$. So if $y = az$, then y^2 commutes with b and y has order 4. Further $(by^{-1})^4 \in Z(\langle b, y\rangle)$ and hence the involution $[[b, y]^2, y] \in Z(\langle b, y\rangle)$ and so $[[b, y]^2, y^2] = 1 = [b, y, y]^2$. Thus we conclude that $[[a, b], [z, b], [z, a]] = 1$ and by symmetry $[[z, a], [z, b], [a, b]] = 1$. Also from $[b, y, y]^2 = 1$ we get $[[z, b, a], [a, b, z]] = 1$. Then $[b, z, a, z] = [[b, z], [a, z]]^a$ is an involution and commutes $[a, b]$. Let $u = [a, b][b, z]$. Since $u^8 = 1$, we deduce that $u^4 \in Z(H)$ and so $[[a, b], [b, z]]^2 = 1$. Therefore $[b, z, a, b]$ commutes with $[a, b]$. Using the Hall–Witt identity we see that $[[b, z, a], [a, b]]$ belongs to $Z(G)$ and so is trivial. But then $[[b, z], [a, b]] \in Z(H)$ and is also trivial. By symmetry $[[a, z], [a, b]] = 1$ also. Finally it now follows that $(abz)^4$ commutes with a, b and hence with z and so $(abz)^4 = 1 = [[a, z], [b, z]]$ and G is nilpotent. Now it is relatively straight forward to determine the complete structure of G. We find that $\Gamma_7(G) = 1$ and $\Gamma_6(G) = \langle [[a, z, b], [b, z, a]] = [[a, z], [b, z]]^2 = [[a, b], [a, z], [b, z]] = [[a, b], [b, z], [a, z]]\rangle$. Moreover,

$$\Gamma_5(G)/\Gamma_6(G) = \langle c1, c2, c3, c4\rangle$$

with

$$\begin{aligned} c1 &= [[a, z, b], [a, b]] = [[a, z, b], [b, z]], \\ c2 &= [[b, z, a], [b, a]] = [[b, z, a], [a, z]], \\ c3 &= [[a, b, z], [a, z]], \\ c4 &= [[a, b, z], [b, z]], \end{aligned}$$

where

$$[[a, z, b], [a, z]] = [a, z, b]^2 = c2c3, [[b, z, a], [b, z]] = [b, z, a]^2 = c1c4,$$

and

$$\Gamma_4(G)/\Gamma_5(G) = \langle [[a,b],[a,z]], [[a,b],[b,z]], [[a,z],[b,z]] \rangle.$$

The relation $[(az)^4, b] = [[a,z]^2, b] = 1$ follows. □

Remark. In the group above if we add the relations $[a,b]^2 = [a,z]^2 = [b,z]^2 = 1$, the relation $[ab,z]^4 = [[a,z],[b,z]]^2$ holds and hence ab does not belong to $L(4,G)$ and this set is not a subgroup. It is easily established that $a,\ b \in L(4,G)$ for a group G satisfying these relations. Finally we show that $L(4,G)$ belongs to the Hirsch–Plotkin radical of G.

Theorem 5. *$L_4(G) = \langle L(4,G) \rangle$ is locally nilpotent.*

Proof. Let $a \in L(4,G)$. Then $a^2 \in B(4,G)$ and so belongs to the Hirsch–Plotkin radical $H_1(G)$. We work modulo this group and show that $a \in H_2(G)$ the second term of the upper radical series. Since a is a right Engel-element of G, it follows by [1] Theorem 7.34 that $a \in H_1(G)$. The group $\langle a^x, a^y, a \rangle$ is generated by three involutions and by the result above we have that $[[a^x,a]^2, a^y] = 1$. Therefore $[a^x,a]^2$ belongs to the centre of normal closure A of $a \in G$. Now $[a^x,a] = [a,x,a] = [a,x]^2$ and so $[a,x]^4 = [a,x^4] \in Z(A)$ for every $x \in G$. In particular $A/Z_2(A)$ has exponent 4 and is consequently locally nilpotent. Therefore A is locally nilpotent and the result is established. □

References

[1] Robinson, D. J. S., Finiteness Conditions and Generalized Soluble Groups, Part 2. Springer, Berlin 1972.

[2] Kappe, L. C., Newell, M. L., On the Levi and Bell Centres of a Group. Rend. Circ. Mat. Palermo (2) 23 (1990), 173–181.

Varieties of Representations of Groups, Quasivarieties and Pseudovarieties of Groups

Boris Plotkin

Abstract. The paper is devoted to the first order calculus in representations of groups. Varieties of representations of groups, quasivarieties, pseudovarieties of representations of groups are considered. We study also classes of groups, connected with these classes of representations. Some problems are formulated.

1991 Mathematics Subject Classification: 20E10

1. Introduction

The main part of this paper relates to the connections between three notions indicated in the title. We consider also other questions, which are united by the general idea of connections between logic and representations of groups. The paper use the books [PV] and [V], for the independence of the account some information from these books is given.

Let us start with some remarks about varieties of representations of groups. We fix a ring K, which is commutative, associative and with the unit. The representation over K is considered as a two-sorted algebra $\rho = (V, G)$, where V is a K-module, G is a group and the action $\circ$ of the group G in V is given. For every $a \in V$ and $g \in G$ we write $a \circ g \in V$. The action $\circ$ is defined through the representation $\rho : G \to \operatorname{Aut} V$. Thus ρ denotes the representation as a homomorphism and as a pair (V, G). In the category of representations, morphisms are homomorphisms changing V and G. This category is simultaneously a variety, and it is easy to list the identities of this variety. We consider subvarieties of this variety.

The free representation in the variety of all representations over the ring K is defined on the pair of sets X and Y and has the form (XKF, F), where $F = F(Y)$ is a free group on the set Y , KF is the group algebra and XKF is a free KF module over the set X. If $w = x_1u_1 + \cdots + x_nu_n \in XKF$, $u_i \in KF$, $x_i \in X$ and $f \in F$, then $w \circ f = x_1(u_1f) + \cdots + x_n(u_nf)$. If (V, G) is an arbitrary representation, then variables from Y are changed in G and variables from X are changed in the module V.

Offprint from
Infinite Groups 94, Eds.: de Giovanni/Newell

In accordance with the construction of the free representation, the identities of representations have the form

$$x_1 \circ u_1 + \cdots + x_n \circ u_n \equiv 0 \tag{1}$$

or

$$f = 1 \tag{2}$$

The identities of the first type are the identities of action while the identities of the second type are the group identities.

We can speak of the identities of individual representations and of the identities of classes of representations. Identities define varieties.

We do not consider arbitrary varieties of representations, but rather only special saturated varieties. A variety $\mathfrak{X}$ is saturated if a representation (V, G) belongs to $\mathfrak{X}$ if and only if the corresponding faithful representation belongs to $\mathfrak{X}$. The identities of such saturated $\mathfrak{X}$ are just the identities of the first type, and it is easy to see that they can be reduced to the identities of the type $x \circ u \equiv 0$, $u = u(y_1, \ldots, y_n) \in KF$. This means that we are interested in the action of the group, but not in the group itself.

The identity $x \circ u \equiv 0$ holds in the representation (V, G) if for every homomorphism $\mu : F \to G$ and every $a \in V$ we have $a \circ u^{\mu} = 0$. Here we take into account that we have also the homomorphism $\mu : KF \to KG$ and together with G, the group algebra KG also acts in V.

2. Some Examples of Identities

Let us consider the identity

2.1. $x \circ (y - 1) \equiv 0.$

If in the representation (V, G) this identity holds, it means that the group G acts trivially in V. We denote the variety of trivial representations by $\mathcal{G}$.

2.2. $x \circ (y_1 - 1) \ldots (y_n - 1) \equiv 0.$

This identity defines the variety $\mathcal{G}^n$ of n-unitriangular representations. This identity is fulfilled in (V, G) if there is a series

$$0 = V_0 \subset V_1 \subset \cdots \subset V_{n-1} \subset V_n = V$$

of G-invariant submodules in V, such that the group G acts trivially in all factors.

If V is a finite dimensional space over the field K, this means that in some base all matrices of elements of G have the triangular form with units on the principal diagonal.

2.3. $x \circ (y - 1)^n \equiv 0.$

The variety defined by such an identity is denoted by $\mathcal{G}_n$. The representation (V, G) belongs to $\mathcal{G}_n$ if all elements from G act in V as n-unipotents. If V is an n-dimensional space over the field K this means that all eigenvalues of elements of G are units.

2.4. $x \circ ([y_1, y_1'] - 1) \dots ([y_n, y_n'] - 1) \equiv 0.$

The variety defined by such identity is the variety of n-triangular representations. This means that (V, G) lies in such a variety if in V there is a series of the same kind as in example 2.2, and the group G acts as an abelian group in all factors. If V is an n-dimension space, then all matrices of elements from G are triangular in some base.

2.5. If G is a group, then the pseudoidentity of G is a formula of the type

$$(f_1 \equiv 1) \vee \cdots \vee (f_n \equiv 1), \quad f_i \in F,$$

which is fulfilled in G. The last means that for every homomorphism $\mu : F \to G$ we have $f_i^{\mu} = 1$ for some $i = 1, 2, \dots, n$.

If such a pseudoidentity takes place in G, then in every representation of this group there exists the identity

$$x \circ (f_1^{y_1} - 1) \dots (f_n^{y_n} - 1) \equiv 0.$$

Here $y_i \in Y, \quad f_i^{y_i} = y_i^{-1} f y_i$.

We note now that to every variety of representations $\mathcal{X}$, the ideal Id $\mathcal{X}=U \subset KF$ is associated.

This ideal is invariant under the action of endomorphisms of the group F. The ideal U consists of all elements $u \in KF$, such that the identity $x \circ u \equiv 0$ takes place in every representation from $\mathcal{X}$.

For the variety $\mathcal{G}$ we have Id $\mathcal{G} = \Delta$, i.e., the augmentation ideal of KF. We have also Id $\mathcal{G}^n = \Delta^n$.

3. Actions with Varieties

We can define the product of varieties, which is coordinated with the product of ideals.

If $\mathcal{X}_1$, and $\mathcal{X}_2$ are two varieties, then the product $\mathcal{X}_1\mathcal{X}_2$ is defined by the rule: $(V, G) \in \mathcal{X}_1\mathcal{X}_2$, if there is a G-invariant submodule $V_1 \subset V$, such that $(V_1, G) \in \mathcal{X}_1$, and $(V/V_1, G) \in \mathcal{X}_2$.

This multiplication leads to a semigroup of varieties, which is denoted by $\mathcal{M} = \mathcal{M}(K)$. In terms of this semigroup, $\mathcal{G}^n$ is the n-th power of the variety $\mathcal{G}$. If K is a field, then $\mathcal{M}$ is a free semigroup [PV].

Now let $\mathcal{X}$ be a variety of representations and Θ a variety of groups. We define the variety of representations $\mathcal{X} \times \Theta$ by the rule: a representation (V, G) belongs to $\mathcal{X} \times \Theta$ if there is a normal subgroup $H \subset G$, such that $(V, H) \in \mathcal{X}$ and $G/H \in \Theta$. Thus, the semigroup $\mathcal{N}$ of group varieties acts in the semigroup $\mathcal{M}$ as a semigroup of endomorphisms.

If K is a field, then this action is free (see [PV]), and it implies the well-known fact that $\mathcal{M}$ is a free semigroup.

From the above remarks we can deduce also that if K is a field, then every variety $\mathcal{X}$ can be uniquely represented in the form

$$\mathcal{X} = (\mathcal{X}_1 \times \Theta_1) \cdot \ldots \cdot (\mathcal{X}_n \times \Theta_n),$$

where $\Theta_1, \ldots, \Theta_n$ are the group varieties and $\mathcal{X}_1, \ldots, \mathcal{X}_n$ are the varieties of representations, irreducible in this sense.

Going back to example 2.4 we can represent the variety from this example as

$$\mathcal{G}^n \times \mathcal{A} = (\mathcal{G} \times \mathcal{A})^n$$

where $\mathcal{A}$ is the variety of all Abelian groups.

The variety $\mathcal{G}_n$ from example 2.3 is an irreducible one.

Comparing $\mathcal{G}_n$ and $\mathcal{G}^n$ we can notice that always $\mathcal{G}^n \subset \mathcal{G}_n$. On the other hand, if K is a field with char $K = 0$, then for every n we can find $m = \varphi(n) > n$, such that $\mathcal{G}_n \subset \mathcal{G}^m$. It follows from the results of E. Zelmanov [Z]. In the case char $K \neq 0$ a similar result is not true.

We can also consider the lattice of varieties of representations. This lattice is well coordinated with the semigroup $\mathcal{M}$ and with the action of the semigroup $\mathcal{N}$. There is no much known about this lattice.

4. Problems

Now, we are going to recall some information from the publications [P1], [P2], [PV], [V] and fix some new problems and results.

Problem 1. Consider the identities for the representations of the type

$$(K^n, GL_n(K))$$

under different conjectures about K and, maybe, n.

We mean that we have to treat the form of the identities and investigate the possibility of finding a finite basis of identities.

Note that when char $K = 0$ the group $GL_n(K)$ contains a free group and it has no nontrivial group identities. But we are looking for the identities of action.

This problem is connected with the results of A. Kemer [K] and A. Jakovlev (unpublished) about associative algebras. However, in the case of groups the problem is more difficult.

On the other hand, for the varieties of representations of Lie algebras, it is known [I] that not only the representation $(K^n, GL_n(K))$ but each finite dimensional representation over a field of zero characteristic has a finite basis of identities. The similar fact for varieties of representations of groups is not known and it makes sense to formulate the particular case of this problem.

Problem 2. Let (V, G) be a finite dimensional representation over a field with a solvable group G. Is it true that such a representation is always finitely based?

This problem has a positive solution up to a subgroup of finite index. Indeed, by the theorem of Kolchin–Mal'cev for some subgroup H of finite index in the solvable group G, the representation (V, H) lies in $\mathcal{G}^n \times \mathcal{A}$. On the other hand, recent results of A. Krasilnikov and A. Shmelkin [Kr] state that under some conditions on K the variety $\mathcal{G}^n \times \mathcal{A}$ and each of its subvarieties are finitely based.

Problem 3. Describe the lattice of varieties of representations of abelian groups over a field of prime characteristic.

In the case of characteristic zero this question has a simple solution, see [PV].

Problem 4. Let a group G act on a set M. We consider the linearization KM and thus arises the linear representation (KM, G). The problem is to describe the identities of this representation. In particular, G can be a full symmetric group S_M with the natural action.

In the last case we can use the pseudoidentities of the group S_M and add to the obtained set of identities the standard identity of Amitsur–Levitsky, but it is not enough to get all the identities.

5. Dimension Subgroup

Let $\mathfrak{X}$ be a variety of representations. Denote by $\vec{\mathfrak{X}}$ the class of all groups G having faithful representations (A, G) in $\mathfrak{X}$. Such a class $\vec{\mathfrak{X}}$ is not always a variety of groups.

Theorem 1. *For every $\mathfrak{X}$ and every K the class $\vec{\mathfrak{X}}$ is a quasivariety of groups* [PV].

Recall that quasivarieties are defined by quasiidentities. For groups, quasiidentities have the form

$$(f_1 \equiv 1) \wedge \cdots \wedge (f_n \equiv 1) \to (f \equiv 1).$$

We see that for every K the class $\vec{\mathfrak{X}}$ has a characterization in terms of group theory. However, we have no general rules how to construct the necessary quasiidentities.

Let now $U = \operatorname{Id} \mathfrak{X}$. For every group G from the given $\mathfrak{X}$ denote by U_G the verbal ideal in KG. The representation

$$(KG/U_G, G)$$

is the free representation of the group G in the variety $\mathcal{X}$. Denote by $D_{\mathcal{X}}(G)$ the kernel of this representation. We call this subgroup a dimension subgroup for given $\mathcal{X}$ and K. For every representation $(V, G) \in \mathcal{X}$ with given G, the subgroup $D_{\mathcal{X}}(G)$ belongs to the kernel of (V, G).

Now, we can say that the inclusion $G \in \vec{\mathcal{X}}$ means that the representation $(KG/U_G, G)$ is faithful or that $D_{\mathcal{X}}(G)$ is the unit in G. In addition, we have $G/D_{\mathcal{X}}(G) \in \vec{\mathcal{X}}$ for every group G, and if for some normal subgroup H holds $G/H \in \vec{\mathcal{X}}$, then $D_{\mathcal{X}}(G) \subset H$. Indeed, if $G/H \in \vec{\mathcal{X}}$ then we have a faithful representation $(V, G/H)$ in $\mathcal{X}$. The corresponding (V, G) also lies in $\mathcal{X}$ and the kernel of this representation is H. Therefore, $D_{\mathcal{X}}(G) \subset H$.

From these remarks we can conclude that the dimension subgroup $D_{\mathcal{X}}(G)$ is well characterized in terms of the quasivariety $\vec{\mathcal{X}}$.

Now, we consider a special case, when $\mathcal{X}$ is the variety $\mathcal{G}^n$. For every K we have $\vec{\mathcal{G}}^n \subset \mathcal{N}_{n-1}$, where $\mathcal{N}_{n-1}$ is the variety of nilpotent groups of the class $n-1$.

Let char $K = 0$. It is easy to prove that in this case the quasivariety $\vec{\mathcal{G}}_n$ is the quasivariety $\mathcal{N}_{n-1,0}$ of all torsion free groups from $\mathcal{N}_{n-1}$. The quasiidentities of this $\vec{\mathcal{G}}_n$ can be written explicitly.

If K is a field of prime characteristic p, then $\vec{\mathcal{G}}_n$ is a variety $\mathcal{N}_{n-1,p}$ of all p-nilpotent groups of class $n-1$. This variety had been studied by Magnus.

The difficult case is $K = \mathbb{Z}$. The conjecture was that $\vec{\mathcal{G}}^n = \mathcal{N}_{n-1}$. For the variety $\vec{\mathcal{G}}_n$ we have Id $\mathcal{G}^n = \Delta^n$, and then $D_{\mathcal{G}^n}(G) = D_n(G)$ is the classical dimension subgroup. Hence, $G \in \vec{\mathcal{G}}_n$ if and only if $D_n(G) = 1$, and the equality $\vec{\mathcal{G}}^n = \mathcal{N}_{n-1}$ means that for every $G \in \mathcal{N}_{n-1}$ there is $D_n(G) = 1$. The last means also that for every G the subgroup $D_n(G)$ coincides with the n-th member of the lower central series of the group G. It is the well-known dimension subgroup problem. So, the conjecture $\vec{\mathcal{G}}^n = \mathcal{N}_{n-1}$ is equivalent to the dimension subgroup problem.

For $n = 1, 2, 3$, it is proved that $\vec{\mathcal{G}}_n = \mathcal{N}_{n-1}$, but for $n = 4$, it is no longer true. In 1969 E. Rips [R] constructed a finite 2-group in $\mathcal{N}_3$ but not lying in $\vec{\mathcal{G}}^4$. The same for every $n > 4$ was done not long ago by N. Gupta [G]. For every $n \geq 4$ he constructed a finite metabelian 2-group G, such that $G \in \mathcal{N}_{n-1}$ and $G \notin \vec{\mathcal{G}}^n$. It is very interesting that if p is prime, $p \neq 2$ and G is a p-group in $\mathcal{N}_{n-1}$, then $G \in \vec{\mathcal{G}}^n$ (see [G] for this result and references).

Thus, we know now that for every $n > 3$ the class $\vec{\mathcal{G}}_n$ does not coincide with $\mathcal{N}_{n-1}$ and that the n-th dimension subgroup is not the n-th member of the lower central series. But what is the description of the class $\vec{\mathcal{G}}^n$ and what is the n-th dimension subgroup in terms of group theory? There arises the following problem, which looks like a natural development of the dimension subgroup problem.

Problem 5. Describe the quasiidentities of the class $\vec{\mathcal{G}}_n$ over $\mathbb{Z}$ for every n. The solution of this problem must be accompanied by searching for some basis of quasiidentities.

This problem was solved by E. Rips [R] for $n = 4$ and thereby he solved the dimension subgroup problem. He constructed an infinite system of quasiidentities, which defines the quasivariety $\vec{\mathcal{G}}^4$. The question of the possibility of reducing this system to some finite one has not been studied. It seems to us that such a finite system does not exist.

We can also consider the question of the description of the quasivariety $\vec{\mathcal{G}}^n$ over some other rings, say over $\mathbb{Z}_n$. For this ring the dimension subgroup problem has been studied.

The classes $\vec{\mathfrak{X}}$ for other interesting $\mathfrak{X}$ also can be investigated. It is natural to study the following problem. Let Θ be a quasivariety of groups. What is the variety $\mathfrak{X}$, such that $\vec{\mathfrak{X}} = \Theta$. The solution of this problem essentially depends on K, and it is known that such a variety of representation $\mathfrak{X}$ does not always exist. In this way we have to pass to the quasivarieties of representations. We will make some remarks on this subject later.

6. Connections with Pseudovarieties of Groups

We turn now to the connections with pseudoidentities and pseudovarieties of groups. For an arbitrary group G denote by $vr_K(G)$ the variety over K, generated by all representations of the group G. In example 2.5 we saw how the pseudoidentities of the group G can be transformed into the identities of all representations of this group, that is into the identities of the variety $vr_K(G)$. We can prove [PK] that if G is a finite group and K is a field with the characteristic which does not divide the order of the group G, then all identities of the variety $vr_K(G)$ can be constructed in such a way. In the modular case, i.e., char K divides $|G|$, it is not true.

Problem 6. Describe the identities of the variety $vr_K(G)$ for the finite group G in the modular case.

Observe here that the variety $vr_K(G)$ is generated by only one regular representation (KG, G). So this problem is the particular case of Problem 4. The problem is of interest also in the case of abelian group G, and, hopefully, the solution is not difficult. Note parenthetically, that pseudoidentities of a finite abelian group is described by its invariants.

Using the results of the book [PV] we can formulate the following two theorems.

Theorem 2. *If groups G_1 and G_2 have the same pseudoidentities, then the varieties $vr_K(G_1)$ and $vr_K(G_2)$ coincide.*

Theorem 3. *If two irreducible, faithful representations (V_1, G_1) and (V_2, G_2) have the same identities, then the group G_1 and G_2 have the same pseudoidentities.*

These theorems are valid over an arbitrary ring K.

It is well- known that if two finite algebras (one-sorted or many-sorted) have the same pseudoidentities then they are isomorphic. Thus, if in Theorem 3 the groups G_1 and G_2 are finite, then they are isomorphic. Moreover, in the conditions of the Theorem 3, for every field K not only the groups G_1 and G_2 but even the representations (V_1, G_1) and (V_2, G_2) are isomorphic [P2], [V].

B. Plotkin and A. Kushkuley have written a yet unpublished paper [PK] on the topic of this section. A. Kushkuley has proved also new results about connections between varieties of representations of groups and pseudovarieties of groups in finite and infinite cases.

Along with pseudovarieties of groups one can study pseudovarieties of representations.

Problem 7. Let (V_1, G_1) and (V_2, G_2) be two finite dimensional representations of finite groups. What can we say about these representations if it is known that they have the same pseudoidentities?

If the field K is finite, then according to the above-mentioned general result, the representations are isomorphic. Is it true when K is infinite?

7. Some Remarks about Terminology

If Θ is a variety of algebras, maybe many-sorted, and W is a free algebra in Θ, then a quasiidentity in Θ is a formula of the form

$$(w_1 \equiv w_1') \wedge \cdots \wedge (w_n \equiv w_n') \to (w \equiv w'),$$

while a pseudoidentity has the form

$$(w_1 \equiv w_1') \vee \cdots \vee (w_n \equiv w_n').$$

Here all w_i and w_i' are elements in W and the symbol "$\equiv$" connects the elements of the same sort.

Pseudoidentities are called also disjunctive identities. In fact, there is no fixed terminology and sometimes the term "pseudoidentity" has another meaning, which we stated previously.

In the case when Θ is the variety of all representations over K, the terms "quasivariety" and "pseudovariety" stand, as a rule, for saturated quasi and pseudovarieties. B. Plotkin and A. Gvaramija [PGG] have proved that a quasivariety of representations is saturated if and only if it can be defined by quasiidentities without group equalities. It is likely that a similar fact is true also for pseudovarieties.

Quasivarieties of representations have been studied by A. Matveev [M]. He proved that the class $\mathcal{X} \times \Theta$ is a quasivariety of representations if and only if $\mathcal{X}$ is a quasivariety of representations and Θ is a variety of groups.

8. Examples of Pseudo- and Quasiidentities of Representations, Additional Remarks

Let us consider a set of quasiidentities

8.1. $x \circ y^n \equiv x \to x \circ y = x, \quad n = 1, 2, \dots$

This set of quasiidentities characterizes special pure representations. Let $\mathcal{X}$ be a quasivariety defined by this set of quasiidentities, and $\vec{\mathcal{X}}$ be a class of groups with faithful representations in $\mathcal{X}$. Then $\vec{\mathcal{X}}$ is the class of torsion free groups. That is, every group G in $\vec{\mathcal{X}}$ is a torsion free group and every torsion free G has a faithful representation in $\mathcal{X}$. On the other hand, there are faithful representations of torsion free groups which do not belong to $\mathcal{X}$.

Not further that not only for varieties of representations but also for quasivarieties $\mathcal{X}$, the class $\vec{\mathcal{X}}$ is a quasivariety of groups. Moreover, if Θ is an arbitrary quasivariety of groups, then there exists a quasivariety of representations $\mathcal{X}$, such that $\vec{\mathcal{X}} = \Theta$.

8.2. The pseudoidentity

$$x \circ y^{n_1} \equiv x \vee x \circ y^{n_2} \equiv x \vee \cdots \vee x \circ y^{n_k} \equiv x$$

holds in the representation (V, G), if for every $a \in V$ and $g \in G$ we can find n_i, $i = 1, 2, \dots, k$ such that $a \circ g^{n_i} = a$.

For saturated pseudovarieties $\mathcal{X}$ we can also consider the classes $\vec{\mathcal{X}}$. What is, in particular, $\vec{\mathcal{X}}$ in example 8.2?

Problem 8. Consider the nature of classes $\vec{\mathcal{X}}$ for saturated pseudovarieties $\mathcal{X}$.

At first glance one might have the impression that $\vec{\mathcal{X}}$ is a pseudovariety of groups. But this is not true.

Now, a few words about another question. We know that in the case of identities it is sufficient to consider identities with one variable x running the module on which a group acts. For quasiidentities and pseudoidentities this is not valid. In examples 8.1 and 8.2 we used only one x and it would be of interest to give well-motivated examples with many x's.

9. Halmos Algebras

In order to consider some other problems, connected with logic in representation theory, we introduce some general notions.

As earlier, let Θ be a variety of algebras (maybe many-sorted), X be a set (also many-sorted in general) and W be a free algebra in Θ over X. With this data we can construct some polyadic Halmos algebra $U = U(X, \Theta)$.

U is a Boolean algebra with additional operations. These operations are connected with elements from the semigroup End W and with existential quantifiers of the form $\exists(Y),\ Y \subset X$. A set of axioms defines the variety of Halmos algebras in given Θ. We will consider these axioms after the definition of quantifiers.

Given a Boolean algebra H, an existential quantifier $\exists$ is a map $\exists : H \to H$ subject to the following conditions.

1. $\exists 0 = 0$,
2. $a \leq \exists a$,
3. $\exists(a \wedge \exists b) = \exists a \wedge \exists b$,

where $a, b \in H$, 0 is a zero element in H.

Every such quantifier is a closure operator in H. Universal quantifiers are defined dually.

Definition. Given a set X, a Halmos algebra H in Θ is:

1.1 H is a Boolean algebra.

1.2 The semigroup End W acts in H as a semigroup of boolean endomorphisms.

1.3 The action of quantifiers of the form $\exists(Y),\ Y \subset X$ is defined.

2. These actions are connected by the following conditions:

2.1 $\exists(\emptyset)$ acts trivially.

2.2 $\exists(Y_1 \cup Y_2) = \exists(Y_1)\exists(Y_2)$.

2.3 $s_1\exists(Y) = s_2\exists(Y)$, if $s_1, s_2 \in$ End W and $s_1(x) = s_2(x)$ for $x \in X \setminus Y$.

2.4 $\exists(Y)s = s\exists(s^{-1}Y)$ for $s \in$ End W, if the following conditions are fulfilled:

2.4.1 $s(x_1) = s(x_2) \in Y$ implies $x_1 = x_2$,

2.4.2 If $x \notin s^{-1}Y$, then $\Delta s(x) \cap Y = \emptyset$.

Here, $s^{-1}Y = \{x | s(x) \in Y\}$ and $\Delta s(x)$ is a support of the element $w = s(x)$ of W, i.e., the set of all $x \in X$, which is involved in the expression of the element w.

Along with Halmos algebras, just defined, we consider also Halmos algebras with equalities. Equality is a function $e : W \times W \to H$, defined for every pair of elements $w, w' \in W$ being of the same sort. Equality $e(w, w')$ can be regarded as a nullary operation on H. This function e should satisfy axioms of equality, see [P3].

The algebra $U = U(X, \Theta)$ mentioned above is a particular example of Halmos algebras. This algebra is constructed from first order logic, connected with given variety Θ. First of all, in this logic the elementary formulas of the type $w \equiv w'$, where $w, w' \in W$ and have the same sort, are considered. From these elementary formulas we can construct arbitrary formulas, using the symbols $\wedge$, $\vee$, $\neg$ and quantifiers $\exists x,\ x \in X$. Elements of the algebra U are these formulas, considered up to some natural equivalence. The equality $e(w, w')$ in the algebra U is the formula $w \equiv w'$. While constructing U we have to assume that every subset of all elements of a fixed sort in the set X is an infinite subset.

The approach developed in this section allows to consider not only a pure logic, but also a logic in the fixed Θ. We can deal with the logic of group theory, logic of ring theory, etc. In our case, Θ is a variety of representations of groups over some K. The transition to Θ-logics was inspired by various applications in algebra and computer science (for more details see the book [P3]).

Let us consider now some other examples of Halmos algebras. Assume that Θ is an arbitrary variety of algebras. Let an algebra $\mathcal{A}$ lie in Θ and let $\mathcal{M}_{\mathcal{A}}$ be a set of all subsets in $\mathrm{Hom}(W, \mathcal{A})$, i.e., $\mathcal{M}_{\mathcal{A}} = \mathrm{Sub}(\mathrm{Hom}(W, \mathcal{A}))$. For every $A \in \mathcal{M}_{\mathcal{A}}$, $s \in \mathrm{End}\, W$, $Y \subset X$ and $\mu : W \to \mathcal{A}$ we define:

1. $\mu \in \exists(Y)A$ if A contains $\nu : W \to \mathcal{A}$, such that $\mu(x) = \nu(x)$ for every $x \in X \setminus Y$.
2. $\mu \in sA$ if and only if $\mu s \in A$.
3. $e(w, w') = \{\mu \mid w^{\mu} = w'^{\mu}\}$.

In this way we define the action of the semigroup $\mathrm{End}\, W$ in $\mathcal{M}_{\mathcal{A}}$, the quantifiers $\exists(Y)$, $Y \subset X$, and the equality. It can be proved that all axioms of Halmos algebras with equality in Θ are fulfilled in $\mathcal{M}_{\mathcal{A}}$.

In the algebra $\mathcal{M}_{\mathcal{A}}$ we can distinguish a subalgebra $V_{\mathcal{A}}$ consisting of all subsets in $\mathrm{Hom}(W, \mathcal{A})$, which can be recognized on finite subsets of the set X.

The following fact plays a crucial role. For every algebra $\mathcal{A} \in \Theta$ there is a canonical homomorphism

$$f_{\mathcal{A}} : U \to V_{\mathcal{A}}.$$

For each formula $u \in U$ the image $f_{\mathcal{A}}(u)$ is the value of u in the algebra $\mathcal{A}$. This value is a subset in $\mathrm{Hom}(W, \mathcal{A})$. If $f_{\mathcal{A}}(u) = 1 = \mathrm{Hom}(W, \mathcal{A})$ then u holds in $\mathcal{A}$.

For example, if u is the formula $w \equiv w'$ then

$$f_{\mathcal{A}}(u) = \{\mu : W \to \mathcal{A} \mid w^{\mu} = w'^{\mu}\}.$$

The formula $w \equiv w'$ is an identity in $\mathcal{A}$ if $f_{\mathcal{A}}(w \equiv w') = \mathrm{Hom}(W, \mathcal{A})$.

The kernel of the homomorphism $f_{\mathcal{A}}$ is a filter in U, consisting of all $u \in U$ with $f_{\mathcal{A}}(u) = 1$. This $\mathrm{Ker} f_{\mathcal{A}}$ is the elementary theory of the algebra $\mathcal{A}$.

All varieties in Θ are controlled in the free algebra W, but for quasivarieties and other classes we need another object. In such a way the algebra U controls all axiomatizable classes in Θ.

10. Algebraic Logic in Representations of Groups

Halmos algebras can be used in the situation when Θ is the variety of representations over a given ring K. Let us take two infinite sets X and Y, where variables of X run the module V, and variables of Y run the acting group G. We have $F = F(Y)$ and $W = (XKF, F)$.

There are two types of elementary formulas:

1) $x_1 \circ u_1 + \cdots + x_n \circ u_n \equiv 0, \quad u_i \in KF$.

2) $f \equiv 1, \quad f \in F$.

The formulas of the first type we called formulas of action. An arbitrary formula of action is constructed from the elementary formulas of action by using the Boolean operations $\vee$, $\wedge$, $\neg$ and quantifiers with variables from the set X. We do not use quantifiers with variables from the acting group. In particular, we can speak about the identities of action, quasiidentities and pseudoidentities of action and so on. In the general case universal formulas have the form:

$$(w_1 \equiv w_1') \vee \cdots \vee (w_n \equiv w_n') \vee (v_1 \not\equiv v_1') \vee \cdots \vee (v_m \not\equiv v_m'),$$

where all $w, v \in W$.

We have mentioned earlier that every saturated variety of representations can be defined by identities of action. The same is valid for quasivarieties and, maybe, for pseudovarieties and universal classes of representations.

Proposition 1. *Let u be an arbitrary formula of action, $\rho = (V, G)$, $\bar{\rho} = (V, \bar{G})$ be a representation and the corresponding faithful representation. Then the formula u holds in ρ if and only if it holds in $\bar{\rho}$.*

This proposition is proved directly,using the homomorphisms

$$f_\rho : U \to V_\rho \quad \text{and} \quad \mathrm{f}_{\bar{\rho}} : \mathrm{U} \to \mathrm{V}_{\bar{\rho}}.$$

Proposition 2. *Let u be a formula of action, let $\rho = (V, G)$ be a representation and let $\rho' = (V, H)$ be its subrepresentation, where H is a subgroup in G. Then if u is fulfilled in ρ, u is also fulfilled in ρ'.*

From these two facts follows

Proposition 3. *Let T be a set of formulas of action and $T' = \mathcal{X}$ be a class of representations defined by T. Then $\mathcal{X}$ is saturated and right hereditary.*

Recall that a class $\mathcal{X}$ is called saturated if the representation $\rho = (V, G)$ belongs to $\mathcal{X}$ if and only if the representation $\bar{\rho} = (V, \bar{G})$ belongs to $\mathcal{X}$. The same can be defined in the following way: two representations (V, G) and (V', G') are called similar, if the corresponding faithful representations are isomorphic. An abstract class $\mathcal{X}$ of representations is saturated if and only if $\mathcal{X}$ is invariant under passing to similar representations.

Right hereditary means that $(V, G) \in \mathcal{X}$ implies $(V, H) \in \mathcal{X}$ if H is a subgroup in G and (V, H) is a subrepresentation in (V, G). Proposition 3 gives sufficient conditions for determining a class by formulas of action. The following question is on necessary and sufficient conditions.

Problem 9. Is it true that an abstract class of representations $\mathcal{X}$ is defined by some set T of formulas of action if and only if $\mathcal{X}$ is axiomatizable, saturated and right hereditary?

This question, as it seems to us, does not appear to be so difficult. We must use the known conditions of axiomatizability and pass to the class of all representations of the free group F of countable rank in the given $\mathcal{X}$.

It may also happen that the property of right hereditary follows from the first two conditions.

A set of formulas $T \subset U$ is called saturated if the class of representations $\mathcal{X} = T'$ is a saturated class.

Problem 10. Is it true that a set T is saturated if and only if it is equivalent to some set T_1 of formulas of actions?

This problem is connected with the previous one. Two sets T and T_1 are equivalent when the classes T' and T_1' coincide. Syntactically it means that T and T_1 generate in U the same filter [P3].

Independently we can speak about a the equivalence of sets of identities, quasiidentities, pseudoidentities and universal formulas. For the conditions of such equivalences in terms of the algebra U, see [P4]. Hence, problem 10 can be specified for such sets of formulas. We can also consider characterization of single saturated formula.

Let us conclude with the following result.

Proposition 4. *If $\mathcal{X}$ is a universal and saturated class of representations over K, then the class of groups $\vec{\mathcal{X}}$ is a universal class.*

This means that such $\vec{\mathcal{X}}$ admits a description by universal formulas in the group theory logic. In particular, it can be applied to the pseudovariety of representations $\mathcal{X}$.

References

[G] Gupta, N., A solution of the dimension subgroup problem. J. Algebra 138 (1991), 479–490.

[I] Il'tjakov, A. V., On varieties of representations of Lie algebras. Preprints 9, 10, Novosibirsk, SOAN, Math. Institute, 1991.

[K] Kemer, A. R., The finite basis property for identities of associative algebras. Algebra i Logika 26 (1987), 597–641.

[Kr] Krasil'nikov, A. N., On the laws on triangulable matrix representations of groups. Trudy Moscow Math. Obsh. 52 (1989), 229–245.

[M] Matveev, M., About quasivarieties of group representations. Izv. Vuz. Math. (1990).

[P1] Plotkin, B. I., Varieties of group representations. Uspekhi Math. Nauk 32 (1977), 3–68.

[P2] —, Varieties in group representations of finite groups. Locally stable representations. Matrix groups and varieties of representations. Uspekhi Math. Nauk 34 (1979), 65–95.

[P3] —, Universal Algebra, Algebraic Logic and Databases. Kluwer Academic Publishers, 1994.

[P4] —, Algebraic logic and some problems in algebra. To appear.

[PGG] —, Greenglaz, L., Gvaramija, A., Algebraic Structures in Automata and Database Theory. World Scientific, Singapore–New-Jersey 1992.

[PK] —, Kushkuley, A., Identities of regular representations of groups. Manuscript, Riga 1980.

[PV] —, Vovsi, S. M., Varieties of Group Representations. General Theory and Applications, Zinatne, Riga 1983.

[R] Rips, E., On the fourth integer dimension subgroup. Israel J. Math. 12 (1972), 342–346.

[V] Vovsi S. M., Topics in Varieties of Group Representations. Cambridge University Press, Cambridge 1991.

[Z] Zelmanov, E. I., On Engel Lie algebras. Sibirski. Math. Zh. 29 (1988), 112–117.

Groups with a Minimality Condition

*Vladimir I. Senashov**

1991 Mathematics Subject Classification: 20F50

1. Introduction

The imposition of restrictions on chains of subgroups has often been used in the investigation concerning the structure of infinite groups. Many authors have in particular considered groups satisfying the minimal condition on all subgroups or the minimal condition on all abelian subgroups. On the other hand, generalizing the concept of a locally finite group, S. P. Strunkov introduced in [13] binary finite groups, and later V. P. Shunkov [9] considered the wider class of conjugately biprimitively finite groups. Here a group G is called ***conjugately biprimitively finite*** if for every prime p and for every finite subgroup H of G any two conjugate elements of order p of $N_G(H)/H$ generate a finite subgroup. One of the main results relating these finiteness conditions is due to A. N. Ostylovskii and V. P. Shunkov [6], and states that a conjugately biprimitively finite group without involutions and satisfying the minimal condition on subgroups is a Chernikov group. Examples of S. P. Novikov, S. I. Adjan [1] and A. Yu. Ol'shanskii [4] show that such result cannot be generalized to arbitrary periodic groups without involutions.

We shall say that a group G satisfies the ***primary minimality condition*** if for each prime p every chain

$$G_1 > G_2 > \cdots > G_n > \cdots$$

of subgroups of G, such that each set $G_n \setminus G_{n+1}$ contains an element g_n with $g_n^{p^{k_n}} \in G_{n+1}$ for some k_n, stops after finitely many steps. An almost locally soluble group satisfying the primary minimality condition will be called a ***generalized Chernikov group***. This name is motivated by the following result: every almost locally soluble group G satisfying the primary minimality condition contains a complete part $\tilde{G}$, the

* Supported by the Krasnoyarsk Region Foundation of Science (3F0194)

Offprint from
Infinite Groups 94, Eds.: de Giovanni/Newell

factor group $G/\tilde{G}$ is locally normal, and every element of G centralizes all but finitely many Sylow subgroups of $\tilde{G}$ (see for instance [7]).

In this article we shall prove the following result:

Theorem. *Let G be a periodic group without involutions. Then G is a generalized Chernikov group if and only if it is conjugately biprimitively finite and the normalizers of all its finite non-trivial subgroups are generalized Chernikov groups.*

The already quoted examples by Novikov, Adjan and Ol'shanskii prove that in our theorem the condition that G is conjugately biprimitively finite cannot be removed.

2. Some Preliminary Results

Lemma 2.1 (see [5]). *Let G be a conjugately biprimitively finite group without involutions. If G satisfies the minimal condition on abelian subgroups, then it is a Chernikov group.*

Lemma 2.2 (see [7]). *Let G be a generalized Chernikov group. Then G has complete part $\tilde{G}$, and the factor group $G/\tilde{G}$ is locally normal. Moreover, every element of G centralizes all but finitely many Sylow subgroups of $\tilde{G}$.*

Lemma 2.3 (see for instance [8]). *Every locally normal group with Chernikov Sylow subgroups is layer-finite.*

Lemma 2.4 (see [2]). *Let G be a periodic locally soluble group such that the normalizer of every finite non-trivial subgroup of G is layer-finite. Then G is layer-finite.*

Lemma 2.5 (see [10]). *Let G be a group, and let A and B locally finite subgroups of G with Chernikov Sylow subgroups. If $A \cap B$ has finite index in A and B, then there exists a subgroup of finite index U of $A \cap B$ such that $N_G(U)$ contains A and B.*

Lemma 2.6 (see [6]). *Let G be a conjugately biprimitively finite group without involutions. If the centralizer of every non-trivial element of G is a Chernikov group, then G is a Chernikov group.*

Lemma 2.7 (see [12]). *Let G be a periodic group without involutions, and let H be a subgroup of G and a an element of H of prime order p. If for every element g of $G \setminus H$ the subgroup $\langle a, g^{-1}ag\rangle$ is finite and $H \cap g^{-1}Hg = 1$, then G is a semidirect product $G = F_p \ltimes H$, where F_p has no elements of order p and $H = N_G(\langle a\rangle)$.*

Lemma 2.8 (see [11], Lemma 4.6). *Let G be a Frobenius group of the form $G = F \ltimes \langle x \rangle$, where x has prime order p. Assume also that the subgroup $\langle x, x^g \rangle$ is finite for every $g \in G$ and that every abelian $\langle x \rangle$-invariant subgroup of prime exponent of F is finite. Then F is a nilpotent group.*

Lemma 2.9 (see [3], Corollary 5.4.1). *Let G be a locally finite group containing an element of prime order with finite centralizer. Then G is nilpotent-by-finite.*

3. Proof of the Theorem

The condition of the theorem is clearly necessary. Thus in the following we will suppose that G is conjugately biprimitively finite and that for every finite non-trivial subgroup X of G the normalizer $N_G(X)$ is a generalized Chernikov group. The proof will be accomplished in a series of lemmas.

Lemma 3.1. *Every Sylow subgroup of G is a Chernikov group.*

Proof. Clearly all primary abelian subgroups of G satisfy the minimal condition on subgroups. Then every Sylow subgroup of G satisfies the minimal condition on abelian subgroups, and so is a Chernikov group by Lemma 2.1. □

Lemma 3.2. *Every locally finite subgroup B of G is generalized Chernikov.*

Proof. The subgroup B is locally soluble by the famous Feit–Thompson theorem; since the Sylow subgroups of G are Chernikov groups by Lemma 3.1, it is well-known that B contains a complete part A. If $A \neq 1$, clearly A contains a finite non-trivial characteristic subgroup, and then B is a generalized Chernikov group by the hypothesis of the theorem. Suppose now that $A = 1$. It follows from Lemma 2.2 and Lemma 2.3 that every generalized Chernikov subgroup of B is layer-finite. In particular, the normalizer of every finite non-trivial subgroup of B is layer-finite, and so B itself is layer-finite by Lemma 2.4. Thus B is contained in the normalizer of some finite non-trivial subgroup, and so it is a generalized Chernikov group. □

Lemma 3.3. *It can be assumed that the locally finite radical of G is trivial.*

Proof. Suppose that the locally finite radical L of G is not trivial. Then L is a generalized Chernikov group, and it follows from Lemma 2.2 and Lemma 2.3 that either L has non-trivial complete part or it is layer-finite. In both cases L contains a finite non-trivial characteristic subgroup, and hence by the hypothesis G is a generalized Chernikov group. □

Lemma 3.4. *Every generalized Chernikov subgroup C of G is contained in a maximal generalized Chernikov subgroup of G.*

Proof. Let

$$C < H_1 < \cdots < H_n < \cdots$$

be a chain of generalized Chernikov subgroups of G containing C. Then the union V of this chain is locally finite, and so it is also a generalized Chernikov subgroup of G by Lemma 3.2. Application of Zorn's Lemma yields that C is contained in a maximal generalized Chernikov subgroup of G. □

Lemma 3.5. *The intersection of two distinct maximal generalized Chernikov subgroups of G is finite.*

Proof. Assume that A and B are distinct maximal generalized Chernikov subgroups of G such that the intersection $D = A \cap B$ is infinite, and suppose first that the complete part $\tilde{D}$ of D is not trivial. Then by hypothesis $H = N_G(\tilde{D})$ is also a generalized Chernikov subgroup of G. Since $\tilde{A}$ and $\tilde{B}$ are abelian and $\tilde{D} \leq \tilde{A} \cap \tilde{B}$, then $\tilde{A}$ and $\tilde{B}$ are contained in H, so that they are also contained in $\tilde{H}$. Clearly $N_G(\tilde{A})$ is a generalized Chernikov group, so that $N_G(\tilde{A}) = A$ and $\tilde{B} \leq \tilde{H} \leq \tilde{A}$. We obtain similarly that $\tilde{A}$ lies in $\tilde{B}$, so that $\tilde{A} = \tilde{B}$ and hence $A = N_G(\tilde{A}) = N_G(\tilde{B}) = B$. This contradiction shows that $\tilde{D} = 1$, and in particular D is layer-finite.

Suppose now that A and B both have non-trivial complete part, and let P and Q be non-trivial Sylow subgroups of $\tilde{A}$ and $\tilde{B}$, respectively. Since P is a Chernikov abelian normal subgroup of A, the factor group $A/C_A(P)$ is finite, so that in particular $C_D(P)$ has finite index in D. Similarly, the subgroup $C_D(Q)$ has finite index in D, and hence $M = C_D(P) \cap C_D(Q)$ is infinite. Moreover, M is layer-finite, so that it contains a finite non-trivial characteristic subgroup, and $N = N_G(M)$ is a generalized Chernikov group. Clearly P is contained in $A \cap N$, and N lies in a maximal generalized Chernikov subgroup of G by Lemma 3.4, so that it follows from the first part of the proof that $N \leq A$. Thus $Q \leq N \leq A$. This contradiction shows that A and B cannot have both non-trivial complete part. Suppose that $\tilde{A} \neq 1$, so that $\tilde{B} = 1$ and B is layer-finite. Let P be a non-trivial Sylow subgroup of $\tilde{A}$, and let b be an element of $B \setminus A$. Clearly the index $|B : C_B(b)|$ is finite, so that $M = C_D(P) \cap C_D(b)$ is infinite and $N = N_G(M)$ is a generalized Chernikov group. As above we obtain that N is contained in A, so that $b \in N \leq A$, a contradiction. Therefore $\tilde{A} = \tilde{B} = 1$, and A and B both are layer-finite.

Let $a \in A \setminus B$ and $b \in B \setminus A$. Then $C_D(\langle a, b\rangle)$ has finite index in D, and so is infinite. Let $d \neq 1$ be an element of $C_D(\langle a, b\rangle)$ and $C = C_G(d)$. Clearly C is contained in $N_G(\langle d\rangle)$, and hence is a generalized Chernikov group. Moreover $E = A \cap C$ has finite index in A, so that E is infinite. As C is not contained in A, it follows from the first part of the proof that $\tilde{C} = 1$, and so also C is layer-finite. In particular $\langle b\rangle^C$ is finite, and E has finite index in $C_1 = E\langle b\rangle^C$. Application of Lemma 2.5 yields that E contains a subgroup of finite index U which is normalized by A and C_1. On the other hand, U is layer-finite, so that it contains a finite non-trivial characteristic subgroup, and $N_G(U)$ is a generalized Chernikov group. Since $A < \langle A, b\rangle \leq N_G(U)$, we obtain a contradiction. The lemma is proved. □

By Lemma 2.6 the group G contains an element $a \neq 1$ such that $C_G(a)$ is not a Chernikov group. Clearly $N = N_G(\langle a\rangle)$ is a generalized Chernikov subgroup of G, and we can consider a maximal generalized Chernikov subgroup H of G containing N.

Lemma 3.6. *Let b be an element of H such that $C_H(b)$ is infinite. Then $C_G(b)$ is contained in H.*

Proof. Let M be a maximal generalized Chernikov subgroup of G containing $C_G(b)$. Then $H \cap M$ is infinite, and so $H = M$ by Lemma 3.5. Therefore $C_G(b)$ is contained in H. □

Lemma 3.7. *There exists an element b of H of prime order such that $C_H(b)$ is finite.*

Proof. Assume that all elements of prime order of H have infinite centralizer in H. Let g be an element of $G \setminus H$, and suppose that $H \cap g^{-1}Hg \neq 1$. If b is an element of prime order in $H \cap g^{-1}Hg$, the centralizer $C_H(b)$ is infinite, and so $C_G(b) \leq H$ by Lemma 3.6. On the other hand, b has infinite centralizer also in $g^{-1}Hg$, so that $C_G(b)$ is contained also in $g^{-1}Hg$. Then $H \cap g^{-1}Hg$ is infinite, and so $g^{-1}Hg = H$ by Lemma 3.5. It follows that $\langle H, g\rangle$ is a generalized Chernikov subgroup of G, a contradiction. Therefore $H \cap g^{-1}Hg = 1$ for every $g \in G \setminus H$. Application of Lemma 2.7 yields that $G = F \ltimes H$, where F and H are coprime and every element of prime order of H acts regularly on F. Application of Lemma 2.8 gives now a contradiction, since the locally finite radical of G is trivial by Lemma 3.3. This contradiction proves the lemma. □

Proof of the Theorem. By Lemma 3.7 the subgroup H contains an element b of prime order such that $C_H(b)$ is finite. Thus the locally finite group H is nilpotent-by-finite by Lemma 2.9. It follows now easily from the structure of generalized Chernikov groups that H is a Chernikov group (see Lemma 2.2). This last contradiction completes the proof of the theorem. □

References

[1] Adjan, S. I., Burnside Problem and Identities in Groups. Nauka, Moscow 1975.

[2] Ivko, M. N., Senashov, V. I., On a new class of infinite groups. Computing Center of RAN (Krasnoyarsk), preprint n.1 (1993), 30–44.

[3] Khukhro, E. I., Nilpotent Groups and their Automorphisms. De Gruyter Exp. Math. 8, Walter de Gruyter, Berlin–New York 1993.

[4] Ol'shanskii, A. Yu., Geometry of Defining Relations of Groups. Nauka, Moscow 1989.

[5] Ostylovskii, A. N., Locally finiteness of some groups with minimality condition for abelian subgroups. Algbra i Logika 16 (1977), 63–73.

[6] —, Shunkov, V. P., On locally finiteness of a class of groups with minimality condition. Research in Group Theory (Krasnoyarsk 1975), 32–48.

[7] Parvluk, I. I., Shafiro, A. A., Shunkov, V. P., On locally finite groups with primary minimality condition for subgroups. Algebra i Logika 13 (1974), 324–336.

[8] Senashov, V. I., Layer-Finite Groups. Nauka, Novosibirsk 1993.

[9] Shunkov, V. P., On one class of p-groups. Algebra i Logika 9 (1970), 484–496.

[10] —, On locally finite groups of finite rank. Algebra i Logika 10 (1971), 199–225.

[11] —, On embedding of prime order elements in a group. Nauka, Moscow 1992.

[12] —, Sozutov, A. I., On infinite groups full by Frobenius subgroups II. Algebra i Logika 18 (1979), 206–223.

[13] Strunkov, S. P., Subgroups of periodic groups. Dokl. Akad. Nauk SSSR 170 (1966), 279–281.

On Groups with Chernikov Centralizer of a 2-Subgroup

Pavel Shumyatsky

Abstract. We show that if G is a locally soluble periodic group having an elementary subgroup A of order 2^n such that $C_G(A)$ is Chernikov then $G/F_n(G)$ is Chernikov.

1991 Mathematics Subject Classification: 20F50

Let G be a locally soluble periodic group, A a subgroup of order p^n of G, where p is a prime, n an integer. Suppose that $C_G(A)$ is Chernikov. There are clear evidences in favor of the following

Conjecture. *With the above assumptions, $G/F_n(G)$ is Chernikov.*

Here $F_n(G)$ means the nth term of the Hirsch–Plotkin series of G, which is defined as follows.

Let $F(X)$ denote the Hirsch–Plotkin radical of a group X. Then

$$F_0(G) = 1, \quad F_{\alpha+1}(G)/F_\alpha(G) = F(G/F_\alpha(G)), \quad F_\mu(G) = \bigcup_{\alpha<\mu} F_\alpha(G)$$

for ordinals α and limit ordinals μ.

In particular the above conjecture is suggested by numerous results on finite groups admitting a group of automorphisms whose set of fixed points is "small" (see for example [6, 12]).

If $|A|$ is a prime then $n = 1$ and the conjecture is true by a result of B. Hartley [5]. The author showed recently that in this case G is even nilpotent-by-Chernikov [10].

The present paper is devoted to the case when A is an elementary 2-group. We prove here

Theorem. *Let G be a periodic locally soluble group containing an elementary 2-subgroup A of rank n such that $C_G(A)$ is Chernikov. Then*

(1) *$G/F_n(G)$ is Chernikov.*

Offprint from
Infinite Groups 94, Eds.: de Giovanni/Newell

(2) *If G is soluble then there exists a series in G*

$$1 = N_0 \leq N_1 \leq \cdots \leq N_n \leq G$$

such that N_i is normal in G, N_{i+1}/N_i is nilpotent for every i and G/N_n is Chernikov.

Our notation is mostly standard. If a group A acts on a group G we write G_A for the subgroup $\bigcap_{a\in A^\#}[G,a]$ of G. If G is a group, π a set of primes we write G_π for a Sylow π-subgroup of G and $G_{\pi'}$ for a Sylow π'-subgroup of G. A Chernikov group is a finite extension of a direct product of finitely many quasicyclic groups. Theorem of Feit and Thompson on solubility of finite groups of odd order [3] will be used without reference.

In the sequel we shall frequently and sometimes without reference use the following well known facts

Lemma 1. *Let A be a finite p-group acting on locally finite p'-group G. Then*

(1) $G = [G, A]C_G(A)$;

(2) $[G, A]$ *is normal in G;*

(3) *if N is an A-invariant normal subgroup of G then $C_{G/N}(A) = C_G(A)N/N$;*

(4) $[G, A, A] = [G, A]$;

(5) *if A is elementary abelian and $\mathcal{M}$ is the set of maximal subgroups of A then $G = \prod_{B\in\mathcal{M}} C_G(B)$.*

Lemma 2. *Let p be a prime, G a locally finite p'-group acted on by an elementary abelian group A of order p^n. Then we have*

(1) *if $C_G(A)$ is Chernikov then G/G_A admits an elementary abelian p-group of operators B of order $\leq p^{n-1}$ such that $C_{G/G_A}(B)$ is Chernikov;*

(2) *if N is any A-invariant normal subgroup of G then $G_AN/N \leq (G/N)_A$.*

Proof. See Lemma 1.9 of [8]. □

Recall that a group G is said to be radical if $G = F_\rho(G)$ for some ordinal ρ.

Lemma 3. *Let G be a locally soluble periodic group admitting a finite p-group A of automorphisms with $C_G(A)$ Chernikov. Then G is radical.*

Proof. By [7, 3.2] $G/O_{p'}(G)$ is Chernikov so we may assume that G is a p'-group. It is sufficient to show that $F(G) \neq 1$. First, suppose that every A-invariant normal subgroup in G has non-trivial intersection with $C_G(A)$. Since $C_G(A)$ satisfies the minimal condition, there exists a minimal subgroup C of $C_G(A)$ such that $C = N \cap C_G(A)$ for some normal A-invariant subgroup N of G. Then the normal closure

$\bar{C}$ of C in G is evidently a minimal normal subgroup in GA. By a theorem of McLain [7, p.11] $\bar{C}$ is abelian, whence $F(G) \neq 1$.

Now, let G contain an A-invariant normal subgroup N such that $C_N(A) = 1$. Then there exists a number $h = h(A)$ depending only on A such that for every A-invariant finite subgroup K of N we have $K = F_h(K)$ [2]. It follows now (see [7, Ch 1, §K]) that $F_h(N) = N$. Then $F(G) \neq 1$ as $F(N) \leq F(G)$. The proof is now completed. □

Lemma 4 (Hartley, [5, Lemma 2.10]). *Let H be a group with Hirsch–Plotkin radical F, and suppose that H/F is locally nilpotent. Then*

(1) *$F = F(X)$ for any $F \leq X \leq H$;*

(2) *if q is a prime, Q is a q-subgroup of H and $F_{q'}$ is the Sylow q'-subgroup of F, then*

$$C_Q(F_{q'}) = Q \cap F.$$

Lemma 5. *Let μ be an ordinal, and let N be a normal Chernikov subgroup of a periodic group G. Then $NF_\mu(G)/N$ is a subgroup of finite index in $F_\mu(G/N)$.*

Proof. Let F be the preimage of $F_\mu(G/N)$ in G. Then $C_F(N)N$ has finite index in F [7, ch. 1, §F]. Clearly, $C_F(N)$ is a normal subgroup of G which is contained in $F_\mu(G)$. Hence, $C_F(N)N \leq F_\mu(G)N$. □

Lemma 6. *Let G be a locally finite $2'$-group admitting a finite elementary 2-group of automorphisms A such that $C_G(A)$ is Chernikov. Let M be a normal abelian divisible subgroup of G. Then $C_G(C_M(A))$ has finite index in G.*

Proof. First, suppose that $A =\langle \phi \rangle$ is of order two. Set $H = [G, \phi]$. Then H' is Chernikov [5]. In particular, $[M, H, H]$ is Chernikov. Let $C = C_H([M, H, H])$. Then $|H : C|$ is finite [7, Theorem 1.F.3]. We have $[M, C, C, C] = 1$, whence $[M, C] = 1$ (Lemma 1), that is, $C_H(M)$ is of finite index in H. Since $C_G(\phi)$ is Chernikov, it is clear that $C_G(C_M(A)) \cap C_G(A)$ has finite index in $C_G(A)$. Since $G = HC_G(A)$, the result follows.

Let now A be of rank $n \geq 2$, and let $A_1, A_2, \ldots, A_{2^n-1}$ be the maximal subgroups of A. Set $G_i = C_G(A_i)$; $i = 1, \ldots, 2^n - 1$. As we have seen, $C_{G_i}(C_M(A))$ has finite index in G_i for every $i \in \{1, \ldots, 2^n - 1\}$. Now arguing as in [1, Lemma 7], we obtain the required conclusion. □

Lemma 7. *Let G be a locally finite $2'$-group acted on by an elementary group A of order 2^n. Suppose Z is a normal A-invariant subgroup of G such that $C_Z(A) = 1$. Then the following conditions hold.*

(1) *if G is soluble and has derived length k then $[Z, \underbrace{G_A, \ldots, G_A}_{s \text{ times}}] = 1$, where $s = s(n, k)$ is a number depending only on n and k;*

(2) *every p-subgroup of G_A centralizes every q-subgroup of Z.*
In particular, $G_A \cap Z \leq F(G)$.

Proof. (1) is an immediate corollary of [9, Lemma 9]. The part (2) follows easily from (1). □

Lemma 8. *Let G be a locally finite $2'$-group acted on by a finite elementary 2-group A. If $C_G(A)$ is Chernikov then $|G_A : F(G_A)|$ is finite.*

Proof. We write F and F_2 for $F_1(G_A)$ and $F_2(G_A)$ respectively. It is sufficient to prove that F_2/F is finite. First, show that for every prime q the Sylow q-subgroup of F_2/F is finite. If not, then for some q, G_A/F contains a countably infinite q-subgroup Q_1/F. Arguing as in [4, p. 492] we can choose a countable q-subgroup Q of G_A such that $QF = Q_1$. Put $F^0 = [F_{q'}, Q]$. Suppose that $C_{F^0}(A)$ is infinite and write M for the normal closure in G of the divisible part of $C_{F^0}(A)$.

We will show that M is hypercentral. Evidently F^0 possesses a normal in G series

$$1 \leq S_1 \leq \cdots \leq S_i \leq \cdots \qquad (*)$$

such that for every i the quotient S_{i+1}/S_i is either the maximal normal in GA/S_i subgroup of F^0/S_i satisfying $C_{S_{i+1}/S_i}(A) = 1$ or a minimal normal in GA/S_i subgroup of F^0/S_i with $C_{S_{i+1}/S_i}(A) \neq 1$. Put

$$\Gamma = \{i;\ C_{S_{i+1}/S_i}(A) = 1\},$$
$$\Delta = \{i;\ C_{S_{i+1}/S_i}(A) \neq 1\}.$$

It is clear from the definition that for each $i \in \Delta$ there exists a prime p such that S_{i+1}/S_i is an elementary abelian p-group. We can view the S_{i+1}/S_i as a GA/S_i-module over the field of p elements. A standard local argument yields that Theorem B of Hartley and Isaacs [6] remains true if we allow G to be a periodic locally soluble group. Since $C_{S_{i+1}/S_i}(A)$ is finite, by Theorem B of [6] the S_{i+1}/S_i is finite. Thus, we see that S_{i+1}/S_i is finite for every $i \in \Delta$. Since M has no subgroup of finite index, M centralizes S_{i+1}/S_i for every $i \in \Delta$.

By Lemma 7 (2) Q centralizes each S_{i+1}/S_i for every $i \in \Gamma$. As $M \leq [F_{q'}, Q]$, we obtain that M centralizes the series $(*)$. It follows now that M is hypercentral. Bearing in mind that M is generated by its quasicyclic subgroups, we conclude that M is abelian [11, Part 2, p.125]. Therefore, by Lemma 6, $C_M(A)$ is contained in some normal Chernikov subgroup, that is, M is Chernikov. By Lemma 5 we may assume that $M = 1$. Then, evidently, Δ is finite. Put

$$R = \bigcap_{i \in \Delta} C_Q(S_{i+1}/S_i).$$

Since S_{i+1}/S_i is finite for every $i \in \Delta$, it follows that R is of finite index in Q. As R is a q-group, by Lemma 7 (ii) R centralizes S_{i+1}/S_i for every $i \in \Delta$. Thus, R centralizes $(*)$. This implies that R centralizes F^0. Since $F^0 \geq [F_{q'}, R]$, it follows that $[F_{q'}, R, R] = 1$, whence $[F_{q'}, R] = 1$. By Lemma 4 $C_Q(F_{q'}) \leq F$, that is, $R \leq F$. As R is of finite index in Q, we deduce that $Q/Q \cap F$ is finite.

Let σ be the finite set of primes s such that $C_G(A)$ contains an element of order s. We show that the Sylow σ'-subgroup H/F of F_2/F is finite. Let F_σ be the Sylow σ-subgroup of F and $C = C_H(F_\sigma)$. Then A acts fixed point freely on $H/F_\sigma \leq (G/F_\sigma)_A$. Therefore H/F_σ is locally nilpotent, and so are $C/C \cap F_\sigma$ and C, since $C \cap F_\sigma \leq Z(F_\sigma)$. Hence $C \leq F$. If H/F is infinite we can choose an A-invariant σ'-subgroup S of H such that $S/S \cap F$ is (countably) infinite. Then $S \cap F \leq C$ and, since $C \leq F$, $S \cap C = S \cap F$. Therefore $S/S \cap F$ acts faithfully on F_σ. Now write $F^1 = [F_{\sigma'}, S]$. We can consider a normal in GA series of F^1 which is similar to the series $(*)$ in F^0. Repeating the argument applied previously to $(*)$, we can show that $S \cap C$ is of finite index in S, which shows that S and, therefore, H/F is finite. The proof is now completed. □

Lemma 9. *Assume hypotheses of Lemma 8. If G is soluble then G_A has a nilpotent subgroup of finite index.*

Proof. We use induction on the derived length of G. Let T be the last term of the derived series of G. By induction we may assume that TG_A/T has a nilpotent subgroup of finite index. It follows from [5, §4] that TG_A/T has a normal in G/T nilpotent subgroup F/T of finite index. Let M be the normal closure in G of the divisible part of $C_T(A)$. Then, by Lemma 6, M is Chernikov. Therefore $C_G(M)$ has finite index in G [7, Theorem 1. F. 3]. Taking $F \cap C_G(M)$ instead of F, we may assume that $F \leq C_G(M)$. Then F contains a normal in G nilpotent subgroup of finite index if and only if F/M does. So, without loss of generality, we may assume that $M = 1$ and use induction on $|C_T(A)|$. Let S be the maximal normal in GA subgroup of G_A such that $C_S(A) = 1$. By Lemma 7 there exists an integer l such that $S \leq Z_l(G_A)$. Therefore G_A contains a nilpotent subgroup of finite index if and only if G_A/S does. Thus, we may assume that $S = 1$. Then, as we have seen, Theorem B of Hartley and Isaacs [6] shows that there exists a normal in G finite subgroup $R \leq G_A \cap T$ such that $C_R(A) \neq 1$. We have $|C_T(A)| > |C_{T/R}(A)|$, whence G_A/R contains a normal in G/R nilpotent subgroup of finite index. Since R is finite, we conclude that G_A possesses a normal nilpotent subgroup of finite index, as required. □

Theorem. *Let G be a periodic locally soluble group containing an elementary 2-subgroup A of rank n such that $C_G(A)$ is Chernikov. Then*

(1) *$G/F_n(G)$ is Chernikov.*

(2) *If G is soluble then there exists a series in G*

$$1 = N_0 \leq N_1 \leq \cdots \leq N_n \leq G$$

such that N_i is normal in G, N_{i+1}/N_i is nilpotent for every i and G/N_n is Chernikov.

Proof. (1) Let H be the maximal normal subgroup of G containing no element of order two. By [7, 3.2 and 3.17] G/H is Chernikov, so it is sufficient to show that $H/F_n(H)$ is Chernikov. We consider A as a group of operators on H. By Lemma 9 $H_A/F(H_A)$ is finite. By Lemma 2 $H_1 = H/H_A$ admits an elementary 2-group of operators B of order at most 2^{n-1} such that $C_{H_1}(B)$ is Chernikov. Arguing by induction n we may assume that $H_1/F_{n-1}(H_1)$ is Chernikov. Thus, if $L = H/F(H_A)$ then L is an extension of a finite group by H_1. Now, using Lemma 5, we see that $L/F_{n-1}(L)$ is Chernikov. Therefore $H/F_n(H)$ is Chernikov, as required.

A proof of (2) can be obtained in a similar way. □

References

[1] Belyaev, V. V., Locally finite groups with Chernikov Sylow p-subgroups. Algebra and Logic 20 (1981), 393–402.

[2] Dade, E. C., Carter subgroups and Fitting heights of finite solvable groups. Illinois J. Math. 13 (1972), 347–369.

[3] Feit, W., Thompson, J. G., Solvability of groups of odd order. Pacific J. Math. 13 (1963), 773–1029.

[4] Hartley, B., The Schur–Zassenhaus Theorem in locally finite groups. J. Australian Math. Soc. (Series A) 22 (1976), 491–493.

[5] —, Periodic locally soluble groups containing an element of prime order with Černikov centralizer. Quart. J. Math. Oxford (2) 33 (1982), 309–323.

[6] —, Isaacs, I. M., On characters and fixed points of coprime operator groups. J. Algebra 131 (1990), 342–358.

[7] Kegel, O. H., Wehrfritz, B. A. F., Locally Finite Groups. North Holland, Amsterdam, 1973.

[8] Shumyatsky, P., Groups with regular elementary 2-groups of automorphisms. Algebra and Logic 27 (1988), 447–457.

[9] —, A four-group of automorphisms with a small number of fixed points. Algebra and Logic 30 (1991), 481–489.

[10] —, On locally finite groups admitting an automorphism of prime order whose centralizer is Chernikov. Houston Math. J. To appear.

[11] Robinson, D. J. S., Finiteness Conditions and Generalized Soluble Groups. Springer-Verlag, Berlin 1972.

[12] Turull, A., Groups of automorphisms and centralizers. Math. Proc. Cambridge Philos. Soc. 107 (1990), 227–238.

On Torsion-Free Groups with Trivial Center

*Andrzej Strojnowski**

Abstract. Let Γ be a Bieberbach group which means that Γ is a torsion free group and it has a free abelian normal subgroup A of finite index. If A is maximal it is called the translation subgroup, and then Γ/A is called the point group of Γ. We will show a series of examples of Bieberbach groups with trivial center and with the translation group A of minimal rank. Such groups are of interest to differential geometers, since they arise as fundamental groups of primitive, compact flat Riemannian manifolds. All these groups are torsion free but not right orderable. Some of them do not satisfy the unique product condition. Hence the group rings over such groups are good for testing Kaplansky's conjecture about units in group rings.

1991 Mathematics Subject Classification: 20F34

1. Introduction

By a primitive manifold X we mean a compact, connected flat Riemannian manifold, such that the first Betti number $b_1(X)$ of X vanishes. Recall the following fundamental result:

Theorem 1.1 (Bieberbach, Charlap, Hiller, Sah, Zassenhaus). *If Γ is the fundamental group of a primitive manifold X, then Γ fits into a short exact sequence*

$$0 \to M \to \Gamma \to H \to 1,$$

where H is the finite holonomy group of X and M is a faithful integral representation of H (an H-lattice). Furthermore Γ is torsion-free and the center $C(\Gamma)$ of Γ is trivial. Conversely given such an H, M and Γ there exists a corresponding flat manifold X.

The main result of [3] is:

Theorem 1.2. *If H is a finite group then H is the holonomy group of some primitive manifold X if and only if no cyclic Sylow subgroup of H has a normal complement.*

* Supported by KBN grant No 2 1115 91 01

Offprint from
Infinite Groups 94, Eds.: de Giovanni/Newell

We construct some examples of Bieberbach groups Γ with trivial center, such that the holonomy group H is p-group. This gives us an estimate for the minimal dimension of a primitive H-manifold for the number of such manifolds (see problem 3 of [3]).

2. Construction

Theorem 2.1. *Let $n > 1$ be an integer. Let $G = A * H$ be a subdirect product of the free abelian group A with a free basis $\{g_j;\ 1 \leq j < n\}$ and of a cyclic group H of order n, generated by h with the following action: $g_j^h = g_{j+1}$ for $j < n-1$ and $g_{n-1}^h = g_1^{-1} \cdot g_2^{-1} \cdot \ldots \cdot g_{n-1}^{-1}$. Let $B = \langle g_1^{t_1} \cdot g_2^{t_2} \cdot \ldots \cdot g_{n-1}^{t_{n-1}} \in A\,;\ n \mid \sum_{i=1}^{n-1} t_i \rangle$ be a subgroup of the group A. Then:*

(1) *B is a free abelian group and the following elements form a basis*

$$\begin{aligned} b_1 &= g_1 \cdot g_2^{-1}, \\ b_2 &= b_1^h = g_2 \cdot g_3^{-1}, \\ &\vdots \\ b_{n-1} &= b_{n-2}^h = g_{n-2} \cdot g_{n-1}^{-1}, \\ b_{n-1} &= b_{n-2}^h = g_1 \cdot g_2 \cdot \ldots \cdot g_{n-2} \cdot g_{n-1}^2. \end{aligned}$$

(2) *The group B is equal to the commutator group $B = [G, G]$.*

(3) *The quotient group G/B is isomorphic to a direct product of two copies of the cyclic group of order n. One of this groups is generated by image of h and the second by image of $g = g_1$.*

(4) *The center $Z = Z(G)$ is trivial.*

(5) *$A = \Delta(G)$ is a maximal FC-subgroup of the group G.*

Proof. (1) It is obvious that b_i belongs to the group B, and that they all generate this group. Furthermore $A^n \subset B$ so torsion-free rank of both of this groups are equal to $n-1$. Hence the set $\{b_1, b_2, \ldots, b_{n-1}\}$ is a basis of B.

(2) Since the basis of B is h-invariant so this group is a normal subgroup of G. It is obvious that the group G is generated by B, g and h. Furthermore $g^h B = g_2 B = \left(g \cdot b_1^{-1}\right) B = gB$. Hence the quotient group is isomorphic to the direct product of two cyclic groups of order n.

(3) By 2) $[G, G] \subset B$. Now we show that b_i's belong to $[G, G]$. In fact:

$$b_i = g_i \cdot g_{i+1}^{-1} = g_i \cdot h^{-1} \cdot g_i^{-1} \cdot h \in [G, G], \text{ for } i = 1, 2, 3, \ldots, n-1.$$

(4) Let $a \in Z(G)$. Then $a \in A$ since $(ch)^g \neq ch$ for each $c \in A$. Put $a = g_1^{t_1} \cdot g_2^{t_2} \cdot \ldots \cdot g_{n-1}^{t_{n-1}}$. Then $a^h = g_1^{-t_{n-1}} \cdot g_2^{t_1 - t_{n-1}} \cdot \ldots \cdot g_{n-1}^{t_{n-2}-t_{n-1}}$, and so

$$\begin{aligned} t_1 &= -t_{n-1}, \\ t_2 &= t_1 - t_{n-1}, \\ &\vdots \\ t_{n-1} &= t_{n-2} - t_{n-1}. \end{aligned}$$

Hence

$$\begin{aligned} t_{n-2} &= 2t_{n-1}, \\ t_{n-3} &= 3t_{n-1}, \\ &\vdots \\ t_2 &= (n-2)\, t_{n-1}, \\ t_1 &= (n-1)\, t_{n-1}. \end{aligned}$$

But $t_1 = -t_{n-1}$ so $0 = n \cdot t_{n-1}$, and consequently

$$0 = t_{n-1} = t_{n-2} = \cdots = t_2 = t_1.$$

Therefore $a = 1$.

(5) G contains the following conjugacy class.

$$\left\{g_1, g_2, \ldots, g_{n-1}, g_1^{-1} \cdot g_2^{-1} \cdot \ldots \cdot g_{n-1}^{-1}\right\}.$$

Hence the subgroup A generated by these elements is contained in $\Delta(G)$. Suppose $a = h^i b \notin A$. Then the conjugacy class of a contains infinitely many elements $a^{g^j} = g^{-j} h^i b g^j = h^i b g_{i+1}^{-j} g_1^j$. So $A = \Delta(G)$. □

Theorem 2.2. *Let $n = p^k$ be a power of a prime number. Let $\Gamma_{t,k}$ be a subgroup of G^{p+t-1} generated by $\Delta = B^{p+t-1}$ and by elements:*

$$\begin{array}{rlllllllllll} y = (h^{p-1}g, & h^{p-2}g, & \ldots, & hg, & g, & h, & h, & h, & \ldots, & 1) \\ x_1 = (h, & h, & \ldots, & h, & h, & \underset{p+1}{g}, & 1, & 1, & \ldots, & 1) \\ x_2 = (h, & h, & \ldots, & h, & h, & h, & \underset{p+2}{g}, & 1, & \ldots, & 1) \\ \vdots & & & & & & & & & \\ x_{t-1} = (h, & h, & \ldots, & h, & h, & h, & h, & h, & \ldots, & \underset{p+t-1}{g}) \end{array}$$

Then $\Gamma_{t,k}$ is a torsion-free group with trivial center. Furthermore Δ is a maximal FC- subgroup of Γ_1 such that the quotient group Γ_1/Δ is an elementary p-group of order p^t.

Proof. Let $\pi_r : G^{p+t-1} \to G$ be a map to the rth component. It is easy to see that $\pi_r(\Gamma_1)$ is onto for each r. Since $Z(G) = \{1\}$ so $Z(\Gamma_1) = \{1\}$. It is obvious that Δ is a normal FC-subgroup of $\Gamma_{t,k}$ of index $p^{t \cdot k}$ and that it is a free abelian group of rank $(p^k - 1) \cdot (p + t - 1)$. Let $u = b \cdot y^{i_o} \cdot \prod_{j=1}^{t-1} x_j^{i_j} \in \Gamma \setminus \Delta$, where $b \in \Delta$ and $0 \le i_j < p$. We will show that u is of infinite order and that it has infinite conjugacy class. Without loss of generality we can assume that $u^p \in \Delta$.

Step 1. Suppose $i_o = 0$, and let $r = \max\{j > 0; i_j \ne 0\}$. Then $\pi_r(u) = \pi_r(b) \cdot g^{i_r}$ belongs to $A \setminus B$. Hence u is not a torsion element. Furthermore one of component of u belongs to $G \setminus A$ so u has infinite conjugacy class.

Step 2. Suppose $i_o \ne 0$. Then first p components of u belong to distinct cosets of A in G. Hence u has infinite conjugacy class. Furthermore one of this component belongs to $G \setminus B$ so u is not a torsion element. □

Theorem 2.3. *Let Γ_2 be the subgroup of $\Gamma_{t,k}$ generated by the elements $y, x_1, x_2, \dots, x_{t-1}$. Then Γ_2 is a torsion-free group with trivial center. Furthermore $\Delta(\Gamma_2)$ is a free abelian group of rank $(p^k - 1) \cdot (p + t - 1)$ and the quotient group $\Gamma_2/\Delta(\Gamma_2)$ is a product of t copies of the cyclic group of order p^k.*

Proof. Since each map $\pi_r(\Gamma_2)$ is onto so $Z(\Gamma_2) = \{1\}$. By the theorem above the group Γ_2 is torsion-free and $\Delta(\Gamma_2) \subset B^{p+t-1}$. To prove that $\Delta(\Gamma_2)$ has rank $(p^k - 1) \cdot (p + t - 1)$ we show that $\Delta(\Gamma_2)$ contains:

$$\left(g_i^{p^k},\ 1,\ 1,\ \dots,\ 1\right) = x_1^{-i+1} (x_1 y)^{p^k} x_1^{i-1},$$

$$\Big(1,\ \dots,\ 1,\ \underset{r}{g_i^{p^k}},\ 1,\ \dots,\ 1\Big) = x_1^{-i+1} \left(x_1^r y\right)^{p^k} x_1^{i-1}, \text{ for } 1 \le r \le p,$$

$$\Big(1,\ \dots,\ 1,\ \underset{r}{g_i^{p^k}},\ 1,\ \dots,\ 1\Big) = y^{-i+1} x_{r-p}^{p^k} y^{i-1}, \text{ for } p+1 \le r \le p+t-1.$$

□

Corollary 2.4. *Let H be a product of t copies of the cyclic group of order p^k. Then $\delta(H) \le (p^k - 1) \cdot (p + t - 1)$, where $\delta(H)$ is the minimal dimension of a primitive H-manifold.*

Remark 2.5. By theorem 1.1. of [4], if H is an elementary abelian p-group of order p^t, then $\delta(H) = (p - 1) \cdot (p + t - 1)$.

Let $D_4 = Z_2 \times Z_2$. Then $\partial(D_4) = 3$.

Theorem 2.6. *All primitive D_4-manifolds are isomorphic.*

Proof. Let Γ be a Bieberbach group of Hirsch rank 3 with trivial center, such that $\Gamma/\Delta(\Gamma) \simeq D_4$. We prove that $\Gamma \simeq \langle x, y;\ x^{-1} y^2 x = y^{-2},\ y^{-1} x^2 y = x^{-2} \rangle = P$.

It is known that $\Delta(P)$ is a free abelian group with basis x^2, y^2 and $(xy)^2$ (see [5], Chapter 13). Let D_4 be a group of linear automorphism of Q^3 which acts faithfully and such that 1 is eigenvalue for each automorphism. Then in a suitable basis we can express theses automorphism by following matrices:

$$\begin{pmatrix} 1 & 0 & 0 \\ 0 & 1 & 0 \\ 0 & 0 & 1 \end{pmatrix}, \begin{pmatrix} 1 & 0 & 0 \\ 0 & -1 & 0 \\ 0 & 0 & -1 \end{pmatrix}, \begin{pmatrix} -1 & 0 & 0 \\ 0 & -1 & 0 \\ 0 & 0 & 1 \end{pmatrix}, \begin{pmatrix} -1 & 0 & 0 \\ 0 & 1 & 0 \\ 0 & 0 & -1 \end{pmatrix}.$$

Let $a, b \in \Gamma$ be elements such that Γ is generated by a, b and $\Delta = \Delta(\Gamma)$. Then a^2, b^2 and $(ab)^2$ form a free basis of some subgroup of Δ of finite index and the subgroup generated by a, b is isomorphic to P. We will write the group Δ additively. Let $\alpha_1, \alpha_2, \alpha_3$ be a basis of Δ. Now there exist integers $r_{i,j}$ such that:

$$\begin{array}{rcl} a^2 & = & (\; r_{1,1} \quad r_{1,2} \quad r_{1,3} \;) \\ b^2 & = & (\; r_{2,1} \quad r_{2,2} \quad r_{2,3} \;) \\ (ab)^2 & = & (\; r_{3,1} \quad r_{3,2} \quad r_{3,3} \;) \end{array}$$

By elementary operations on the basis we can find a new basis $\beta_1, \beta_2, \beta_3$, in which:

$$\begin{array}{rcl} a^2 & = & (\; s_{1,1} \quad 0 \quad 0 \;) \\ b^2 & = & (\; s_{2,1} \quad s_{2,2} \quad 0 \;) \\ (ab)^2 & = & (\; s_{3,1} \quad s_{3,2} \quad s_{3,3} \;) \end{array}$$

Now $\beta_1^a = \beta_1$ and $\beta_1^b = \beta_{1.}^{-1}$ Suppose $s_{1,1} = 2t$. Then $\left(a\beta_1^{-t}\right)^2 = a^2\beta_1^{-2t} = 1$. A contradiction. Hence $s_{1,1} = 2t + 1$ is an odd number and $\left(a\beta_1^{-t}\right)^2 = \beta_1$. Suppose $s_{2,2} = 2k$. Then $\left(b\beta_2^{-k}\right)^2 = (c\ 0\ 0)$. A contradiction. Hence $s_{2,2} = 2k + 1$ is an odd number and $\left(b\beta_2^{-k}\right)^2 = (c\ 1\ 0)$. We have got the following new basis of Δ : $\gamma_1 = a_1^2, \gamma_2 = b_1^2$ and $\gamma_3 = \beta_3$, where $a_1 = a\beta_1^{-t}$ and $b_1 = b\beta_2^{-k}$. Put $(a_1b_1)^2 = (m, n, w)$. Since

$$\begin{aligned} (w \cdot \gamma_3)^{a_1b_1} &= \left((a_1b_1)^2 - m \cdot (a_1)^2 - n \cdot (b_1)^2\right)^{a_1b_1} \\ &= \left((a_1b_1)^2 + m \cdot (a_1)^2 + n \cdot (b_1)^2\right) = (2m.2m, w), \end{aligned}$$

we have $\gamma_3^{a_1b_1} = \left(\frac{2m}{w}, \frac{2n}{w}, 1\right)$. Suppose $w = 2r$. Then $\left(a_1b_1\gamma_3^{-r}\right)^2 = (0\,0\,0) = 1$. A contradiction. Hence $w = 2r + 1$ is an odd number and

$$\left(a_1b_1\gamma_3^{-r}\right)^2 = \left(\frac{m}{w}, \frac{n}{w}, 1\right) = \eta_3.$$

Hence $\gamma_1, \gamma_2, \eta_3$ form a basis. But $\eta_3 = \left(a_1b_1\gamma_3^{-r}\right)^2 = \left(a_1b_1\gamma_1^i\gamma_2^j\eta_3^q\right)^2 = \left(a_1b_1\eta_3^q\right)^2$, and $\left(b_1\eta_3^q\right)^2 = b_1^2$. It follows that the map $\varphi : P \to \Gamma$, given by $\varphi(x) = a_1$ and $\varphi(y) = b_1\eta_3^q$, is an isomorphism. □

Theorem 2.7. *Let $p > 2$ be a prime and let $H = Z_p \times Z_p$. Then there exists at least 2 non isomorphic primitive H-manifolds.*

Proof. Let $\Gamma_1 = \Gamma_{2,1}$ from the previous section and Γ_2 be its subgroup generated by y and $x = x_1$. We show that these groups have non isomorphic abelianizations.

Put $C = \{(c_1, c_2, \dots, c_{p+1}) \in \Gamma_1 \; ; \; c_i \in [B, G]\}$. Then $[\Gamma_1, \Gamma_1]$ is generated by C and all conjugates of $u = [x, y]$. But $u^p \in C, u \notin C$ and $u^\alpha = u \cdot [u, \alpha] \in \langle u, C \rangle$ for $\alpha \in \Gamma_1$. Hence the index of C in $[\Gamma_1, \Gamma_1]$ is equal to p. Since $[\Gamma_1 : B] = p^2$ and $[B, C] = p^{p+1}$, the index $[\Gamma_1, \Gamma_1]$ in Γ_1 is equal to p^{p+2}.

Let us look for Γ_2. $\Delta = \Delta(\Gamma_2)$ is generated by $[\Gamma_2, \Gamma_2]$, x^p and y^p. Hence the index of $[\Gamma_2, \Gamma_2]$ in Δ is less then p^2 (it is easy to prove that it is equal to p^2). We have $[\Gamma_2 : [\Gamma_2, \Gamma_2]] \leq p^4$. It follows that $\Gamma_1$and Γ_2 are not isomorphic. □

References

[1] Bieberbach, L., Über die Bewegungsgruppen der Euklidischen Räume I. Math. Ann. 72 (1912), 400–412.

[2] Charlap, L., Compact flat Riemannian manifolds I. Ann. Math. 81 (1965), 15–30.

[3] Hiller, H., Sah, C.-H., Holonomy of flat manifolds with $b_1 = 0$. Quarterly J. Math. 37 (1986), 177–187.

[4] Hiller, H., Marciniak, Z., Sah, C.-H., Szczepański, A., Holonomy of flat manifolds with $b_1 = 0$, II. Quart. J. Math. 38 (1987), 213–220.

[5] Passman, D. S., The Algebraic Structure of Group Rings. Wiley Interscience, New York 1977.

[6] Zassenhaus, H., Beweis eines Satzes über diskrete Gruppen. Abh. Math. Semin. Univ. Hamburg 21 (1938), 289–312.

Groups with Pronormal Subgroups of Proper Normal Subgroups

Igor Ya. Subbotin

1991 Mathematics Subject Classification: 20F16

1. Introduction

A subgroup H of a group G is said to be ***pronormal*** if for every element x of G the subgroups H and H^x are conjugate in $\langle H, H^x \rangle$ (see [9]). Finite groups in which all subgroups are pronormal have been studied in [7]. The structure of locally solvable groups with this property was obtained in [6]. The groups in which all subgroups of each proper normal subgroup are pronormal (***NP-groups***) are a natural generalization of groups in which all subgroups are pronormal (***KS-groups***), and obviously any *NP*-group G satisfies the transitivity condition for normal subgroups (i.e. G is a ***T-group***). Solvable T-groups have been studied in [8], and we will use the well-known results contained there.

The aim of this article is to obtain a complete description of locally solvable *NP*-groups (Theorem 2). We will also describe locally solvable groups in which all non-invariant cyclic primary subgroups are abnormal (Theorem 3). This is a generalization of the main results from [3].

2. Some Preliminaries

We will use the following properties of pronormal subgroups.

Lemma 1 (see [9] and [6]). *Let H be a subgroup of a group G. Then the following statements hold:*

(1) *If H is pronormal in G, then H is pronormal also in every subgroup of G containing it.*

Offprint from
Infinite Groups 94, Eds.: de Giovanni/Newell

(2) *Let N be a normal subgroup of G contained in H. Then H/N is a pronormal subgroup of G/N if and only if H is a pronormal subgroup of G.*

(3) *If N is a normal subgroup of G and H is pronormal in G, then HN is pronormal in G.*

(4) *If H is pronormal and subnormal in G, then H is normal in G.*

(5) *If H is pronormal in G and N is a normal subgroup of G containing H, then $G = N \cdot N_G(H)$, where $N_G(H)$ is the normalizer of H in G.*

(6) *If H is pronormal in G, then $N_G(H) = N_G\big(N_G(H)\big)$.*

A subgroup A is said to be ***quasicentral*** in a group G if all subgroups of A are normal in G. The following lemma is the main result of [6].

Lemma 2. *A non-periodic locally solvable group in which all subgroups are pronormal is abelian. Moreover, a periodic locally graduated group G has all its subgroups pronormal if and only if it satisfies the following conditions:*

(1) *$G = A \leftthreetimes B$, where A is an abelian quasicentral Hall subgroup without involutions and B is a Dedekind group.*

(2) *The derived subgroup $G' = [G, G]$ is the direct product of A and the derived subgroup B' of B.*

(3) *Every Sylow $\pi(B)$-subgroup of G is a complement to A in G.*

The groups in which all subgroups of the derived subgroup are invariant in the group (***KI-groups***) have been studied in [10] and [11]. These groups are generalizations of solvable groups in which normality is a transitive relation (***solvable T-groups***) (see [4] and [8]). Another generalization of T-groups can be found in [5].

Lemma 3 (see [10]). *The derived subgroup of any KI-group is abelian.*

Recall that the automorphism s of the group G is called a ***power automorphism*** if $s(A) = A$ for every $A \leq G$. Power automorphisms have been studied in [2] and [8]. In particular:

Lemma 4. *The group of all power automorphisms of any group is abelian.*

The following properties of T-groups have been obtained in [8].

Lemma 5.

(i) *The lower central series of a T-group G terminates with $A = [G', G]$, i.e. $[A, G] = A$; also $|G' : A| = 1$ or 2.*

(ii) *The odd component of the last term N of the lower central series of a periodic solvable T-group is an abelian Hall quasicentral subgroup of G, and the factor group G/N is a direct product of primary groups (s-group);*

(iii) *A primary solvable T-group is abelian if $p \neq 2$, and if $p = 2$ it is a group of one of the following types:*
  (0) *Dedekind group;*
  (1) $G = (C \leftthreetimes \langle z \rangle) \times D$;
  (2) $G = (C \leftthreetimes \langle z \rangle) \times D$;
  (3) $G = (C \leftthreetimes \langle z, s \rangle) \times D$;
  (4) $G = (C \leftthreetimes K\langle z \rangle) \times D$;

  where C is a divisible abelian 2-group, and $C \neq 1$ in cases (1)–(3), $\exp D = 2$, *the element z has order* 2 *in case* (1) *and order* 4 *in cases* (2)–(4), $\langle z, s \rangle$ *is a quaternion group, K is a Prüfer* 2*-subgroup,* $[K, C] = 1$, $z^2 \in K$ *and* $t^z = t^{-1}$ *for any* $t \in C \times \langle z^2 \rangle$ *in cases* (1)–(2), $t \in C \times \langle z^2, s \rangle$ *in case* (3), *and* $t \in C \times K$ *in case* (4).

The following theorem is interesting in connection with [10] and [11].

Theorem 1. *Let G be a locally solvable group such that all subgroups of the derived subgroup G' are pronormal in G. Then all subgroups of G' are normal in G (G is a KI-group).*

Proof. Let G be a locally solvable group in which all subgroups of G' are pronormal. By Lemma 2, G is a solvable group. The derived subgroup G' is a group in which all subgroups are pronormal (KS-group). If G' is not periodic, then G' is abelian (by Lemma 2). All subgroups of G' are pronormal and subnormal in G in this case. By point (4) of Lemma 1, all of them are normal in G.

Let G' be a periodic group. By Lemma 2, $G' = A \leftthreetimes B$, where A is an abelian Hall quasicentral subgroup of G' without involutions, and B is a Dedekind group. It is easy to show that the derived subgroup of the factor group $\bar{G} = G/A$ coincides with $G'/A \simeq B$, and it is a Dedekind group whose subgroups are all pronormal in $\bar{G}$. By point (4) of Lemma 1, the group $\bar{G}$ is a KI-group. By Lemma 3, the derived subgroup of $\bar{G}$ is abelian. Then B is abelian. Since the subgroup A is quasicentral in G, then by Lemma 4 the factor group $G/C_G(A)$ is abelian and $C_G(A) \geq G'$. Thus $G' = A \times B$ is an abelian group and all its subgroups are subnormal in G. It follows that all subgroups of G' are normal in G. □

3. *NP*-Groups

Lemma 6. *Let G be a locally solvable NP-group. Then the derived subgroup G' of G is abelian.*

Proof. Let C be an arbitrary finitely generated subgroup of the group G. Then C is a solvable group in which all subgroups of the derived subgroup C' are pronormal. By Theorem 1, C is a *KI*-group. By Lemma 3, the derived subgroup of C is abelian. Let g_1 and g_2 be two arbitrary elements of G'. We can write each of them as finite product of commutators of elements of G. Therefore g_1 and g_2 belong to the derived subgroup of a finitely generated subgroup of G. Hence $[g_1, g_2] = 1$, and G' is abelian. □

Theorem 2.

(a) *A periodic locally solvable group G is an NP-group if and only if $G = A \leftthreetimes B$, where A is a quasicentral abelian Hall subgroup without involutions and B is a group of one of the following types:*

(1) *B is a Dedekind group;*

(2) *$B = C \leftthreetimes \langle z \rangle$, C is a non-trivial divisible abelian 2-subgroup, z is an element of order 2 or 4, $z^2 \in Z(G)$, $c^z = c^{-1}$ for each $c \in C$;*

(3) *$B = C \leftthreetimes K\langle z \rangle$, where C is a divisible abelian 2-subgroup, K is a Prüfer 2-subgroup, z is an element of order 4, $z^2 \in K$, $t^z = t^{-1}$ for each $t \in C \times K$. Moreover, the derived subgroup G' is the direct product of A and the derived subgroup B' of B, and any Sylow $\pi(B)$-subgroup of G is a complement to A in G.*

(b) *A non-periodic non-abelian locally solvable group G is an NP-group if and only if $G = A\langle b \rangle$, where $A \neq 1$ is an abelian non-periodic quasicentral subgroup, b is an element of order 2 or 4, and also $A^2 = A$, $a^b = a^{-1}$ for each $a \in A$.*

Proof. Case 1. *G is periodic.*

Necessity. First of all note that the group G is solvable by Lemma 6. Obviously, by point (4) of Lemma 1, G satisfies the transitivity condition for normal subgroups. By points (1)–(2) of Lemma 5, there exists an odd component A of the last term of the lower central series of G, and it is an abelian Hall subgroup without involutions, and also the factor group G/A is an s-group.

Let B be an arbitrary Sylow $\pi(G/A)$-subgroup of G, and $G_1 = A \leftthreetimes B$. We will show that $G_1 = G$. Assume $G_1 \neq G$. Then G_1 is a proper normal subgroup of the *NP*-group G. By definition, all subgroups of G_1 are pronormal in G. By point (5) of Lemma 1, $G = G_1 N_G(B) = A N_G(B)$. Obviously, there is no $\pi(A)$-element in G/G_1. Moreover,

$$G/G_1 = AN_G(B)/G_1 \simeq N_G(B)/N_G(B) \cap G_1.$$

Since B is the unique Sylow $\pi(G/A)$-subgroup of its normalizer $N_G(B)$ and $B \leq G_1 \cap N_G(B)$, then G/G_1 contains neither $\pi(A)$ nor $\pi(G/A)$-elements. It follows that $G/G_1 = 1$ and $G = G_1$. Hence $G = A \leftthreetimes B$, where B is an arbitrary Sylow $\pi(G/A)$-subgroup of G. Obviously B is an s-group, and all its Sylow p-subgroups for $p > 2$ are abelian, and the Sylow 2-subgroup B_2 of B either is a Dedekind group, or a group of one of the types (1)–(4) of point (3) of Lemma 5. Let $B_{2'}$ be the complement to B_2 in B. If $B_{2'} \neq 1$, then $A \leftthreetimes B_2$ is a proper normal subgroup of G, and all subgroups of $A \leftthreetimes B_2$ are pronormal in G. By Lemma 2 the group B_2 is a Dedekind group. But in this case B is a Dedekind group, and we obtain a group of type (1) of our theorem. If $D \neq 1$, there is a proper normal subgroup $\tilde{G}$ of G such that $G/\tilde{G} \simeq D$, and in the same way we obtain that G is a group of the required form. We have now to consider groups in which $D = B_{2'} = 1$. If $G = A \leftthreetimes (C \leftthreetimes \langle z, s\rangle)$, then $A \leftthreetimes (C \leftthreetimes \langle z\rangle)$ is a proper normal subgroup of G, and all its subgroups are pronormal in G. Hence all subgroups of $C \leftthreetimes \langle z\rangle$ are pronormal. This is a contradiction to Lemma 2. By point (3) of Lemma 5, G is a group of one of the types (2)–(3) of our theorem. The necessity has been proved in this case.

Sufficiency. If G is a group of type (1) of the theorem, our statement follows from Lemma 2.

If G is a group of types (2)–(3) of the theorem, we will prove that all proper normal subgroups of G are contained in the subgroup $Q = A \times M$, where M is the product of the divisible part of B by $\langle z^2\rangle$. Let g be an arbitrary element of some normal subgroup N of G. We can write $g = amz^t$, where $a \in A$, $m \in M$, $z^t \in \langle z\rangle$. If $\langle z^t\rangle = \langle z\rangle$, then for any $x \in G'$ we have $g^x = amz^{tx} = amz^t x^k$. Since $\langle x\rangle = \langle x^k\rangle$, then G' is contained in N, and in this case $N = G$. Hence, if N is a proper normal subgroup of G, we have $t \equiv 0 (\mathrm{mod} 2)$ and $N \leq Q$. It is easy to show that all subgroups of Q are normal in G. The sufficiency of the conditions of our theorem has been proved.

Case 2. *G is a non-periodic group.*

Necessity. Let G be a non-periodic locally solvable non-abelian group in which all subgroups of each proper normal subgroup are pronormal. Let x be an element of infinite order of G. By Lemma 6 the derived subgroup G' of G is a proper abelian normal subgroup, and all its subgroups are normal in G. If G' is a periodic group, we can write $G_2 = G' \leftthreetimes \langle x^2\rangle$, $G_3 = G' \leftthreetimes \langle x^3\rangle$. Then G_2 and G_3 are proper normal subgroups of G, and all subgroups of G_2 and G_3 are pronormal. By Lemma 2, this means that G_2 and G_3 are abelian. Hence $[x^2, G'] = [x^3, G'] = 1$ and $x = x^3 x^{-2} \in C_G(G')$. Then all elements of infinite order of G centralize G', and $C = C_G(G')$ is an abelian group with elements of infinite order. Since C is a proper abelian normal subgroup of G, then all subgroups of C are normal in G. For any element $b \in G \setminus C$, we have $x^b = x^{-1}$. In fact, if $x^b = x$ and y is an arbitrary element of G', then $y^b = y^t$ and $(xy)^b = xy^t = (xy)^k$, where $k = 1$ or -1. If $k = 1$, then $xy^t = xy$, and $y^t = y$, and $b \in C$. But $b \notin C$, and hence $k = -1$, $xy^t = x^{-1}y^{-1}$, $y^{b+1} = x^2$. But $|x| = \infty$ and $|y| < \infty$. This means that G' is not periodic.

If G' is not periodic, let x be an element of infinite order of G', and let b be an element of the set $G' \setminus C$, where $C = C_G(G')$. then $G'\langle b\rangle$ is a normal subgroup of G. If $G'\langle b\rangle \neq G$, then it is a *KS*-group with an element of infinite order, and $G'\langle b\rangle$ is abelian by Lemma 2. But $b \notin C$, and hence $G = G'\langle b\rangle$. If $x^b = x$, then for any $y \in G'$ we have $(xy)^b = (xy)^k = xy^t$. If $|y| < \infty$, then $(xy)^k = xy$. If $|y| = \infty$, and $y^b = y^{-1}$, then $(xy)^k = xy^{-1}$. If $|xy| < \infty$, then $(xy)^b = xy$, and we have a contradiction. If $|xy| = \infty$, then $k = 1$ or $k = -1$. But both cases are impossible. So, if $x^b = x$, then $b \in C$, a contradiction. Hence $x^b = x^{-1}$. Then $x^{b^2} = x$, $b^2 \in C$, and $g^b = g^{-1}$ for any $g \in G'$. Therefore $R = G'\langle b^2\rangle$ is abelian, and it is a proper normal subgroup of G. Hence all subgroups of R are normal in G, and $g^b = g^{-1}$ for any $g \in R$. Then $(b^2)^b = b^{-2}$, so that $b^4 = 1$. Thus $G = G'\langle b\rangle$, b has order 4 or 2 and $g^b = g^{-1}$ for any $g \in G'$. Then $[G', b] = G' = (G')^2$. The necessity of the conditions of the theorem has been proved.

Sufficiency. Let G be a group satisfying the conditions of the second part of the statement. Let N be a proper normal subgroup of G. Let $n = gb_1 \in N$, $g \in G'$, $b_1 \in \langle b\rangle$, $a \in G' \setminus N$, $a^b = a^{-1}$. Then

$$n^a = (gb_1)^a = ga^{-1}b_1a = gb_1b_1^{-1}a^{-1}b_1a = gb_1[b_1, a].$$

It follows that $[b_1, a]$ belongs to N. If $[b_1, a] = 1$, then $b_1 \in \langle b^2\rangle$ and $n \in G'\langle b^2\rangle$. If $[b_1, a] \neq 1$, then $[b_1, a] = a^2$ and $a^2 \in N$. So, for any $d \in G'$, $d^2 \in N$. But $(G')^2 = G'$, and this means that $G' \leq N$. This argument shows that all proper normal subgroups of G are contained in $G'\langle b^2\rangle$. Hence the sufficiency of the theorem in Case 2 is proved, and our theorem is completely proved. □

Corollary 1. *A solvable non-KS-group G is an NP-group if and only if it has an abelian quasicentral subgroup C of index 2 such that $[G, C] = C$. Moreover, in such a case the 2-component of C is divisible and every element of C whose order either is infinite or a power of 2 is mapped by conjugation with the elements of G into its inverse.*

Proof. The necessity follows from Theorem 2.

Sufficiency. Let N be a proper normal subgroup of G which is not contained in C. Then $G = NC$ and $C = [G, C] \leq G' \leq N \cap C$, a contradiction. Furthermore it is also clear that C/C^2 is a central section, and hence $C = C^2$. The corollary is proved. □

4. Groups with Abnormal Non-Normal Primary Cyclic Subgroups

A subgroup D is called an ***abnormal subgroup*** of a group G if $x \in \langle D, D^x\rangle$ for every $x \in G$ (see for instance [9]). Finite groups in which all non-normal subgroups are abnormal have been studied in [3]. The following corollary of Theorem 2 is a generalization of [3].

Corollary 2. *A non-periodic locally solvable group in which all non-normal subgroups are abnormal is abelian. Moreover, a locally graduated periodic group G has the property that all its non-normal subgroups are abnormal if and only if $G = A \leftthreetimes \langle b \rangle$, $A = G'$, $2 \notin \pi(A)$, $\pi(A) \cap \pi(G/A) = \emptyset$, all subgroups of A are normal in G and $b^p \in Z(G)$.*

The next theorem is a generalization of this corollary.

Theorem 3.

(1) *A periodic locally graduated non-Dedekind group G has the property that all its non-normal cyclic primary subgroups are abnormal if and only if $G = A \leftthreetimes \langle b \rangle$, where A is an abelian Hall subgroup without involutions, $A = G'$, all subgroups of A are normal in G, $\langle b \rangle$ is a cyclic p-subgroup and $b^p \in Z(G)$.*

(2) *A non-periodic locally solvable group G has the property that all its non-normal cyclic primary subgroups are abnormal if and only if G has a periodic part that is a quasicentral subgroup of G.*

Proof. Case 1. G is a periodic locally graduated group.

Necessity. Let G be a periodic locally graduated group in which all non-normal cyclic primary subgroups are abnormal. Hence all primary cyclic subgroups of G are pronormal. By Theorems 2 and 3 of [13], G is a solvable T-group. By Theorem 6.1.1 of [8] G is an extension of an abelian Hall quasicentral subgroup A without involutions by a Dedekind group B. If all p-subgroups of G are normal, where p is any prime in $\pi(B)$, then all primary subgroups of G are normal (A is a Hall subgroup of G), and G is a Dedekind group. Suppose that for some prime $p \in \pi(B)$ there exists in G a cyclic non-normal p-subgroup $\langle b \rangle$. Then $\langle b \rangle$ is abnormal in G. Since the factor group G/A is a Dedekind group, then the subgroup $A \leftthreetimes \langle b \rangle$ is normal in G. Therefore $G = A \leftthreetimes \langle b \rangle$. By Theorem 6.1.1 of [8] the subgroup A is the last member of the lower central series of G. Hence $A = G'$ (G/A is cyclic). Assume that $b^p \notin Z(G)$. then $\langle b^p \rangle$ is an abnormal subgroup of G, and $G = A \leftthreetimes \langle b^p \rangle$. This contradiction shows that $b^p \in Z(G)$. The necessity of the conditions of our theorem in Case 1 has been proved.

Sufficiency. Let $\langle x \rangle$ be a cyclic q-subgroup of G (q prime). If $q \in \pi(A)$, then $\langle x \rangle$ is normal in G. Suppose that $q \notin \pi(A)$, so that $q = p$. If $x \in Z(G)$, obviously $\langle x \rangle \lhd G$. Let x be a non-central element of G, and consider the subgroup $G_0 = A \leftthreetimes \langle x \rangle$. If $b \in G_0$, then $G = G_0$. Let $b \notin G_0$. If g_0 is any element of G, then $g_0 = a_0 b^t$, where $a_0 \in A$, $b^t \in \langle b \rangle$. If $(t, p) = 1$, then $b^t \in G_0$ and $b \in G_0$, that is impossible. Hence $(t, p) = p$. But in this case $[a, b^t] = 1$ for every $a \in A$, and G_0 is abelian. Write $x = a_x b^y$, where $a_x \in A$, $b^y \in \langle b \rangle$, $[b^y, A] = 1$ and $(y, p) = p$. Hence $x^b = a_x^b b^y = a_x^z b^y$. Since $(|a_x|, |b^y|) = 1$, $[b^y, a_x] = 1$, then there exists a number k such that $x^b = x^k = (a_x b^y)^k$. Therefore $\langle x \rangle \lhd G$. Let $G = A \leftthreetimes \langle x \rangle$. In this case

$[A, \langle x\rangle] = A$ $(A = G')$. By [12] $A \cap Z(G) = 1$. Let g be an arbitrary element of G. Then $g = x^n a$, where $x^n \in \langle x\rangle$, $a \in A$, and $x^g = x^{x^n a} = x^a$. Since $\langle a\rangle \cap Z(G) = 1$, then $[x, a] = a^m$, $(m, |a|) = 1$. But $[x, a] \in \langle\langle x\rangle, \langle x^g\rangle\rangle$. Hence $a \in \langle\langle x\rangle, \langle x^g\rangle\rangle$, i.e. $g = ax^n \in \langle\langle x\rangle, \langle x^g\rangle\rangle$, and $\langle x\rangle$ is abnormal in G. The first part of our theorem has been proved.

Case 2. G is a non-periodic locally solvable group.

The sufficiency of the conditions of the theorem is evident in this case.

Necessity. If all primary cyclic subgroups of G are normal, then our theorem is evident. Let $\langle a\rangle$ be an abnormal cyclic p-subgroup (p prime) of G. Then $\langle a\rangle = N_G(\langle a\rangle)$ and each subgroup of G containing $\langle a\rangle$ is abnormal in any subgroup of G in which is contained. We will show that in this case $\langle a\rangle \neq N_G(\langle a\rangle)$. Since $\langle a^p\rangle \neq N_G(\langle a^p\rangle)$, then $\langle a^p\rangle \lhd G$. Obviously the subgroup $\langle a\rangle/\langle a^p\rangle$ is an abnormal subgroup of $G/\langle a^p\rangle$. Hence, without loss of generality it can be assumed that $|a| = p$. Let g be an element of infinite order of G, and put $A = \langle a, g\rangle$. Since G is a locally solvable group, there exists in A a proper normal subgroup of finite index N. Since A is a finitely generated group and N has finite index in A, then N is a finitely generated group. Since $\langle a\rangle$ is an abnormal subgroup of A, then $a \notin N$. It is also clear that every cyclic primary subgroup of N is normal in G. Therefore N has a periodic part $t(N)$, and all subgroups of $t(N)$ are normal in G. Since $N_G(\langle a\rangle) = \langle a\rangle$, then a induces a regular automorphism on N. If $t(N) = 1$, then N is a solvable torsion-free group, and without loss of generality it can be assumed that g belongs to an abelian subgroup A_1 of N which is normal in A. Let M be the normal closure of $\langle g\rangle$ in A, and let $G_1 = M \rtimes \langle a\rangle$. Obviously M is a free abelian group of finite rank. Then $M^p \lhd G_1$ and M/M^p is a finite non-trivial p-subgroup. Hence G_1/M^p is a finite p-group. It follows that $\langle M^p, a\rangle/M^p$ is properly contained in its normalizer in G/M^p, that is impossible. The case when $t(N) \neq 1$ can be reduced to the above situation by considering the groups $\bar{N} = N/t(N)$ and $\bar{A} = A/t(N)$. The theorem has been proved. □

From Theorem 2 of [1] and our theorem we obtain the following result.

Corollary 3. *The class of locally finite groups in which all cyclic primary subgroups are pronormal coincides with the class of locally finite groups in which all abelian subgroups which are not self-normalizing are normal.*

5. Appendix

We prove now the following result.

Theorem 4. *Let A be some subgroup property. Then the following conditions are equivalent for solvable groups:*

(1) *each non-A subgroup supplements the derived subgroup of the group;*

(2) *each subgroup of each proper normal subgroup of the group is an A-subgroup.*

Proof. Let G be a solvable group with the property (1), and let N be an arbitrary proper normal subgroup of G. If some subgroup M of N is not an A-subgroup, then the subgroup $G'M$ coincides with G, and in particular $G = G'N$. Clearly $D = G' \cap N$ is a normal subgroup of G, and $G/D = G'/D \times N/D$. Since N/D is abelian, it follows that $(G/D)' = (G'/D)'$. But G is a solvable group, so that $G'/D = 1$ and $G' = D$. Then $G' \leq N$ and $N = G$. This contradiction proves that the condition (2) follows from the condition (1).

Let now G be a solvable group with the property (2), and let B be a non-A subgroup of G. If $N = G'B$ is a proper normal subgroup of G, then all subgroups of N are A-subgroups. But B is not an A-subgroup, so that $G = N = G'B$. The theorem is proved. □

Corollary 4. *Let G be a locally solvable group. Then every non-pronormal subgroup of G supplements the derived subgroup G′ if and only if G is a group of one of the types of Theorem* 2.

References

[1] Chernikov, N. S., Groups in which all abelian subgroups which are different from their normalizers are invariant. In: Studies in Group Theory, 117–127, Nauka Dumka, Kiev, 1978.

[2] Cooper, C. D. H., Power automorphisms of a group. Math. Z. 107 (1968), 335–356.

[3] Fattahi, A., Groups with only normal and abnormal subgroups. J. Algebra 28 (1974), 15–19.

[4] Gaschütz, W., Gruppen in denen das Normalteilersein transitiv ist. J. Reine Angew. Math. 198 (1957), 87–92.

[5] de Giovanni, F., Franciosi, S., Groups in which every infinite subnormal subgroup is normal. J. Algebra 96 (1985), 566–580.

[6] Kuzennyi, N. F., Subbotin, I. Ya., Groups in which all subgroups are pronormal. Ukrain. Math. J. 39 (1987), 251–254.

[7] Peng, T. A., Finite groups with pro-normal subgroups. Proc. Amer. Math. Soc. 20 (1969), 232–234.

[8] Robinson, D. J. S., Groups in which normality is a transitive relation. Proc. Cambridge Philos. Soc. 60 (1964), 21–38.

[9] Shemetkov, L. A., Formations of Finite Groups. Nauka, Moscow 1978.

[10] Subbotin, I. Ya., Infinite *KI*-groups with Chernikov part of the commutant. In: Investigations of Groups with Given Properties of Systems of Subgroups, 79–93, Kiev 1981.

[11] —, On the hypercentral coradical of a *KI*-group. Ukrain. Math. J. 34 (1982), 532–536.

[12] —, Groups decomposable into a quasicentralizer product. Ukrain. Math. J. 37 (1985), 529–531.

[13] —, Kuzennyi, N. F., Locally soluble groups in which all infinite subgroups are pronormal. Soviet Math. 32 (1988), 126–131.

Some Examples of Factorized Groups and Their Relation to Ring Theory

Yaroslav P. Sysak

1991 Mathematics Subject Classification: 20D40, 20F19

1. Introduction

Let G be a group, R a commutative ring with 1 and let M be a right RG-module. A map $d : G \to M$ is called a derivation from G to M if $d(gh) = d(g)h + d(h)$ for all g and h in G. We will say that the module M is radical if there exists a surjective derivation d from G onto M. Natural examples of radical modules can be obtained as follows.

Let J be a radical ring in the sense of Jacobson [Ja, p. 4]. Then J forms a group under the operation $a \circ b = a + b + ab$ for all a and b in J. This group is called the adjoint group J° of J. The additive group J^+ of J becomes a $\mathbb{Z}J^\circ$-module via the rule $a^b = a + ab$ for all a and b in J. The identity map $J \to J$ is a bijective derivation from J° onto J^+. Therefore, the $\mathbb{Z}J^\circ$-module J^+ is radical, and thus we can consider radical modules in analogy to radical rings.

The next theorem describes such modules from two standpoints, that of the ring theory and that of the group theory, and motives their study.

Theorem 1. *Let G be a group and let M be a right RG-module. Then the following statements are equivalent:*

(1) *the module M is radical;*

(2) *the module M as a right RG-module is isomorphic to the module $\Delta_R(G)/\rho$, where $\Delta_R(G)$ is the augmentation ideal of the group ring RG and ρ is a right ideal of RG such that*

$$\Delta_R(G) = G - 1 + \rho;$$

Offprint from
Infinite Groups 94, Eds.: de Giovanni/Newell

(3) *in the semidirect product $M \rtimes G$ of the module M and the group G there exists a subgroup H such that*

$$M \rtimes G = M \rtimes H = GH.$$

Moreover, if d is a surjective derivation from G onto M, then the right ideal ρ and the subgroup H can be chosen such that

$$\operatorname{Ker} d = G \cap (1 + \rho) = G \cap H.$$

Of course, if the group G is abelian, then ρ is an ideal of RG and the factor ring $\Delta_R(G)/\rho$ is therefore radical. In general, the right ideal ρ however need not be an ideal.

The basic problem, which is considered here, refers to studying the connections between the structure of a group G and that of a radical RG-module M. In the case of a soluble or even hyperabelian group G these connections can be very tight. The book [AFG] contains a wealth of material on this subject in the language of triply factorized groups. In this connection, we cite the following results.

Theorem 2. *Let G be a hyperabelian group and let M be a radical RG-module. If G is*

(1) *a p-group or*

(2) *a group of finite abelian section rank or*

(3) *a group of finite torsion-free rank,*

then M regarded as a group has the same property respectively.

Recall that G is a group of finite abelian section rank if every elementary abelian section of G is finite; and the group G has finite torsion-free rank if it has a finite series whose factors are either periodic or infinite cyclic.

Recall also that the statements (1) and (2) of Theorem 2 were proved by N. S. Chernikov [Ch] and J. Wilson [Wi], respectively, and independently by myself in [S1] and [S2]. The statement (3) has been recently obtained in [S3].

In the general case, where the group G is not hyperabelian, the statements (1) and (2) of Theorem 2 do not hold. The first such examples were actually given in [S1]. This will be seen also from the following slightly more general result.

Theorem 3. *There exist a periodic locally soluble and residually finite group G and a radical RG-module M such that*

(1) *the additive group of M is an infinite elementary abelian q-subgroup for a prime q, and*

(2) *G has no elements of order q.*

Moreover, we can arrange that either

(3) *G is a p-group, where p is any odd prime if $q = 2$, and $p = 2$ if $q = 3$ or*

(4) *the Sylow p-subgroups of G are finite abelian for all odd primes p.*

From Theorem 1 and Theorem 3 we immediately obtain the next two results about triply factorized groups.

Corollary 1. *For each prime p there exists a locally finite group G of the form*

$$G = M \rtimes A = M \rtimes B = AB,$$

where A and B are residually finite p-subgroups of G and M is a normal elementary abelian q-subgroup with $q \neq p$. In particular, the group G is radical and it is not a p-group.

Recall that a group is radical if it has an ascending normal series with locally nilpotent factors.

Note that V. Sushchanskii [Su] also proved that, for each prime p, there exists a residually finite group G of the form $G = AB$ with locally finite p-subgroups A and B, which is not a p-group. But his group is non-periodic and, moreover, it is neither radical nor even locally soluble.

Corollary 2. *There exists a periodic locally soluble group G of the form*

$$G = M \rtimes A = M \rtimes B = AB$$

with an infinite normal elementary abelian 2-subgroup M and subgroups A and B of finite abelian section rank. In particular, the group G has infinite abelian section rank.

As was stated above, an arbitrary group of the form $G = AB$ with periodic subgroups A and B need not be periodic even if A and B are locally finite. The first such example was given by Suchkov [AFG, Theorem 3.1.2]. On the other hand, if the group G is radical and the subgroups A and B are hyperabelian, then G is periodic [AFG, Theorem 3.2.6]. Therefore, the following question is of interest: is every radical product of two periodic subgroups periodic? It is the part 1 of Question 6 from [AFG]. The positive answer to this question follows directly from the next our result.

Theorem 4. *Let the group $G = AB$ be the product of two locally finite subgroups A and B. Then the Hirsch–Plotkin radical of G is periodic.*

This theorem is indeed easy to deduce from our main result about radical modules over periodic groups.

Theorem 5. *Let G be a group in which every two-generated subgroup is finite and let M be a radical $\mathbb{Z}G$-module. Then the additive group of M is periodic.*

We remark that no example seems to be known of a radical $\mathbb{Z}G$-module M such that G is periodic but the additive group of M is not periodic.

The following result about unitary matrices over the field $\mathbb{C}$ of complex numbers will play a decisive role in the proof of Theorem 5.

Proposition. *Let A and B be two unitary $m \times m$ matrices over $\mathbb{C}$ which have no common non-zero fixed vector. Then the matrix $B - nA + (n-1)E$ is nonsingular for every $n > 1$.*

Here, E denotes the identity $m \times m$ matrix.

One further application of Proposition is the next statement about the augmentation ideal $\Delta_{\mathbb{Q}}(G)$ of the group algebra $\mathbb{Q}G$ of a finite group G over the field $\mathbb{Q}$ of rational numbers.

Corollary 3. *Let $G = \langle g, h \rangle$ be a finite group with generators g and h. Then*

$$\Delta_{\mathbb{Q}}(G) = (h - ng + n - 1)\mathbb{Q}G$$

for all $n > 1$.

By means of Proposition we also get the following results.

Theorem 6. *Let M be an abelian group and let G be a group which acts on M such that M is partitioned into finitely many orbits under G. If every two-generated subgroup of G is finite, then the group M is periodic.*

Before formulating our last theorem we recall that a transitive permutation group G on a set X is a group of finite permutation rank if the stabilizer of a point in G has only finitely many orbits on X.

Theorem 7. *Let G be a transitive permutation group of finite permutation rank on a set X and let H be the stabilizer of a point x of X in G. If every two-generated subgroup of H is finite, then the Fitting subgroup of G is periodic. Moreover, if H is locally finite, then the Hirsch–Plotkin radical of G is also periodic.*

From Theorem 7 we deduce

Corollary 4. *Let G be a radical group of the form $G = ABA$ with periodic subgroups A and B. If at least one of these subgroups is finite, then the group G is periodic.*

We do not know whether the statement of Corollary 4 holds if the subgroups A and B are both infinite.

Our notation is standard as in [Ro].

2. Proof of Theorem 1

First, we will prove that the statement (2) follows from (1). Let M be a radical RG-module and let $d : G \to M$ be a surjective derivation from G onto M. Since the augmentation ideal $\Delta_R(G)$ is a free R-module with the free generating set $\{g-1 \mid g \in G\}$, the map $g-1 \mapsto d(g)$ of this set onto the R-module M can be uniquely extended to an R-homomorphism $\widehat{d} : \Delta_R(G) \to M$ of $\Delta_R(G)$ onto M. Furthermore, for all g and h of G, we have $\widehat{d}((g-1)h) = \widehat{d}((gh-1)-(h-1)) = \widehat{d}(gh-1) - \widehat{d}(h-1) = d(gh) - d(h) = d(g)h = \widehat{d}(g-1)h$. Therefore, $\widehat{d}$ is a homomorphism of the right RG-module $\Delta_R(G)$ onto the right RG-module M. Consequently, the kernel $\operatorname{Ker}\widehat{d}$ of $\widehat{d}$ is a right ideal of the group ring RG contained in $\Delta_R(G)$ and the RG-modules $\Delta_R(G)/\operatorname{Ker}\widehat{d}$ and M are isomorphic.

Let x be an arbitrary element of $\Delta_R(G)$. The derivation d being surjective, we have $\widehat{d}(x) = d(g) = \widehat{d}(g-1)$ for an element g of G whence $\widehat{d}(x-g+1) = 0$. But then $x-g+1$ belongs to $\operatorname{Ker}\widehat{d}$ and hence x is contained in the set $G-1+\operatorname{Ker}\widehat{d}$. Therefore, $\Delta_R(G) = G-1+\operatorname{Ker}\widehat{d}$ and obviously $G \cap (1+\operatorname{Ker}\widehat{d}) = \operatorname{Ker} d$. Thus, setting $\rho = \operatorname{Ker}\widehat{d}$ we obtain the statement (2).

Next we will prove that the statement (3) follows from (2). Let the RG-module M be isomorphic to the RG-module $\Delta_R(G)/\rho$, where ρ is a right ideal of RG such that $\Delta_R(G) = G-1+\rho$. We use the multiplicative notation for the operation on M, and if m is an element of M and g is an element of RG, then we write m^g for the image of m under g. Let $S = M \rtimes G$ be the semidirect product of M by G. Denote by H the subset $\{gm_g \mid g \in G, m_g \in M\}$ of S where $m_g = g-1+\rho$. Since for all g and h of G we have $gm_g \cdot hm_h = (gh)(m_g^h m_h) = (gh)m_{gh}$ because $m_g^h m_h = (g-1+\rho)h+(h-1+\rho) = gh-1+\rho = m_{gh}$, the subset H is a subgroup of S and it is immediately verified that $M \cap H = 1$ and $H \cap G = G \cap (1+\rho)$.

Let gm be an arbitrary element of S, where $g \in G$ and $m \in M$. As $\Delta_R(G) = G-1+\rho$, we have $m = h-1+\rho = m_h$ for some $h \in G$. Hence $gm = gm_h = (gm_g)(m_g^{-1}m_h)$ belongs to $M \rtimes H$, because $gm_g \in H$ and $m_g^{-1}m_h \in M$. On the other hand, $gm = gm_h = (gh^{-1})(hm_h)$ belongs to GH because $gh^{-1} \in G$ and $hm_h \in H$. Consequently, $S = M \rtimes H = GH$ and thus we obtain the statement (3).

Finally, we will show that the statement (1) follows from (3). For convenience, we continue to use the multiplicative notation for the operation on M.

In the semidirect product $S = M \rtimes G$ of the RG-module M by the group G, let there be a subgroup H such that $S = M \rtimes H = GH$. Then, for each element g of G, there exists a unique element m of M such that gm belongs to H. Define a map $d : G \to M$ by $d(g) = m$. If h is an element of G and $d(h) = n$, then hn is contained in H and therefore $(gm)(hn) = (gh)(m^h n)$ belongs to H. Hence $d(gh) = d(g)^h d(h)$ and thus d is a derivation of G into M. Next, if m is an arbitrary element of M, then $m = g^{-1}h$ for certain elements g of G and h of H. Therefore, $gm = h$ belongs to H and consequently $d(g) = m$. Hence the derivation d is surjective, and thus the module M is radical. Moreover, since each element h of H can be written in the form $h = gm$

with g of G and m of M and, by definition, $m = d(g)$, we have $H = \{gd(g) \mid g \in G\}$ and $G \cap H = \operatorname{Ker} d$. This proves the theorem.

3. Proof of Theorem 3

We begin with an exposition of a general construction.

Let R be a finite homomorphic image of the ring $\mathbb{Z}$ and let $C_1, C_2, \dots, C_n, \dots$ be an infinite sequence of finite groups satisfying, for each $n \geq 1$, the following condition:

($*$) in the group ring RC_n there exist linearly independent elements x_n and y_n such that $x_n + y_n R$ is contained in $x_n C_n$.

Denote by A_n the standard wreath product $(\cdots (C_n \wr C_{n-1}) \cdots \wr C_2) \wr C_1$. For each $n \geq 1$ we will construct a suitable RA_n-module M_n as follows.

Let $A_{n0} = C_n$ and $A_{ns} = A_{ns-1} \wr C_{n-s}$ for $1 \leq s \leq n-1$. Obviously $A_{nn-1} = A_n$. Denote by B_{ns} the base group of the wreath product A_{ns}. We will first construct certain RA_{ns}-modules M_{ns} by induction.

If $s = 0$, then we put $M_{n0} = x_n RC_n$. Suppose that we have already constructed the RA_{ns-1}-module M_{ns-1}. We can consider this module as an RB_{ns}-module if we assume that, for every non-identity element c of C_{n-s}, the subgroup A_{ns-1}^c of B_{ns} acts trivially on M_{ns-1}. Then the induced module $M_{ns-1} \otimes_{RB_{ns}} RA_{ns}$ (see [Ro, p. 231]) is the required RA_{ns}-module M_{ns}. We take now the module M_{nn-1} as an RA_n-module M_n .

If we denote by T_n the subset $C_n C_{n-1} \dots C_1$ of A_n, then it is easy to see that the RA_n-module M_n can be written in the form

$$M_n = \sum_{c \in T_n} (x_n R)c.$$

We put $u_n = \sum_{c \in T_{n-1}} x_n c$ and $V_n = y_n RA_{n-1}$.

Lemma 1. *For each $n \geq 1$, we have*

(1) $u_n + V_n$ *is contained in* $u_n A_n$ *and*

(2) $u_{n+1}(A_n - 1) = 0$.

Proof. (1) Let v be an arbitrary element of V_n. We have to prove that $u_n + v$ belongs to $u_n A_n$. Since V_n lies in M_n, we get $V_n = y_n RA_{n-1} = \sum_{c \in T_{n-1}} (y_n R)c$. Therefore, $v = \sum_{c \in T_{n-1}} r_c y_n c$ for some elements r_c of R. Hence $u_n + v = \sum_{c \in T_{n-1}} x_n c + \sum_{c \in T_{n-1}} r_c y_n c = \sum_{c \in T_{n-1}} (x_n + r_c y_n)c$. But, by the condition ($*$), $x_n + r_c y_n = x_n a_c$ for an element a_c of C_n. Consequently, $u_n + v = \sum_{c \in T_{n-1}} (x_n a_c)c = \sum_{c \in T_{n-1}} (x_n c)a_c^c$. In accordance with the construction of the module M_n we have $(x_n c)a_b^b = x_n c$ if

$b \neq c$. Therefore, $u_n + v = \sum_{c \in T_{n-1}} (x_n c) \left(\prod_{c \in T_{n-1}} a_c^c \right) = u_n \left(\prod_{c \in T_{n-1}} a_c^c \right)$ belongs to $u_n A_n$, as required.

(2) It suffices to show that $u_{n+1} a = u_{n+1}$ for each a of C_{i+1}, $0 \leq i \leq n-1$. We put $T_{ni} = C_n \dots C_{i+1}$ for $1 \leq i \leq n-1$. Then $T_n = T_{ni} T_i$ and $u_{n+1} = \sum_{c \in T_n} x_n c = \sum_{c \in T_i} \left(\sum_{b \in T_{ni}} x_{n+1} b \right) c$. Therefore, $u_{n+1} a = \sum_{c \in T_i} \left(\sum_{b \in T_{ni}} x_{n+1} b \right) ca = \sum_{c \in T_i} \left(\sum_{b \in T_{ni}} x_{n+1} b \right) (cac^{-1}) c$. But in case $c \neq 1$, then, as we already know, $\left(\sum_{b \in T_{ni}} x_{n+1} b \right) (cac^{-1}) = \sum_{b \in T_{ni}} x_{n+1} b$. If $c = 1$, we have $\left(\sum_{b \in T_{ni}} x_{n+1} b \right) a = \sum_{b \in T_{ni}} x_{n+1} ba = \sum_{b \in T_{ni}} x_{n+1} b$ because $T_{ni} a = T_{ni}$ for each a of C_{i+1}. Thus $\left(\sum_{b \in T_{ni}} x_{n+1} b \right) (cac^{-1}) = \sum_{b \in T_{ni}} x_{n+1} b$ and hence $u_{n+1} a = u_{n+1}$, which is what had to be proved. □

Next, the mapping $x_n \to y_{n+1}$ determines an isomorphism from the RA_n-module M_n into the RA_n-module V_{n+1} and defines therefore an RA_n-isomorphism $\varphi_n : M_n \to M_{n+1}$ with $M_n^{\varphi_n} \leq V_{n+1}$.

Finally, let M be the direct limit

$$M = \lim_{\rightarrow} (M_n, \varphi_n)$$

of the modules M_n and their isomorphisms φ_n, and let A be the union

$$A = \bigcup_{n=1}^{\infty} A_n$$

of the groups A_n. Clearly, A is a locally finite and residually finite group and M is an RA-module.

Theorem 3A. *The RA-module M is radical.*

Proof. We put $v_1 = u_1$ and $v_{n+1} = u_{n+1} + v_n^{\varphi_n}$ for every $n \geq 1$. Since $M_n^{\varphi_n}$ lies in V_{n+1}, we obtain, by Lemma 1, that $v_{n+1} + M_n^{\varphi_n}$ is contained in $v_{n+1} A_{n+1}$ and $(v_{n+1} - v_n^{\varphi_n})(A_n - 1) = 0$. Now, for an arbitrary element a of A, we take the least number n such that a belongs to A_n and we define a function $d : A \to M$ via $d(a) = v_n(a-1)$. We will show that d is a derivation from A onto M. Note that if $m \geq n$, then $v_m(a-1) = v_n(a-1)$ for every a of A_n. This follows from the equality $(v_{n+1} - v_n^{\varphi_n})(A_n - 1) = 0$ observed above. Therefore, if b is an element of A and $d(b) = v_m(b-1)$, then denoting the largest of the integers m and n by l, we get $d(ab) = v_l(ab-1) = v_l((a-1)b + (b-1)) = v_l(a-1)b + v_l(b-1) = d(a)b + d(b)$. Hence d is a derivation from A into M. Let x be an arbitrary element of M. Then x belongs to $M_n^{\varphi_n}$ for a certain $n \geq 1$. As was stated above, $v_{n+1} + M_n^{\varphi_n}$ is contained in $v_{n+1} A_{n+1}$. Therefore, $v_{n+1} + x = v_{n+1} a$ for a certain a of A_{n+1}. This implies that $x = v_{n+1} a - v_{n+1} = v_{n+1}(a-1) = d(a)$ and thus d is a surjective derivation from A onto M. The theorem is proved. □

We consider now some special cases of Theorem 3A.

Examples. 1. Let R be the field F_2 of two elements and, for every $n \geq 1$, let C_n be the cyclic group of prime order $p \neq 2$. If we take any nontrivial element c_n in C_n and put $x_n = 1$ and $y_n = c_n - 1$, then the condition $(*)$ holds trivially. Therefore, the group A is a p-group and M is an infinite radical F_2A-module.

2. Let R be the field F_3 of three elements and, for every $n \geq 1$, let C_n be the cyclic group of order 8 with a generator c_n. We put

$$x_n = -1 + c_n^2 + c_n^3 + \cdots + c_n^7 \quad \text{and} \quad y_n = -1 + c_n^2.$$

It is easily verified that the condition $(*)$ is again valid. Hence, the group A is a 2-group and M is an infinite radical F_3A-module.

3. Let $R = F_2$ and let C_n be the cyclic group of order $2^{t_n} - 1$ where $t_1 = 2$ and $t_n = t_1 \dots t_{n-1} + 1$ for all $n \geq 2$. We put $x_n = 1$ and $y_n = c_n - 1$ for any nontrivial element c_n of C_n. Once again the condition $(*)$ is valid. But now the group A is locally soluble with only finite abelian Sylow p-subgroup for all odd primes p, and M is an infinite radical F_2A-module.

It immediately follows from Examples 1–3 that Theorem 3 is proved.

4. Proposition and Its Applications

Proof of Proposition. Since each unitary matrix over the field $\mathbb{C}$ is conjugate by certain unitary matrix to a diagonal matrix, we can assume without any restriction of generality that the matrix A is diagonal. Let $A = \operatorname{diag}(a_1, \dots, a_m)$. Then $|a_s| = 1$ and the matrix $D = nA - (n-1)E$ is diagonal with nonzero elements $na_s - n + 1$ on the main diagonal, $1 \leq s \leq m$.

Suppose that the matrix $B - D$ is singular. Then there exists a nonzero vector $x = (x_1, \dots, x_m)$ such that $x(B - D) = 0$ or $xB = xD$. Applying to the last equality the complex conjugate transpose, which is denoted by $*$, we obtain $B^*x^* = D^*x^*$. Since the matrix B is unitary and the matrix D is diagonal, we get $B^* = B^{-1}$ and $D^* = \overline{D}$, where the bar denotes the complex conjugate. Therefore, $B^{-1}x^* = \overline{D}x^*$ and hence $xB \cdot B^{-1}x^* = xD \cdot \overline{D}x^*$. A straightforward calculation yields the equality

$$|x_1|^2 + \cdots + |x_m|^2 = |na_1 - n + 1|^2|x_1|^2 + \cdots + |na_m - n + 1|^2|x_m|^2.$$

Now we will show that this equality holds provided that $|na_s - n + 1|^2 = 1$ for each s with $x_s \neq 0$. Obviously, it suffices to show that $|na_s - n + 1|^2 \geq 1$ for every s.

Since $|a_s| = 1$, we have $a_s = \cos\varphi + i\sin\varphi$ for an argument φ, where $0 \leq \varphi < 2\pi$. Therefore, $na_s - n + 1 = (n\cos\varphi - n + 1) + in\sin\varphi$ and hence $|na_s - n + 1|^2 = (n\cos\varphi - n + 1)^2 + n^2\sin^2\varphi = (2n^2 - 2n + 1) - (2n^2 - 2n)\cos\varphi \geq (2n^2 - 2n +$

$1) - (2n^2 - 2n) = 1$. Moreover, if $|na_s - n + 1|^2 = 1$, then $\cos\varphi = 1$ whence $\varphi = 0$. Consequently, in the last case $a_s = 1$.

Thus, if $x_s \neq 0$, then $a_s = 1$. We can assume, without any loss of generality, that $x_1, \ldots, x_r$ are nonzero and $x_{r+1} = \cdots = x_m = 0$ for certain r, $1 \leq r \leq m$. Then $a_1 = \cdots = a_r = 1$ and hence $A = \mathrm{diag}(1, \ldots, 1, a_{r+1}, \ldots, a_m)$ and $D = \mathrm{diag}(1, \ldots, 1, na_{r+1} - n + 1, \ldots, na_m - n + 1)$. But then $xA = xD = x$ whence $xB = x$. Therefore, x is a common nonzero fixed vector for the matrices A and B, which contradicts the condition of Proposition. This completes the proof. □

Lemma 2. *Let G be a group in which every two-generated subgroup is finite and let M be a $\mathbb{Z}G$-module with torsion-free additive group. If x and y are linearly independent elements of M, then, for each $n > 1$, at least one of the elements $x + y$ or $x + ny$ is not contained in the set xG.*

Proof. Suppose the contrary and let g and h be elements in G such that $x + y = xg$ and $x + ny = xh$ for certain $n > 1$. We put $H = \langle g, h\rangle$, and let U be a $\mathbb{Z}H$-submodule in M generated by the elements x and y. Since the subgroup H is finite, the submodule U, regarded as a group, is finitely generated. Therefore, the tensor product $V = U \otimes_{\mathbb{Z}} \mathbb{C}$ is a $\mathbb{C}H$-module of finite dimension over the field $\mathbb{C}$. As well known, every representation of a finite group over $\mathbb{C}$ is unitary. Hence, we can consider V as a vector space with a scalar product and H as a group of unitary operators of this space. By Maschke's theorem, the module V is the direct sum $V = V_1 \oplus \cdots \oplus V_m$ of irreducible $\mathbb{C}H$-modules V_i, $1 \leq i \leq m$. Evidently, if x_i and y_i are components of the elements x and y in the submodule V_i, then $x_i + y_i = x_i g$ and $x_i + ny_i = x_i h$. Therefore, without any loss of generality, we can assume that the module V is irreducible. But then the generators g and h of the group H cannot have common non-zero fixed vectors in V, and in view of Proposition, the operator $h - ng + n - 1$ is nonsingular. On the other hand, the equalities $x + y = xg$ and $x + ny = xh$ imply the equality $x(h - ng + n - 1) = xh - nxg + (n - 1)x = 0$, from which it follows that the operator $h - ng + n - 1$ is singular. This contradiction completes the proof. □

Theorem 5A. *Let G be a group and $\Delta_{\mathbb{Z}}(G) = G - 1 + \rho$ for a right ideal ρ of the group ring $\mathbb{Z}G$. If every two-generated subgroup of G is finite, then the additive group of the $\mathbb{Z}G$-module $\Delta_{\mathbb{Z}}(G)/\rho$ is periodic.*

Proof. Assume that the theorem is false and the additive group $\Delta_{\mathbb{Z}}(G)/\rho$ is nonperiodic. Let τ/ρ be the maximal periodic subgroup of $\Delta_{\mathbb{Z}}(G)/\rho$. Then τ is a right ideal in $\mathbb{Z}G$ and the additive group of the $\mathbb{Z}G$-module $\Delta_{\mathbb{Z}}(G)/\tau$ is torsion-free. Therefore, the additive group V of the $\mathbb{Z}G$-module $\mathbb{Z}G/\tau$ is also torsion-free. Let g be an element of G such that $g - 1$ is not contained in τ. We put $x = 1 + \tau$ and $y = g - 1 + \tau$. Then the elements x and y are linearly independent in V and $x + y = xg$. Since $\Delta_{\mathbb{Z}}(G) = G - 1 + \rho = G - 1 + \tau$, we have $2(g - 1) = h - 1 + u$ for certain elements h of G and u of τ. But then $x + 2y = xh$ and hence the elements $x + y$ and $x + 2y$ lie both in the set xG. In view of Lemma 2, this is impossible. This contradiction proves the theorem. □

Clearly, Theorem 5 directly follows from Theorem 1 and Theorem 5A.

Proof of Corollary 3. We put $\rho = (h - ng + n - 1)\mathbb{Q}G$ and assume that $\Delta_{\mathbb{Q}}(G) \neq \rho$. Then $g - 1$ is not contained in ρ. Therefore, if V denotes the $\mathbb{Q}G$-module $\mathbb{Q}G/\rho$ and if $x = 1 + \rho$ and $y = g - 1 + \rho$, then the elements x and y are linearly independent in V. Furthermore, $x + y = g + \rho = (1 + \rho)g = xg$ and $x + ny = 1 + n(g - 1) + \rho = (h - ng + n - 1) + 1 + n(g - 1) + \rho = h + \rho = (1 + \rho)h = xh$, contrary to Lemma 2. Thus $\Delta_{\mathbb{Q}}(G) = \rho$ and the corollary is proved. □

Corollary 3A. *Let $G = \langle g, h\rangle$ be a finite group. Then, for each $n > 1$, the right ideal $(h - ng + n - 1)\mathbb{Z}G$ has a finite index in $\Delta_{\mathbb{Z}}(G)$.*

Proof. According to Corollary 3 the factor group $\Delta_{\mathbb{Q}}(G)/(h - ng + n - 1)\mathbb{Z}G$ is periodic and its subgroup $\Delta_{\mathbb{Z}}(G)/(h - ng + n - 1)\mathbb{Z}G$ is finitely generated, so this subgroup is finite, as required. □

5. Proof of Theorem 4

We need the following lemma about locally nilpotent groups with a locally finite operator group.

Lemma 3. *Let G be a nonperiodic locally nilpotent group and A be a locally finite group, which acts on G. Then G contains certain A-invariant normal subgroups F and H such that $F < H$ and the factor group H/F is an abelian torsion-free group.*

Proof. Let g be an element of infinite order of G. Then there exists an A-invariant normal subgroup F of G such that $F \cap \langle g\rangle = 1$ and F is maximal with respect to these properties. We put $H = \langle g^{xa} \mid x \in G, a \in A\rangle F$. Obviously, H is also an A-invariant normal subgroup in G and the factor group H/F is nontrivial and torsion-free. We will show that this factor group is abelian. Evidently, we can assume without any loss of generality that $F = 1$.

Suppose that the subgroup H is not abelian. Then it has elements u and v such that $w = [u, v] \neq 1$ and $w^u = w^v = w$. We put $N = \langle w^{xa} \mid x \in G, a \in A\rangle$. Clearly, N is a nontrivial A-invariant normal subgroup of G contained in H. Therefore, $N \cap \langle g\rangle \neq 1$ and hence the factor group H/N is periodic. In particular, there are positive integers m and n such that the subgroup $\langle u^m, v^n\rangle$ lies in N. Consequently, there exist a finite subset X of G and a finite subgroup B of A such that $\langle u^m, v^n\rangle$ lies in the subgroup $K = \langle w^{xa} \mid x \in X, a \in B\rangle$. But then $w^{mn} = [u^m, v^n]$ belongs to the derived subgroup K' of K, and hence the factor group K/K' is periodic. On the other hand, K is a finitely generated nilpotent torsion-free group, and thus the factor group K/K' cannot be periodic. This contradiction proves the lemma. □

It is easy to see that Theorem 4 can be now readily deduced from Lemma 3 and Theorem 5.

Indeed, let N be the Hirsch–Plotkin radical of a group $G = AB$ with locally finite subgroups A and B. If N is non-periodic, then, by Lemma 3, there exist A-invariant normal subgroups M and L of N such that $L < M$ and M/L is an abelian torsion-free group. We put $B_1 = AM \cap B$ and $A_1 = B_1M \cap A$. Then $MA_1 = MB_1 = A_1B_1$ and L is normal in MA_1. Passing to the factor group MA_1/L, we can assume that $L = 1$. Then obviously $M \cap A_1 = M \cap B_1 = 1$, and we have the semidirect product $M \rtimes A_1 = M \rtimes B_1 = A_1B_1$ with the normal abelian torsion-free subgroup M and locally finite subgroups A_1 and B_1. By Theorem 1, M is a radical $\mathbb{Z}A_1$-module, and by Theorem 5, M is therefore periodic, contrary to the assumption. Thus, N is periodic, as required.

6. On Groups of Finite Permutation Rank

Proof of Theorem 6. Assume that the group M is non-periodic and let T be the torsion part of M. Then the factor group M/T is torsion-free and satisfies the condition of this theorem. Therefore, without any loss of generality, we can suppose that M is torsion-free. Clearly, M cannot be any locally cyclic group. Hence, M contains at least two linearly independent elements, say u and v. Consider the set of all elements of the form $u + nv$ where n runs over the positive integers. This set falls into finitely many subsets, each of which is contained in a certain orbit of M under G. But if the set of all positive integers is partitioned into finitely many parts, then at least one of these parts contains an arithmetic progression of length three (see, for instance, [Wa]). Therefore, there exist positive integers r and s such that elements $u+rv$, $u+(r+s)v$ and $u + (r + 2s)v$ lie in the same orbit mG for an element m of M. But then, for certain elements a, b, c of G, we have $u + rv = ma^{-1}$, $u + (r + s)v = mb$ and $u + (r + 2s)v = mc$. This implies that $ma^{-1} + sv = mb$ and $ma^{-1} + 2sv = mc$, whence $m + s(va) = m(ba)$ and $m + 2s(va) = m(ca)$. Since the elements m and va are also linearly independent, by Lemma 2 at least one of these elements is not contained in the set mG. This contradiction proves the theorem. □

Proof of Theorem 7. Let A be an arbitrary H-invariant subgroup of G. We put $Y = xA$. Then Y is a subset of X and the subgroup AH acts transitively on Y. It is clear that Y is partitioned by the orbits of H on Y. Since G is a group of finite permutation rank, the number of orbits of H on Y is finite. Let $y_1, \dots, y_n$ be a set of representatives for these orbits. Then, for each element a of A, there exist a representative y_i and an element h of H such that $xa = y_i h$. As $y_i = xa_i$ for some a_i of A, $1 \leq i \leq n$, we have $xa = xa_i h$, and therefore $a_i ha^{-1}$ belongs to H. But then $a_i^h a^{-1}$ is contained in $H \cap A$ and hence a lies in the coset $(H \cap A)a_i^h$. Let B be a normal H-invariant subgroup of A containing $H \cap A$. Then a belongs to the coset Ba_i^h and thus the factor group A/B falls into finitely many conjugacy classes under H.

Assume that every two-generated subgroup of H is finite and the factor group A/B is abelian. Then, by Theorem 6, A/B is periodic. In particular, if A is an abelian normal subgroup of G and $B = A \cap H$, then A/B and therefore A is periodic. Consequently, in the case under consideration, the Fitting subgroup of G is periodic.

Consider now the case where H is a locally finite subgroup and let N be the Hirsch–Plotkin radical of G. If N is a non-periodic subgroup, then by Lemma 3 it contains normal H-invariant subgroups L and M such that $L < M$ and the factor group M/L is an abelian torsion-free group. Then $M \cap H$ is contained in L and, for this reason, as it was stated above, M/L is periodic. This contradiction proves the theorem. □

Proof of Corollary 4. If the subgroup A is finite, then by a result due to B. H. Neumann [AFG, Lemma 1.2.4], the subgroup B has finite index in G and thus G is periodic.

Let us now consider the case where the subgroup B is finite. Since each factor group of $G = ABA$ possesses the same property, it suffices to prove that the Hirsch–Plotkin radical of G is periodic. Without any loss of generality we can assume that the largest normal subgroup of G that is contained in A is 1. Then the representation of G on the cosets of A by right multiplication is a faithful transitive permutation representation of G. As a permutation group, G has the locally finite stabilizer A and finite permutation rank not exceeding the order of B. Therefore, by Theorem 7, the Hirsch–Plotkin radical of G is periodic, as required. □

References

[AFG] Amberg, B., Franciosi, S., de Giovanni, F., Products of Groups. Clarendon Press, Oxford 1992.

[Ch] Chernikov, N. S., Products of groups of finite free rank. In: Groups and Systems of their Subgroups, 42–56, Kiev 1983.

[Ja] Jacobson, N., Structure of Rings. AMS Colloq. Publ. 37. Providence R. I. 1964.

[Ro] Robinson, D. J. S., A Course in the Theory of Groups. Springer, New York-Berlin-Heidelberg 1982.

[Su] Sushchanskii, V. I., Wreath products and general products of groups. Soviet Math. Dokl. 43 (1991), 239–242.

[S1] Sysak, Ya. P., Products of infinite groups. Preprint 82.53. Akad. Nauk Ukrain. Inst. Mat. Kiev 1982.

[S2] —, Radical modules over groups of finite rank. Preprint 89.18. Akad. Nauk. Ukrain. Inst. Mat. Kiev 1989.

[S3] —, Radical modules over hyperabelian groups of finite torsion-free rank. To appear.

[Wa] van der Waerden, B. L., How the proof of Baudet's conjecture was found. In: Studies in Pure Mathematics (ed. by L. Mirsky), 251–260. Academic Press, London 1971.

[Wi] Wilson, J. S., Soluble groups which are products of groups of finite rank. J. London Math. Soc. (2) 40 (1989), 405–419.

FC-Groups: Recent Progress

M. J. Tomkinson

1991 Mathematics Subject Classification: 20F24

1. Introduction

Our notes on FC-groups [30] which appeared in 1984 included a number of open questions. Over the past ten years further progress has been made, some of the questions being answered completely, partial results being obtained in other cases. More importantly, some of the developments have changed the emphasis in current work.

The intention here is to give a brief account of the results obtained in this period indicating, in particular, which of the questions in [30] have been answered and mentioning some of the open problems which arise from this more recent work. As far as possible we have tried to order this survey according to the layout of [30] although it will be seen that there are places where the inter-relationship between the topics has changed since [30] was written.

2. Embedding Questions

If $G_i, i \in I$, are finite groups with centres Z_i, then the centrally restricted product of the groups G_i is defined to be the subgroup of $\prod_{i \in I} G_i$ consisting of all elements of finite order with only finitely many components not belonging to Z_i. The periodic residually finite FC-groups are then completely characterized as subgroups of centrally restricted products of finite groups. [30, Theorem 2.24].

Using this characterization many sufficient conditions for a residually finite periodic FC-group to be an SD$\mathfrak{F}$-group (i.e. a subgroup of a direct product of finite groups) can be easily obtained.

Offprint from
Infinite Groups 94, Eds.: de Giovanni/Newell

It is known that if G is a residually finite FC-group then $G/Z(G) \in \mathrm{SD}\mathfrak{F}$ and $G' \in \mathrm{SD}\mathfrak{F}$ [30, Corollaries 2.26 and 2.27] and it seems that it is mainly bad abelian sections which prevent residually finite periodic FC-groups from being embeddable in direct products of finite groups. In Question 2D we asked whether a periodic residually finite FC-group G with G/G' a direct product of cyclic groups must be an $\mathrm{SD}\mathfrak{F}$-group.

This is true if G/G' has finite exponent [30, Corollary 2.29] and a further case has been dealt with by Kurdachenko.

Theorem 2.1 ([15]). *Let G be a residually finite periodic FC-group with G/G' countable. Then G' is a direct product of countable groups and hence $G \in \mathrm{SD}\mathfrak{F}$.*

In an FC-group, $G/Z(G)$ is always periodic and residually finite and so a number of embedding questions are concerned with conditions for $G/Z(G)$ to be an $\mathrm{SD}\mathfrak{F}$-group. The known examples with $G/Z \notin \mathrm{SD}\mathfrak{F}$ have C_{p^∞}-subgroups contained in G' which suggested the question (2A) of whether an FC-group G with $G' \in \mathrm{SD}\mathfrak{F}$ must have $G/Z(G) \in \mathrm{SD}\mathfrak{F}$. Again Kurdachenko has made some progress with this question.

Theorem 2.2. *Let G be an FC-group with G' a countable $\mathrm{SD}\mathfrak{F}$-group. Then $G/Z(G) \in \mathrm{SD}\mathfrak{F}$.*

It is also known that the general form of Question 2A can be reduced to the case in which G is a metabelian p-group with G' having finite exponent.

In one other question which remains unsolved in its general form the metabelian case has been dealt with. This is the question (2F) of whether or not a residually finite periodic FC-group with countable centre is an $\mathrm{SD}\mathfrak{F}$-group.

Theorem 2.3 ([15]). *If G is a residually finite periodic FC-group with countable centre and G is metabelian then $G \in \mathrm{SD}\mathfrak{F}$.*

Kurdachenko [14] has also obtained a positive answer to this question if one has the condition that $Z(G)$ has finite Sylow subgroups.

The other embedding results obtained by Kurdachenko concern nonperiodic FC-groups. It is easy to see that every FC-group can be embedded in the direct product of a periodic FC-group and a torsion-free abelian group. In the 1970's, Kurdachenko began a series of articles considering the general question of when an FC-group can be embedded in the direct product of finite groups and a torsion-free abelian group. A very satisfactory answer to this question was eventually obtained, the main theorems being given in [16].

Let $t(G)$ denote the torsion subgroup of the FC-group G. For G to be embeddable it is clear that $t(G)$ must be an $\mathrm{SD}\mathfrak{F}$-group. But if T is an $\mathrm{SD}\mathfrak{F}$-group with non-trivial centre, there is an FC-group G with $t(G) \simeq T$ and G not embeddable [30, Theorem 2.31]. Kurdachenko's main result determines the embeddable FC-groups G having $t(G) \simeq T$. This comes from considering the interaction of the centre of $t(G)$ and the torsion-free abelian factor group $G/t(G)$.

Given an SD$\mathfrak{F}$-group T, we define the class $\mathfrak{A}(T)$ of torsion-free abelian groups by saying that $A \in \mathfrak{A}(T)$ if and only if every FC-group G with $t(G) \simeq T$ and $G/t(G) \simeq A$ is embeddable in a direct product of finite groups and a torsion-free abelian group. Kurdachenko's result gives a complete description of the class $\mathfrak{A}(T)$.

Theorem 2.4 ([13], [16]). *Let $T \in$ SD$\mathfrak{F}$, $Z = Z(T)$, $\pi = \pi(Z)$ and let A be a torsion-free abelian group.*

(I) *If Z has finite exponent, then $A \in \mathfrak{A}(T)$ if and only if $|A/A^p| < \infty$, for all $p \in \pi$.*

(II) *If Z has infinite exponent and π is finite, then $A \in \mathfrak{A}(T)$ if and only if $|A/A^p| < \infty$, for all $p \in \pi$, and A is countable.*

(III) *If π is infinite, then $A \in \mathfrak{A}(T)$ if and only if $A = \bigcup_{i=1}^{\infty} A_n$ is the union of an ascending chain of pure subgroups A_n with $r(A_{n+1}/A_n) = 1$, $\pi \cap \mathrm{Sp}(A_{n+1}/A_n)$ is finite for each n, and, for each $p \in \pi$, there is an integer $n(p)$ such that $p \in \mathrm{Sp}(A_{n+1}/A_n)$ for all $n \geq n(p)$.*

In part III, $\mathrm{Sp}(A_{n+1}/A_n)$ denotes the spectrum of A_{n+1}/A_n; that is, the set of primes p such that A_{n+1}/A_n has a C_{p^∞} as a factor.

A dual result to the above is given in [17]. There the torsion-free abelian group A is fixed and the class $\mathfrak{T}(A)$ of all SD$\mathfrak{F}$-groups such that any FC-group G with $G/t(G) \simeq A$ and $t(G) \in \mathfrak{T}(A)$ is embeddable. The class $\mathfrak{T}(A)$ is completely determined.

3. Sections of Direct Products of Finite Groups

It is clear that every QSD$\mathfrak{F}$-group is a periodic FC-group and it was shown by P. Hall [30, Corollary 3.3] that every countable periodic FC-group is a QSD$\mathfrak{F}$-group. Hall [30, Example 3.8] gave an example of an extraspecial 2-group which is not in the class QSD$\mathfrak{F}$ and all known examples of periodic FC-groups which are not in QSD$\mathfrak{F}$ involve an extraspecial section which is not in QSD$\mathfrak{F}$. Most of the progress which has been made on understanding groups which are not in QSD$\mathfrak{F}$ has therefore concentrated on extraspecial p-groups and the construction of examples with various bad properties.

If E is an extraspecial group with centre $Z = \langle z \rangle$ of order p and E/Z is an elementary abelian p-group, then E/Z may be written additively and considered as a vector space V over $GF(p)$. If $x, y \in E$, then $[x, y]$ depends only on the cosets $\bar{x} = xZ$ and $\bar{y} = yZ$; so we can define a mapping $\phi : V \times V \to GF(p)$ by $\phi(\bar{x}, \bar{y}) = k$, where $[x, y] = z^k$. The map ϕ is an alternate bilinear map so that V becomes a symplectic space.

It is well known that a finite dimensional symplectic space is an orthogonal sum of hyperbolic planes $V = H_1 \oplus^{\perp} \cdots \oplus^{\perp} H_n$, where H_i is a two-dimensional subspace with basis $\{x_i, y_i\}$ and $\phi(x_i, y_i) = 1$. This leads to the usual description of finite extraspecial groups.

It is also true that a symplectic space of countable dimension is an orthogonal sum of hyperbolic planes [30, Theorem 3.9]. Again it is possible to obtain a description of all countable extraspecial p-groups. Unfortunately the formulation we gave in [30, Corollary 3.10] is incorrect. A correct description is given by J. I. Hall [8, (4.6)] although his methods are rather different from those we were using. The following argument corrects the error made in [30, Corollary 3.10] and gives a construction for the groups involved.

Theorem 3.1. *If $p > 2$ there are exactly two nonisomorphic countable extraspecial p-groups of exponent p^2.*

Proof. Let X denote the extraspecial group of order p^3 and exponent p and y the extraspecial group of order p^3 and exponent p^2.

If E is a countable extraspecial p-group then it follows from the structure of symplectic spaces of countable dimension that E is the direct product with amalgamated centre of groups isomorphic to either X or Y. We write

$$E = \left(\Upsilon_{i\in I} X_i\right) \Upsilon \left(\Upsilon_{j\in J} Y_j\right)$$

with $X_i \simeq X$ and $Y_i \simeq Y$ (We use Υ to denote the direct product with amalgamated centre). It is known that $Y_1 \Upsilon Y_2 \simeq Y_1 \Upsilon X_2$ so if J is finite a simple induction argument shows that $E \simeq \left(\Upsilon_{n\in\mathbb{N}} X_n\right) \Upsilon Y$. If J is infinite then $\Upsilon_{j\in J} Y_j \simeq \Upsilon_{j\in J} (X_j \Upsilon Y_j)$ so that $E \simeq \left(\Upsilon_{n\in\mathbb{N}} X_n\right) \Upsilon \left(\Upsilon_{n\in\mathbb{N}} Y_n\right) \simeq \Upsilon_{n\in\mathbb{N}} (X_n \Upsilon Y_n) \simeq \Upsilon_{n\in\mathbb{N}} Y_n$.

We note that the two groups $E_1 = \left(\Upsilon_{n\in\mathbb{N}} X_n\right) \Upsilon Y$ and $E_2 = \Upsilon_{n\in\mathbb{N}} Y_n$ are not isomorphic. In E_1 there is a finite subgroup Y whose centralizer has exponent p, whereas in E_2 any finite subgroup is contained in $Y_1 \Upsilon \ldots \Upsilon Y_k$ and so its centralizer contains $\Upsilon_{n>k} Y_n$ which has exponent p^2.

The importance of extraspecial groups in testing whether a group is in QSD$\mathfrak{F}$ arises in the following way. Define classes $\mathfrak{Y}$ and $\mathfrak{Z}$ of periodic FC-groups by

$$G \in \mathfrak{Z} \Leftrightarrow |G : C_G(H)| \leq |H|, \text{ for all infinite subgroups } H \text{ of } G,$$

$$G \in \mathfrak{Y} \Leftrightarrow |G : N_G(H)| \leq |H|, \text{ for all infinite subgroups } H \text{ of } G,$$

It is clear that

$$\text{QSD}\mathfrak{F} \subseteq \mathfrak{Z} \subseteq \mathfrak{Y} \subseteq \text{ periodic } FC\text{-groups}. \qquad (*)$$

In constructing his example of an extraspecial group which is not a QSD$\mathfrak{F}$-group, P. Hall showed that it was not a $\mathfrak{Z}$-group. In fact it is not a $\mathfrak{Y}$-group and the significance of this is indicated by the following theorem.

Theorem 3.2 ([30, Theorems 3.14, 3.15]).

(i) *Let G be a periodic FC-group. Then $G \in \mathfrak{Y} \Leftrightarrow$ every extraspecial section of G is a $\mathfrak{Y}$-group.*

(ii) *Let E be an extraspecial group. Then $E \in \mathfrak{Y} \Leftrightarrow$ for every infinite $H \leq E$ and every maximal abelian subgroup A of H, $|A| = |H|$.*

Under the assumption that $2^\lambda = \lambda^+$, Ehrenfeucht and Faber [30, Example 3.12] constructed an extraspecial p-group E with $|E| = \lambda^+$ but every abelian subgroup having cardinality at most λ.

Questions 3C, 3I, 3J in [30] concerned the necessity of the assumption $2^\lambda = \lambda^+$ and whether one could construct many non-isomorphic extraspecial p-groups E of given cardinality with $E \notin \mathfrak{Y}$. These questions were answered by Shelah and Steprāns as follows.

Theorem 3.3 ([28]).

(i) *Let λ be the successor of a regular cardinal. Then there are 2^λ pairwise nonisomorphic extraspecial p-groups of cardinality λ which are not $\mathfrak{Y}$-groups.*

(ii) *If λ is a weakly compact cardinal then every extraspecial group of cardinality λ is a $\mathfrak{Y}$-group.*

A cardinal λ is ***regular*** if it can not be expressed as a sum of a smaller set of smaller cardinals: $\lambda = \sum_{i \in I} \lambda_i$ with $\lambda_i < \lambda$ and $|I| < \lambda$. In particular ω is regular and so (i) applies when $\lambda = \omega_1$.

A ***weakly compact*** cardinal λ may be defined as a strictly inaccessible cardinal such that there are no λ-Aronszajn trees. Thus the existence of such cardinals does require some additional set-theoretic hypothesis. The important property required for (ii) is that weakly compact cardinals satisfy a very strong partition property, namely that if the complete graph on λ vertices is coloured with less than λ colours then it contains a complete monochrome subgraph of size λ.

Shelah and Steprāns reduced the construction of the counterexamples of 3.3(i) to a purely combinatorial set-theoretic result. As many of the constructions in this area depend on set-theoretic ideas we shall describe the construction of one example to indicate where the combinatorial set theory becomes involved.

We have already described the connection between extraspecial p-groups and symplectic spaces. To construct an example of a symplectic space it is sufficient to give a basis $\{x_i : i < \lambda\}$ and to specify $\phi(x_i, x_j)$, for $i < j$. That is, ϕ arises linearly from a mapping Φ from the set $[\lambda]^2$ of 2-element subsets of λ to $p = \{0, \dots, p-1\}$. To obtain interesting examples of extraspecial groups it is necessary to translate the condition on the group into a condition on the mapping Φ.

If $\Phi : [\lambda]^2 \to p$ and X, Y are two k-element subsets of λ, then $M^\Phi_{X,Y}$ is the $k \times k$ matrix with ij-entry $\Phi(X(i), Y(j))$. The result used by Shelah and Steprāns can then be formulated as follows.

Theorem 3.4 ([28]). *Let λ be the successor of a regular cardinal and k a natural number. Then there is a mapping $\Phi : [\lambda]^2 \to p$ satisfying the following condition.*

If $\{X_\alpha : \alpha < \lambda\}$ is any sequence of disjoint k-subsets of λ then

(a) *there is a $k \times k$ matrix $M = (M_{ij})$ and a λ-subset $\Lambda \subseteq \lambda$ such that, for all $\alpha, \beta \in \Lambda$*

$$M^{\Phi}_{X_\alpha, X_\beta}(i, j) = M_{ij}, \quad \text{whenever } i \neq j,$$

(b) *for any k-vector $d = (d_1, \dots, d_k)$, there are $\alpha, \beta \in \Lambda$ such that*

$$M^{\Phi}_{X_\alpha, X_\beta}(i, j) = \begin{cases} M_{ij}, & i \neq j \\ d_i, & i = j. \end{cases}$$

We indicate why the extraspecial p-group with generators $\{x_i : i < \lambda\}$ and $[x_i, x_j] = z^{\Phi(i,j)}$, for $i < j$, has no abelian subgroups of cardinality λ. To simplify notation, we work in the symplectic space $V = E/Z$ but write $[x_i, x_j]$ for $\phi(\bar{x}_i, \bar{x}_j)$.

Suppose that there are λ elements f_α which commute and put $Y_\alpha = \operatorname{supp} f_\alpha$. Then the Y_α are finite sets and since λ is uncountable and regular, we can assume that they all have the same size. Using the Δ-system lemma we may assume that there is a finite set R such that $Y_\alpha \cap Y_\beta = R$, whenever $\alpha \neq \beta$, and that each f_α has the same projection on R. Considering the elements $f_\alpha - f_\beta$ we now obtain a set of λ commuting elements g_α with $\operatorname{supp} g_\alpha = X_\alpha$ of size k and $X_\alpha \cap X_\beta = \emptyset$, whenever $\alpha \neq \beta$. If $X_\alpha = \{x_{\alpha 1}, x_{\alpha 2}, \dots, x_{\alpha k}\}$, then $g_\alpha = \gamma_1 x_{\alpha 1} + \dots + \gamma_k x_{\alpha k}$ and we can further assume that each g_α has the same coefficient vector $(\gamma_1, \dots, \gamma_k)$ associated with it, and that $x_{\alpha i} \leq x_{\beta j}$ for all i, j, whenever $\alpha < \beta$.

We apply Theorem 3.4 to these $\{X_\alpha : \alpha < \lambda\}$; so there is a subset Λ of λ and a $k \times k$ matrix $M = (M_{ij})$ such that $M^{\Phi}_{X_\alpha, X_\beta}(i, j) = M_{ij}$, whenever $i \neq j$. Now

$$\begin{aligned} [g_\alpha, g_\beta] &= \sum_{i,j} \gamma_i \gamma_j \Phi(X_\alpha(i), X_\beta(j)) \\ &= \sum_{i \neq j} \gamma_i \gamma_j M_{ij} + \sum_i \gamma_i^2 \Phi(X_{\alpha(i)}, X_{\beta(i)}). \end{aligned}$$

If $\sum_{i \neq j} \gamma_i \gamma_j M_{ij} \neq 0$ then let $d = (0, \dots, 0)$; if $\sum_{i \neq j} \gamma_i \gamma_j M_{ij} = 0$ then let $d = (1, 0, \dots, 0)$. By Theorem 3.4, there are $\alpha, \beta \in \Lambda$ such that

$$[g_\alpha, g_\beta] = \sum_{i \neq j} \gamma_i \gamma_j M_{ij} + \gamma_1^2 d_1 \neq 0$$

contrary to the g_α's commuting.

To obtain 2^α nonisomorphic groups, Shelah and Steprāns carried out this construction with the basis elements being taken from λ-subsets X_ξ of $\{x_i : i < \lambda\}$ with $|X_\xi \backslash X_\eta| = \lambda$.

The proof of the nonisomorphism of these groups also depends on making appropriate use of Theorem 3.4.

In [30, Example 3.16] we gave a rather complicated example of an extraspecial $\mathfrak{Z}$-group E which can not be embedded in a direct product with amalgamated centre of groups of order p^3. The intention behind this example was the belief that E was not in the class QSD$\mathfrak{F}$ thus sharpening one of the inclusions in (*). Specifically we asked in Question 3G whether an extraspecial QSD$\mathfrak{F}$-group is always embeddable in a direct product with amalgamated centre of groups of order p^3.

Kurdachenko [18] has answered Question 3G positively making use of a theorem of Gorchakov [30, Theorem 2.4]. Hence QSD$\mathfrak{F}$ is a proper subclass of $\mathfrak{Z}$.

A much simpler example of an extraspecial $\mathfrak{Z}$-group not in QSD$\mathfrak{F}$ has been given by J. Brendle [3].

Theorem 3.5 ([3]). *The extraspecial p-group*

$$E = \langle z, x_i (i < \omega_1) : z^p = x_i^p = 1, \ z \text{ central}, \ [x_i, x_j] = z \text{ if } i < j \rangle$$

is a $\mathfrak{Z}$-group not in the class QSD$\mathfrak{F}$.

If H is a countable subgroup of E, then $H \leq \langle x_i : i < \beta \rangle$, for some $\beta < \omega_1$. Thus $C_E(H) \geq \langle x_i x_j^{-1} : \beta \leq i < j < \omega_1 \rangle$ and so $|E : C_E(H)|$ is countable. Hence $E \in \mathfrak{Z}$.

The classes $\mathfrak{Y}$ and $\mathfrak{Z}$ were introduced to test for groups not being in QSD$\mathfrak{F}$. Brendle observed that there is a further condition satisfied by QSD$\mathfrak{F}$-groups. That is, in a direct product of finite groups one can not have an uncountable set of pairwise noncommuting elements. This property clearly also passes over to QSD$\mathfrak{F}$-groups. However the extraspecial group E above clearly has such a set, namely $\{x_i : i < \omega_1\}$.

Define the class $\mathfrak{P}$ by $G \in \mathfrak{P}$ if and only if G is a periodic *FC*-group and every set of pairwise noncommuting elements in G is countable. One might ask whether this property is close to QSD$\mathfrak{F}$. For example, one might hope that an extraspecial $\mathfrak{P}$-group is a QSD$\mathfrak{F}$-group or that an *FC*-group in which each extraspecial section is a $\mathfrak{P}$-group is itself a $\mathfrak{P}$-group. Unfortunately, neither of these hopes are fulfilled. Brendle [1] constructed an extraspecial $\mathfrak{P}$-group E such that $E \times E \notin \mathfrak{P}$ and so the extraspecial group E is a $\mathfrak{P}$-group but is not QSD$\mathfrak{F}$. A slight variation in Brendle's construction answers both questions.

Let p be an odd prime and let $\Phi : [\omega_1]^2 \to p$ be the mapping given by Theorem 3.4 and define $\Phi' : [\omega_1]^2 \to 2$ by

$$\Phi'(i, j) = \begin{cases} 1, & \text{if } \Phi(i, j) = 0, \\ 0, & \text{if } \Phi(i, j) \neq 0. \end{cases}$$

Then it is clear that Φ' also satisfies the conditions of Theorem 3.4 for the prime 2. We now define an extraspecial p-group E_p generated by $x_i (i < \omega_1)$ with $[x_i, x_j] = \Phi(i, j)$, for $i < j$, and an extraspecial 2-group E_2 generated by $y_i (i < \omega_1)$ with $[y_i, y_j] = \Phi'(i, j)$, for $i < j$. Using the type of argument given after Theorem 3.4 it is fairly easy to see that both E_p and E_2 are $\mathfrak{P}$-groups; that is, they have no uncountable set of pairwise noncommuting elements. But in $E_p \times E_2$ the elements

$x_i y_i (i < \omega_1)$ clearly form an uncountable pairwise noncommuting set. In particular $E_p \times E_2 \notin \mathrm{QSD}\mathfrak{F}$ and hence at least one of E_p, E_2 is an extraspecial $\mathfrak{P}$-group not in $\mathrm{QSD}\mathfrak{F}$. Also $G = E_p \times E_2$ is an FC-group not in $\mathfrak{P}$ but each extraspecial section of G is in $\mathfrak{P}$.

The various results given above show that

$$\mathrm{QSD}\mathfrak{F} \subset \mathfrak{Z} \subseteq \mathfrak{Y} \subset \text{ periodic } FC\text{-groups}$$

$$\mathrm{QSD}\mathfrak{F} \subset \mathfrak{P} \subset \text{ periodic } FC\text{-groups} .$$

The rather more delicate question of whether the classes $\mathfrak{Z}$ and $\mathfrak{Y}$ are distinct has also been considered by Brendle [3].

We remarked in [29, Theorem D(i)] that an extraspecial $\mathfrak{Y}$-group of cardinality ω_1 is a $\mathfrak{Z}$-group. Brendle improves this result to finite-by-abelian groups.

Theorem 3.6 ([3]). *If G is a finite-by-abelian $\mathfrak{Y}$-group of cardinality ω_1, then $G \in \mathfrak{Z}$.*

This theorem makes use of the fact that any countable finite-by-abelian group can be generated by finitely many abelian subgroups [3, Theorem 2.2] and requires the following important step.

Theorem 3.7 ([3]). *Let A be a countable periodic abelian group. Let G be a group of FC-automorphisms of A with $cf|G| > \omega$. Then there is a subgroup B of A such that $|\{B\phi : \phi \in G\}| = |G|$.*

Here an automorphism ϕ of A is an FC-automorphism if $\{a^{-1}a^{\phi} : a \in A\}$ is finite. Theorem 3.6 could be proved for all $\mathfrak{Y}$-groups of cardinality ω_1 if Theorem 3.7 could be improved to A being a countable periodic FC-group. This is Question 1″ in [3].

Brendle uses a forcing argument to show that a result of this type is true under additional set-theoretic hypotheses and so obtains the required generalization of 3.6 in this setting.

Theorem 3.8 ([3]). *Let $\mathcal{V} \models ZFC$. In the model obtained by adding ω_2 Cohen reals to $\mathcal{V}$ any $\mathfrak{Y}$-group of cardinality ω_1 is a $\mathfrak{Z}$-group.*

Thus it is consistent with *ZFC* that the classes $\mathfrak{Y}$ and $\mathfrak{Z}$ coincide for FC-groups of cardinality ω_1. There is no construction at the moment to show that the above is not, in fact, a theorem of *ZFC*.

This question of the relationship between $\mathfrak{Y}$ and $\mathfrak{Z}$ is closely related to questions concerning FC-groups in which $|G : N_G(U)| < \alpha$, for all $U \leq G$, or $|G : C_G(U)| < \alpha$, for all $U \leq G$. We discuss these in the next section.

The importance of extraspecial groups or finite-by-abelian groups in the construction of counterexamples suggested the question (3H): is there a periodic FC-group which is not in the class $\mathrm{QSD}(\mathfrak{F}\,\mathfrak{A})$? This was answered by Kurdachenko who extended a result of Gorchakov to obtain

Theorem 3.9. *If* $G \in \mathrm{QSD}(\mathfrak{F}\,\mathfrak{A})$ *then* $G/Z(G) \in \mathrm{SD}\mathfrak{F}$.

A proof of this result for nilpotent groups of class two appears in [18] but the general result has now been obtained. The example of Gorchakov [30, Example 2.9] then demonstrates the existence of a periodic FC-group which is not in the class $\mathrm{QSD}(\mathfrak{F}\,\mathfrak{A})$.

4. Bounds on Infinite Conjugacy Classes of Subgroups

Most of the results of this section are due to Brendle and relate to Question 7A of [30]. In the 1950's, B. H. Neumann characterized centre-by-finite groups in the following way.

Theorem 4.1. *For a group G the following conditions are equivalent:*

1. *G is centre-by-finite,*
2. $|G : N_G(U)| < \omega$, *for all (abelian) subgroups U of G,*
3. $|U/U_G| < \omega$, *for all (abelian) subgroups U of G.*

It is natural to ask whether this result can be generalized to higher cardinalities and so characterize periodic FC-groups with $|G/Z(G)| < \lambda$. In the example of Shelah and Steprāns given earlier every subgroup U either contains G' and so is normal in G or $U \cap G' = 1$ so that U is abelian and $|U| < \lambda$. So in this example $|U/U_G| < \lambda$ for each subgroup but $|G/Z(G)| = |G| = \lambda$. Thus condition (iii) can not be generalized. However there are abelian subgroups U of G with $N_G(U) = C_G(U) = UZ$ so that $|G : N_G(U)| = |G|$ so that it is still reasonable to attempt to generalize condition (ii).

Brendle concentrates on groups in which $|G : N_G(A)| < \lambda$ for all *abelian* subgroups A and whether these must have $|G/Z(G)| < \lambda$. One way to construct a counterexample would be to construct an extraspecial group of cardinality λ in which each maximal abelian subgroup has index less than λ. But for $\lambda = \omega_1$, we showed in [29, Theorem 5.3] that no such example exists and, in fact, Brendle improved this result to

Theorem 4.2 ([3]). *Let G be a finite-by-abelian group. If* $|G : N_G(A)| \leq \omega$ *for all abelian subgroups A of G then* $|G/Z(G)| \leq \omega$.

The proof of this result depends on Theorem 3.7. If one moves away from finite-by-abelian groups then again a positive result can be obtained under additional set-theoretic hypotheses.

Theorem 4.3 ([3]). *Let $\mathcal{V} \models ZFC$. In the model obtained by adding ω_2 Cohen reals to $\mathcal{V}$ an FC-group G with $|G : N_G(A)| \leq \omega$ for all abelian subgroups A satisfies $|G/Z(G)| \leq \omega$.*

However this result needs to be read in conjunction with

Theorem 4.4 ([3]). *Assuming CH, there is an FC-group G with $|G/Z(G)| = \omega_1$, but $|G : N_G(A)| \leq \omega$, for all abelian subgroups A of G.*

This example is the extension G of a group $H = \mathrm{Dr}\, H_n$, where H_n is a finite p-group of order $p^{n+\binom{n}{2}}$ with $Z(H_n) = H_n'$ having order $p^{\binom{n}{2}}$, by a group of automorphisms whose definition depends on an enumeration of the maximal abelian subgroups of H.

The group G contains a subgroup U such that $|G : N_G(U)| = \omega_1$ and so does not answer the question about FC-groups G in which $|G : N_G(U)| \leq \omega$, for all $U \leq G$. It is still possible that it is a theorem of *ZFC* that $|G/Z(G)| \leq \omega$ in these groups.

When we move up to the next cardinal ω_2 then the situation changes. There is no equivalent of Theorem 4.2 and it is now possible to consider the construction of extraspecial groups in which all the maximal abelian subgroups have small index. Again the construction depends on the underlying set theory in a very precise but rather different way from that above.

To describe the results adequately we need to say a little about the combinatorial objects involved in the construction.

An uncountable cardinal κ is ***inaccessible*** if it is regular and, for all cardinals $\lambda < \kappa$, we have $2^\lambda < \kappa$. If I represents "there is an inaccessible cardinal" then the consistency of $ZFC + I$ can not be deduced from the consistency of *ZFC*. It can be proved that the consistency of $ZFC + \neg I$ follows from that of *ZFC*.

A ***Kurepa tree*** is a tree of height ω_1 with ω_2 uncountable branches such that all the levels are countable. If K represents "there is a Kurepa tree" then the consistency of $ZFC + K$ follows from that of *ZFC*; $ZFC + \neg K$ is equiconsistent with $ZFC + I$. The answer to the question posed now depends on the existence or otherwise of Kurepa trees.

Theorem 4.5 ([3]). *Assume there is a Kurepa tree. Then there is an extraspecial p-group G with $|G| = \omega_2$ but $|G : A| \leq \omega_1$ for all maximal abelian subgroups A of G.*

Since every subgroup of an extraspecial group is either normal or abelian it follows that in this example $|G : N_G(U)| \leq \omega_1$ for all subgroups U of G but $|G/Z(G)| = \omega_2$.

On the other hand

Theorem 4.6 ([3]). *Assuming the consistency of $ZFC + I$, it is consistent that in any FC-group G and for any $\kappa \leq \omega_2$, if $|G : N_G(A)| < \kappa$ for all abelian subgroups A of G then $|G/Z(G)| < \kappa$.*

The construction of the group for Theorem 4.5 is as follows. In the Kurepa tree T there is a family $\{B_\alpha : \alpha < \omega_2\}$ of branches such that $B_\alpha \cap B_\beta$ is countable for all

$\alpha, \beta < \omega_2$. Now form the symplectic space V with basis $X \cup Y$, where $X = \{x_i : i < \omega_1\}$ and $Y = \{y_\alpha : \alpha < \omega_2\}$. The alternative product $\phi : V \times V \to GF(p)$ is defined on X as in the Shelah–Steprāns construction so that every isotropic (= abelian) subspace of $\langle X \rangle$ is countable. We then define

$$\phi(y_\alpha, y_\beta) = 0,$$

$$\phi(x_i, y_\alpha) = \begin{cases} 1, & \text{if } i \in B_\alpha, \\ 0, & \text{if } i \notin B_\alpha. \end{cases}$$

Roughly speaking, if A is an isotropic subspace of V then its projection P on $\langle X \rangle$ is countable. Then $P^\perp \cap \langle Y \rangle$ has codimension ω_1. If Q is the projection of A on $\langle Y \rangle$ then clearly $Q^\perp$ has codimension at most ω_1. It follows that the codimension of $A^\perp$ is at most ω_1 and so in the associated extraspecial group E, $|E : C_E(A)| \leq \omega_1$, for each abelian subgroup A. Hence $|E : N_E(A)| \leq \omega_1$, for each (abelian) subgroup A.

The relationship of the question considered here to the existence of $\mathfrak{Y}$-groups which are not $\mathfrak{Z}$-groups can be seen from Theorem 7.20 of [30] where it is shown that if $|G : C_G(U)| < \lambda$ for all $U \leq G$ with $|U| < \lambda$ and if $|G : N_G(A)| < \lambda$ for all abelian subgroups A of G, then $|G/Z(G)| < \lambda$. Thus the examples above must have a subgroup $U \leq G$ with $|U| < \lambda$ but $|G : C_G(U)| \geq \lambda$; i.e. $G \notin \mathfrak{Z}$. (In fact, a subgroup U corresponding to $\langle X \rangle$ satisfies $|U| = \omega_1$ but $|G : C_G(U)| = \omega_2$). The example also has $|E : N_E(U)| \leq \omega_1$ for all subgroups U and so is very close to being a $\mathfrak{Y}$-group. It seems likely that some variation on this construction could lead to examples of $\mathfrak{Y}$-groups which are not $\mathfrak{Z}$-groups.

5. Minimal Non-*FC*-Groups

A minimal non-*FC*-group is a group G which is not an *FC*-group but in which every proper subgroup is an *FC*-group.

A minimal non-*FC*-group with $G' \neq G$ has a non-trivial finite factor group and these groups were described in some detail by Belyaev and Sesekin. To avoid the Tarski groups which are clearly minimal non-*FC*-groups, Bruno and Phillips considered locally graded minimal non-*FC*-groups showing that these are locally finite. Belyaev showed that a perfect locally finite minimal non-*FC*-group is either a simple group or a p-group for some prime p.

This has now been improved by Kuzucuoglu and Phillips [21] who proved.

Theorem 5.1. *There is no simple locally finite minimal non-FC-group.*

In the proof of this theorem it is shown that a counterexample G would either be linear or contain an element $x \neq 1$ with $C_G(x)$ involving an infinite nonlinear simple group. Since $C_G(x)$ can not then be an *FC*-group, G must be linear. A linear simple locally finite group is generated by two proper subgroups and this leads to a final contradiction.

The question of whether there is a perfect locally finite p-group which is a minimal non-FC-group remains open.

6. Finiteness Conditions on Automorphism Classes

The FC-condition can be defined in two ways in terms of inner automorphisms; either

(a) the (Inn G)-orbits of G are all finite, or

(b) each $\theta \in \operatorname{Inn} G$ is ***virtually trivial*** in the sense that $|G : C_G(\theta)| < \infty$.

Stronger conditions can be imposed on G by requiring that (A) the (Aut G)-orbits of G are all finite, or (B) each $\theta \in \operatorname{Aut} G$ is virtually trivial.

Conditions related to the first of these were considered by Robinson and Wiegold [27] who imposed the stronger condition that the (Aut G)-orbits (or ***automorphism classes***) of elements of G are boundedly finite. If conjugacy classes of elements of G are boundedly finite then it is a well known result of B. Neumann that G is finite-by-abelian.

Theorem 6.1 ([27]). *Let T be the torsion subgroup of $Z(G)$. Then G has boundedly finite automorphism classes if and only if T is finite and* Aut G *induces a finite group of automorphisms on G/T. In particular, G/T is centre-by-finite.*

A similar result of B. Neumann already referred to in Section 4 says that a group G is centre-by-finite if and only if conjugacy classes of subgroups are (boundedly) finite. Robinson and Wiegold considered groups in which automorphism classes of subgroups are boundedly finite and in a later article Lennox, Menegazzo, Smith and Wiegold were able to remove the bounded condition. Thus we have

Theorem 6.2 ([27], [22]). *The group G has finite automorphism classes of subgroups if and only if either*

(a) Aut G *is finite, or*

(b) $G = G_1 \times G_2$, *where G_1 is a locally cyclic torsion group and G_2 is a finite central extension of $C_{p_1^\infty} \times \cdots \times C_{p_n^\infty}$ and $\pi(G_1) \cap \pi(G_2) = \emptyset$.*

Condition (B) was considered by Menegazzo and Robinson [23] who called groups with every automorphism virtually trivial ***VTA-groups***. They showed that a *VTA*-group must satisfy some strong structural conditions.

Theorem 6.3 ([23]). *Let G be a VTA-group. Then*

1. *G' is finite, $Z(G)$ is reduced with finite primary components,*
2. *if G is periodic, then $Z(G)$ is finite.*

They do, in fact, give a necessary and sufficient condition for G to be a *VTA*-group in terms of the cohomology class of the extension of $Z(G)$ by G/Z.

An even stronger restriction was the subject of [24]. The question was raised by Lubotzky and Mann as to whether there exist infinite groups G such that every isomorphism $\theta : H \to K$ between subgroups H and K having finite index in G is virtually trivial in the sense that $|H : C_H(\theta)| < \infty$. It was shown that such a group G contains a subgroup L of finite index such that L is nilpotent of class two with finite derived subgroup and L is generated by elements of order two.

In [24], two examples of infinite 2-groups with this property were constructed. One, assuming CH, is an extraspecial 2-group of cardinality 2^{ω}. The second example does not require any set-theoretic assumptions but leads to a group in which G' is a four-group; again $|G| = 2^{\omega}$. Neither construction leads to examples of higher cardinality, nor is it known whether countable examples exist.

7. CC-Groups and Other Generalizations

During the last few years many of the results on FC-groups have been successfully extended to the class of CC-groups. A group G has Chernikov conjugacy classes (or is a ***CC-group***) if $G/C_G(x^G)$ is a Chernikov group for each $x \in G$. This class was first introduced by Polovickiĭ in the 1960's and he proved some of the basic results. Recently a programme undertaken mainly by J. Otal and his colleagues in Zaragoza has shown that a large part of the theory of FC-groups can be carried out (with suitable modifications) in the class of CC-groups.

We have not the space to give all the results but mention the areas that have been considered with a single reference for each

Sylow theory [25]

Fitting classes [4]

Locally inner automorphisms [26]

Minimal non-CC-groups [9]

Gorchakov's theorem and other embedding results [7]

It is, of course, possible to consider groups in which $G/C_G(x^G)$ belongs to any prescribed class $\mathfrak{X}$ and so obtain $\mathfrak{X}C$-groups. Some results have been obtained when $\mathfrak{X}$ is the class of polycyclic-by-finite groups [6] and further cases are considered in [5].

References

[1] Brendle, J., Cardinal invariants of infinite groups. Arch. Math. Logic 30 (1990), 155–170.

[2] —, Set-theoretic aspects of nonabelian groups. Israel Math. Conf. Proc., 91–105. Bar-Ilan University, Ramat-Gan 1993.

[3] —, Set-theoretic aspects of periodic *FC*-groups - extraspecial groups and Kurepa trees. J. Algebra 162 (1993), 259–286.

[4] Dixon, M. R., Fitting classes of *CC*-groups. Proc. Edinburgh Math. Soc. 31 (1988), 475–479.

[5] Franciosi, S., de Giovanni, F., Kurdachenko, L.A., The Schur property and groups with uniform conjugacy classes. J. Algebra. To appear.

[6] Franciosi, S., de Giovanni, F., Tomkinson, M.J., Groups with polycyclic-by-finite conjugacy classes. Boll. Un. Mat. Ital. (7) 4-B (1990), 35–55.

[7] Gonzalez, M., Incrustaciones de *CC*-grupos. Thesis, Zaragoza 1991.

[8] Hall, J. I., The number of trace-valued forms and extraspecial groups. J. London Math. Soc. 37 (1988), 1–13

[9] Hartley, B., Otal, J., Peña, J. M., Locally graded minimal non *CC*-groups are *p*-groups. Arch. Math. 57 (1991), 209–211.

[10] Kurdachenko, L. A., On the structure of *FC*-groups whose periodic part can be embedded in a direct product of finite groups. Mat. Zametki 25 (1979), 15–25.

[11] —, On conditions for embeddability of an *FC*-group into the direct product of finite groups and an abelian torsion-free group. Mat Sbornik 114 (1981), 566–582.

[12] —, On certain nonperiodic *FC*-groups. In-t. Mat. AN USSR. Kiev 1983

[13] —, *FC*-groups with bounded periodic part. Ukr. Mat. Zh. 35 (1983), 374-378

[14] —, Centrally factorized groups. In-t. Mat. AN USSR. Kiev 1984

[15] —, Residually finite *FC*-groups. Mat. Zametki 39 (1986), 494–506.

[16] —, Nonperiodic *FC*-groups and associated classes of locally normal groups and torsion-free abelian groups. Sib. Mat. Zh. 27 (1986), 104–116.

[17] —, *FC*-groups of countable free rank. Mat. Zametki 40 (1986), 16–30.

[18] —, Two-step nilpotent *FC*-groups, Ukrain. Mat. Zh. 39 (1987) 329–335.

[19] —, On certain classes of *FC*-groups. In-t. Mat. AN USSR. Kiev 1988.

[20] —, On some classes of *FC*-groups. Contemp. Math. 131 (1992), 499–509.

[21] Kuzucuoglu, M., Phillips, R. E., Locally finite minimal non-*FC*-groups. Math. Proc. Cambridge Philos. Soc. 105 (1989), 417–420.

[22] Lennox, J. C., Menegazzo, F., Smith, H., Wiegold, J., Groups with finite automorphism classes of subgroups. Rend. Sem. Mat. Univ. Padova 79 (1988), 87–96.

[23] Menegazzo, F., Robinson, D.J.S., A finiteness condition on automorphism groups. Rend. Sem. Mat. Univ. Padova 78 (1987), 267–277.

[24] Menegazzo, F., Tomkinson, M. J., Groups with trivial virtual automorphism group. Israel J. Math. 71 (1990), 297–308.

[25] Otal, J., Peña, J. M., Sylow theory of *CC*-groups: a survey. Groups St. Andrews 1989, LMS Lecture Notes 160 (1991), 400–407.

[26] —, —, Tomkinson, M. J., Locally inner automorphisms of *CC*-groups. J. Algebra 141 (1991), 382–398.

[27] Robinson, D. J. S., Wiegold, J., Groups with boundedly finite automorphism classes. Rend. Sem. Mat. Univ. Padova 71 (1984), 273–286.

[28] Shelah, S., Steprāns, J., Extraspecial *p*-groups. Ann. Pure Appl. Logic 34 (1987), 87–97.

[29] Tomkinson, M. J., Extraspecial sections of periodic *FC*-groups. Compositio Math. 31 (1975), 285–302.

[30] —, *FC*-Groups, Research Notes in Mathematics 96, Pitman, Boston 1984).

A Remark on Representations of Lattices by Subgroups

Jiří Tůma

Abstract. This paper contains a short proof that every lattice can be embedded in the subgroup lattice of some infinite group. The proof is based on the normal form theorem for free products with amalgamated subgroups.

1991 Mathematics Subject Classification: 20E15, 06B15

P. M. Whitman proved in [Wh] by a rather complicated combinatorial argument that every lattice can be embedded into the partition lattice $\Pi(A)$ on an infinite set A. Using observation made by G. Birkhoff in [Bi] that the partition lattice $\Pi(A)$ can be embedded into the subgroup lattice of the full symmetric group on A by assigning to every partition $\pi \in \Pi(A)$ the subgroup stabilizing each block of π he obtained as a corollary the following theorem.

Theorem 1. *Every lattice can be embedded into the subgroup lattice of an infinite group.*

The author of this note gave in [Tů] a direct proof of the theorem based on Britton's Lemma [Br]. In this note we present a short proof of the theorem using the normal form theorem for free products with amalgamated subgroups.

Let $\mathbf{L}$ be a lattice. We may assume without any loss of generality that the lattice $\mathbf{L}$ has a least element 0. If it is not the case, we simply add a new element 0 to $\mathbf{L}$. By $\mathbf{S}_L$ we denote the restricted symmetric group on L. It consists of all permutations of L fixing all but finitely many elements of L. For a subset $X \subseteq L$ we denote by $\mathbf{S}_X$ the subgroup of $\mathbf{S}_L$ that is the point-wise stabilizer of the complement of X in L.

Let $0 < x < y$ be two different elements of L. Let $\mathbf{Z}_{x,y}$ be the two-element group and $\mathbf{t}_{x,y}$ be its non-zero element. By $\mathbf{T}_{x,y}$ we denote the semidirect product of $\mathbf{S}_{\{0,x,y\}}$ and $\mathbf{Z}_{x,y}$ where $\mathbf{t}_{x,y}$ acts on $\mathbf{S}_{\{0,x,y\}}$ as the inner automorphism defined by the transposition $(0, x)$. The group $\mathbf{S}_{\{0,x,y\}}$ will be referred to as the ***domain*** of $\mathbf{t}_{x,y}$. We define the group $\mathbf{G}_{x,y}$ as the free product of $\mathbf{S}_L$ and $\mathbf{T}_{x,y}$ with the amalgamated

Offprint from
Infinite Groups 94, Eds.: de Giovanni/Newell

subgroup $\mathbf{S}_{\{0,x,y\}}$. The normal form of each element of $\mathbf{G}_{x,y}$ requires to choose coset representatives of the amalgamated subgroup $\mathbf{S}_{\{0,x,y\}}$ in both groups. In $\mathbf{T}_{x,y}$ we choose the elements $1, \mathbf{t}_{x,y}$. In $\mathbf{S}_L$ we choose as the coset representative of a given left coset of $\mathbf{S}_{\{0,x,y\}}$ the unique permutation p of the coset containing the elements $0, x, y$ in different cycles. Then every element of $\mathbf{G}_{x,y}$ can be expressed uniquely in the form

$$p_0\mathbf{t}_{x,y}p_1\mathbf{t}_{x,y}\dots p_k\mathbf{t}_{x,y}p_{k+1}h, \tag{1}$$

where every p_i is a coset representative, $p_i \neq 1$ if $0 < i < k+1$ and $h \in \mathbf{S}_{\{0,x,y\}}$. Note that $\mathbf{S}_L$ is a subgroup of $\mathbf{G}_{x,y}$ and its coset representatives may be chosen in the form (1), where $p_{k+1} = h = 1$.

Similarly, if $x, y \in L$ are non-comparable, then we fix the order of the two elements, say x, y and define analogously an element $\mathbf{t}_{x,y}$ as the only non-zero element of the two-element group $\mathbf{Z}_{x,y}$. This time we choose as the domain of $\mathbf{t}_{x,y}$ the subgroup $\mathbf{S}_{\{x,y,x+y\}}$. We let the element $\mathbf{t}_{x,y}$ act on the domain as the inner automorphism defined by the transposition $(x, x+y)$. The semidirect product of $\mathbf{Z}_{x,y}$ and $\mathbf{S}_{\{x,y,x+y\}}$ will be denoted by $\mathbf{T}_{x,y}$. We choose the coset representatives of $\mathbf{S}_{\{x,y,x+y\}}$ in both groups $\mathbf{T}_{x,y}$ and $\mathbf{S}_L$ in the same way as in the previous paragraph. We define the group $\mathbf{G}_{x,y}$ as the free product of $\mathbf{S}_L$ and $\mathbf{T}_{x,y}$ with the amalgamated subgroup $\mathbf{S}_{\{x,y,x+y\}}$. The normal form for elements of $\mathbf{G}_{x,y}$ is provided by (1) and the coset representatives of $\mathbf{S}_L$ in $\mathbf{G}_{x,y}$ may be chosen in the same way as in the case $x < y$.

We are going to prove that there is a canonical embedding of the lattice $\mathbf{L}$ to the subgroup lattice of the free product of all groups $\mathbf{G}_{x,y}$ with the amalgamated subgroup $\mathbf{S}_L$, where the product is taken over all pairs $x, y \in L$ such that $0 \neq x \not> y$. Let us denote the group by $\mathbf{G}$. Combining the normal forms (1) for different $\mathbf{G}_{x,y}$, we get the following normal form result for the group $\mathbf{G}$.

Proposition 1. *Every element of* $\mathbf{G}$ *can be expressed in a unique way in the form*

$$p_0\mathbf{t}_0p_1\mathbf{t}_1\dots p_k\mathbf{t}_kh, \tag{2}$$

where $\mathbf{t}_i$ *is an abbreviation for* $\mathbf{t}_{x_i,y_i}$, *every* p_i *is a coset representative of the domain of* $\mathbf{t}_i$ *in* $\mathbf{S}_L$, $p_i \neq 1$ *if* $\mathbf{t}_{i-1} = \mathbf{t}_i$, *and* $h \in \mathbf{S}_L$.

Now if $u \in L$, then take the interval $[0, u]$ in $\mathbf{L}$ and define the mapping Φ from $\mathbf{L}$ to the subgroup lattice of $\mathbf{G}$ as follows: $\Phi(u)$ is the subgroup of $\mathbf{G}$ generated by $\mathbf{S}_{[0,u]}$ and all the elements $\mathbf{t}_{x,y}$. The proof that Φ is an embedding of $\mathbf{L}$ into the subgroup lattice of $\mathbf{G}$ will follow from the following sequence of easy claims.

Claim 1. If $q \in \mathbf{S}_{[0,u]}$, $\mathbf{t}_{x,y}$ is arbitrary, and p is the coset representative of the coset of the domain of $\mathbf{t}_{x,y}$ containing q, then $p \in \mathbf{S}_{[0,u]}$.

Proof. We shall restrict ourselves to the case $x < y$. The case x, y non-comparable is even easier. The permutation q can contain two elements from the set $\{0, x, y\}$ in the same cycle only if $\{0, x, y\}$ intersects the interval $[0, u]$ in more than one element. If the intersection contains exactly two elements, then these elements are $0, x$. Only these two elements can be possibly contained in the same cycle of q. If it is the case, then the transposition $(0, x)$ is in both $\mathbf{S}_{\{0,x,y\}}$ and $\mathbf{S}_{[0,u]}$, hence the coset representative $p = q(0, x)$ belongs to $\mathbf{S}_{[0,u]}$. If $0 < x < y < u$, then $\mathbf{S}_{\{0,x,y\}} \subseteq \mathbf{S}_{[0,u]}$, hence $p \in \mathbf{S}_{[0,u]}$ also in this case. In the case x, y non-comparable the only possibility to consider is the latter one.

Claim 2. The mapping Φ is order preserving.

Claim 3. The mapping Φ preserves joins.

Proof. Let $u, v \in L$ be non-comparable. In view of Claim 2 it is sufficient to prove that $\Phi(u + v) \subseteq \Phi(u) + \Phi(v)$. We may assume that u, v is the chosen order for the pair u and v. Since $(0, u) \in \mathbf{S}_{[0,u]} \subseteq \Phi(u)$ and similarly $(0, v) \in \Phi(v)$, we get $(u, v) \in \Phi(u) + \Phi(v)$. Then also $\mathbf{t}_{v,u+v}\mathbf{t}_{u,v}(u, v)\mathbf{t}_{u,v}\mathbf{t}_{v,u+v} = (0, u + v)$ belongs to $\Phi(u) + \Phi(v)$. Now take an arbitrary $z \in L$ such that $0 < z < u + v$. Then also $\mathbf{t}_{z,u+v}(0, u + v)\mathbf{t}_{z,u+v} = (z, u + v) \in \Phi(u) + \Phi(v)$. Since $\mathbf{S}_{[0,u+v]}$ is generated by the transpositions $(z, u + v)$, $z < u + v$, we get $\mathbf{S}_{[0,u+v]} \subseteq \Phi(u) + \Phi(v)$. Thus also $\Phi(u + v) \subseteq \Phi(u) + \Phi(v)$.

Claim 4. The mapping Φ preserves meets.

Proof. Let $u, v \in L$ be non-comparable. In view of Claim 2 it is sufficient to prove that $\Phi(u)\Phi(v) \subseteq \Phi(uv)$. So take any $w \in \Phi(u)\Phi(v)$. Since $\Phi(u)$ is generated by $\mathbf{S}_{[0,u]}$ and the elements $\mathbf{t}_{x,y}$, we can express w in the form

$$w = p_0\mathbf{t}_0 p_1\mathbf{t}_1 \dots p_k\mathbf{t}_k p_{k+1}, \tag{3}$$

where $p_i \in \mathbf{S}_{[0,u]}$ for every i. By Claim 1 we can assume that p_i is a coset representative of the domain of $\mathbf{t}_i$ for every $i < k + 1$, hence (3) is the normal form of w.

Similarly, from $w \in \Phi(v)$ we conclude that

$$w = q_0\mathbf{t}_0 q_1\mathbf{t}_1 \dots q_l\mathbf{t}_l q_{l+1}, \tag{4}$$

where $q_j \in \mathbf{S}_{[0,v]}$ is a coset representative of the domain of $\mathbf{t}_j$ for every $j < l + 1$, hence (4) is also the normal form of w. It follows that $k = l$ and $p_i = q_i \in \mathbf{S}_{[0,uv]}$ for every $i \leq k + 1$. Thus $w \in \Phi(uv)$.

Claim 5. The mapping Φ is injective.

Proof. Since we already know that Φ is a lattice morphism it is sufficient to prove that $\Phi(u) \neq \Phi(v)$ whenever $u < v$. In this case the transposition (u, v) is the normal form and belongs to $\Phi(v)$ but not to $\Phi(u)$.

This completes the proof of the theorem. □

References

[Bi] Birkhoff, G., On the structure of abstract algebras. Proc. Cambridge Philos. Soc. 31 (1935), 433–454.

[Br] Britton, J. L., The word problem. Ann. Math. 77 (1963), 16–32.

[Tů] Tůma, J., A new proof of Whitman's embedding theorem. J. Algebra. To appear.

[Wh] Whitman, P., Lattices, equivalence relations, and subgroups. Bull. Amer. Math. Soc. 52 (1946), 507–522.

A Note on Commutator Laws in Groups

Arthur Weinberger

Abstract. I. D. Macdonald and B. H. Neumann consider an axiomatic treatment of commutator laws in groups. This note answers an open question raised by them concerning the interdependence of certain laws.

1991 Mathematics Subject Classification: 20A05, 08A99

In [1] and [2], I. D. Macdonald and B. H. Neumann began an axiomatic treatment of commutator laws in groups to investigate the completeness and interdependence of these laws. In [2] it is shown that a certain set of laws, called the standard set of laws, together with a Jacobi–Witt–Hall type identity is complete. They discuss the interdependence of these laws by constructing various models, satisfying some of the laws, but not others. This note answers one of the open questions raised in [2] on the interdependence of two of these laws by constructing a suitable model.

In order to make this note self-contained we start with some of the notations and definitions of [1] and [2]. Let G be a group. For $x, y \in G$ we denote with $[x, y] = x^{-1}y^{-1}xy$, the commutator of x and y, and with $x^y = y^{-1}xy$ the conjugate of x by y. The identity of G is denoted by e. Let κ be a binary operation on the elements in post-fix notation, i.e. $xy\kappa$ is κ applied to x and y, where $x, y \in G$. We write $xy\sigma$ as shorthand for $x \cdot xy\kappa$. The κ-operation is called kappatator operation and mimics commutation. The σ-operation mimics conjugation.

In this note we consider the following subset of the laws in [2]:

A1	$xe\kappa = e;$
A3	$xx\kappa = e;$
A4	$xy\kappa \cdot yx\kappa = e;$
I1	$xy\kappa z\kappa = [xy\kappa, z];$
I6	$(xy\kappa)^z = x^z y^z \kappa;$
I7	$xy\kappa z\sigma = (xz\sigma)(yz\sigma)\kappa.$

* The author wants to thank Professor Luise-Charlotte Kappe for her guidance in preparing this note.

Offprint from
Infinite Groups 94, Eds.: de Giovanni/Newell

Following Macdonald and Neumann, we will always assume A1, A3 and A4 for the κ-operation.
In [2] the question is raised if I7 in the presence of I1 implies I6 or not. In this note we construct a model that satisfies A1, A3, A4 together with I1 and I7, but fails to satisfy I6. Thus I1 and I7 do not imply I6.

We record here the following result from [2]:

Proposition 1. *Let G be a free nilpotent group of class 2 and rank ≥ 2 and let $m : G \times G \to \mathbb{Z}$, satisfying*

$$m(x, y) = 0 \text{ for } x \in G' \text{ or } y \in G'; \tag{1.1}$$

$$m(y, x) = m(x, y) \text{ for all } x, y \in G. \tag{1.2}$$

Define κ by

$$xy\kappa = [x, y]^{1+m(x,y)}, \tag{1.3}$$

then

(i) A1, A3, A4 *and* I1 *are satisfied;*

(ii) I7 *is satisfied if and only if*

$$m(xz\sigma, yz\sigma) = m(x, y) \text{ for all } x, y, z \in G; \tag{1.4}$$

(iii) I6 *is satisfied if and only if*

$$m(x^z, y^z) = m(x, y) \text{ for all } x, y, z \in G. \tag{1.5}$$

Remark. We mention here that if there exist $g, h \in G\backslash G'$ such that $m(g, h) \neq 0$ then the κ-operation is different from the commutator operation.

We state now the main result of this note.

Theorem 2. *The laws* I1 *and* I7 *in the presence of* A1, A3 *and* A4 *do not imply* I6.

Proof. Let G be the free nilpotent group of class 2 with two generators given by

$$G = \{(a, b, c) \mid a, b, c \in \mathbb{Z}\}$$

with multiplication

$$(a, b, c)(a', b', c') = (a + a', b + b', c + c' + ab').$$

We note that $Z(G) = G' = \langle (0, 0, 1) \rangle$. We define now a function $f : G \to \mathbb{Z}$ by

$$f((a, b, c)) = \begin{cases} 0 & \text{if } a = b = 0 \\ 2(c \mod 2) + 1 & \text{otherwise.} \end{cases}$$

(Here c mod $2 = 0$, if c is even, and c mod $2 = 1$, if c is odd). For $x, y \in G$ we define $m : G \times G \to \mathbb{Z}$ as

$$m(x, y) = f(x)f(y).$$

Clearly, $m(x, y) = 0$ if $x \in G'$ or $y \in G'$, and $m(x, y) = m(y, x)$, hence (1.1) and (1.2) are satisfied. If we define $xy\kappa = [x, y]^{1+m(x,y)}$, it follows by Proposition 1 that A1, A3, A4 and I1 hold for this κ-operation.

To verify I7, we have to show that (1.4) holds, i.e. $m(xz\sigma, yz\sigma) = m(x, y)$. First let $x \in G'$, then $m(x, y) = 0$ and $xz\kappa = [x, y]^{1+0} = e$, hence $xz\sigma = x \in G'$. It follows $m(xz\sigma, yz\sigma) = m(x, yz\sigma) = 0$, and hence (1.4) holds. Similarly, if $y \in G'$ we obtain $m(x, y) = m(xz\sigma, yz\sigma)$. If $z \in G'$, then $xz\sigma = x \cdot xz\kappa = x[x, z]^{1+0} = x$, and likewise $yz\sigma = y$. Thus again $m(x, y) = m(xz\sigma, yz\sigma)$.

Now assume none of x, y, z is in G'. Let $x = (a, b, c)$, $y = (i, j, k)$ and $z = (r, s, t)$, where $a^2 + b^2$, $i^2 + j^2$ and $r^2 + s^2$ are all non-zero. It suffices to show that $f(x) = f(xz\sigma)$. We have

$$\begin{aligned} xz\sigma &= (a, b, c)[(a, b, c), (r, s, t)]^{1+m(x,z)} \\ &= (a, b, c)(0, 0, c^*)^{1+m(x,z)} \\ &= (a, b, c + c^*(1 + m(x, z))) \end{aligned}$$

for some $c^* \in \mathbb{Z}$. We observe that $f(g) \in \{0, 1, 3\}$ for any $g \in G$. Thus $m(g, h)$ belongs to $\{0, 1, 3, 9\}$, and in particular $m(g, h)$ is odd, if $g, h \in G \backslash G'$. Hence $m(x, z)+1$ is even. Thus

$$f(xz\sigma) = 2((c + c^*(m(x, z) + 1)) \bmod 2) + 1 = 2(c \bmod 2) + 1 = f(x).$$

Similarly, we obtain $f(yz\sigma) = f(y)$. Thus $m(xz\sigma, yz\sigma) = m(x, y)$ also in this case. We conclude that I7 is satisfied.

Finally, we show that I6 fails to hold. By Proposition 1.iii we have that I6 is equivalent to $m(x^z, y^z) = m(x, y)$ for all $x, y, z \in G$. Let $x = (0, 1, 1)$, $z = (1, 0, 1)$ and $y = (1, 0, 0)$. Now $y^z = y$, thus $f(y^z) = f(y) = 2(0 \bmod 2) + 1 = 1$. But $x^z = z^{-1}xz = (-1, 0, -1)(0, 1, 1)(1, 0, 1) = (0, 1, 0)$. Thus

$$f(x^z) = 2(0 \bmod 2) + 1 = 1 \neq 3 = 2(1 \bmod 2) + 1 = f(x).$$

Therefore $1 = m(xz\sigma, yz\sigma) \neq m(x, y) = 3$, and I6 fails to hold. □

References

[1] Macdonald, I. D., and Neumann, B. H., On commutator laws in groups. J. Austral. Math. Soc. Ser. A 43 (1988), 95–103.

[2] —, —, On commutator laws in groups II. Contemp. Math. 109 (1990), 113–129.

Some Finiteness Conditions in Groups

James Wiegold

1991 Mathematics Subject Classification: 20F24, 20F30

The starting-point of this article is the beautiful paper [5], in which B. H. Neumann established some of the earliest very general results on finiteness conditions on groups. The specific result of central interest here was mentioned in a lecture given by J. T. Buckley at Cardiff in 1991, when Neumann was a Visiting Fellow:

Let $\mathfrak{X}$ be the class of groups G such that $|H^G : H| < \infty$ for all subgroups H of G. Then a group G is in $\mathfrak{X}$ if and only if G' is finite, so that $|H^G : H| \leq |G'| < \infty$ for all subgroups H of G.

Thus, if all subgroups of a group have finite indices in their normal closures, there is global bound on these indices. Neumann and J. C. Lennox idependently proposed dualising the class $\mathfrak{X}$ in the following way, and much of this article deals with results obtained on this dual class and to appear in [1], [3], [4] and [7].

Definition 1. A group G is said to be ***core-finite***, *CF* for short, if $|H : \mathrm{Core}_G(H)| < \infty$ for all subgroups H.

Since the $\mathfrak{X}$-groups are exactly the finite-by-abelian groups, an aesthetically pleasing result that might be expected is that *CF*-groups are abelian-by-finite. Alas, this is far from the case, since periodic Tarski-type monsters (see [6]) are certainly *CF* but not abelian-by-finite. Incidentally, these examples illustrate a distinction between $\mathfrak{X}$-groups and their duals the *CF*-groups, namely that proofs of results about $\mathfrak{X}$-groups are usually relatively short and easy, while those for *CF* are less so. A further property of $\mathfrak{X}$-groups that does not carry over to *CF*-groups is as follows. Recall that finiteness of the indices $|H^G : H|$ yields bounded finiteness. Thus we define:

Offprint from
Infinite Groups 94, Eds.: de Giovanni/Newell

Definition 2. A group G is said to be ***boundedly core-finite***, *BCF* for short, if there is an integer n such that $|H : \mathrm{Core}_G(H)| \leq n$ for all subgroups H of G. If n is the least integer with this property, then G is said to be ***core-n***.

CF does not imply *BCF*, and this time we do not need to resort to difficult examples.

Example 1. Set $G = \langle C, g\rangle$, where $C \simeq C_{p^\infty}$ for some prime p, g is of infinite order and $c^g = c^{-1}$ for all c in C. Then G is in $CF \setminus BCF$.

The proof is easy, but perhaps it is worth pointing out that $\langle cg^2\rangle$ has core $\langle g^{2p^n}\rangle$ if c is an element of C of order p^n (at least when p is odd:the case $p = 2$ is slightly different, as usual). It is the element g of infinite order that causes the difficulty, but in general elements of infinite order in *CF*-groups cause little problem, and for many purposes they can be safely ignored. To see this, let a be an element of infinite order in a *CF*-group. Some power of a generates an infinite cyclic normal subgroup, and it is useful to look at the subgroup A of G generated by all the infinite cyclic normal subgroups. A very easy calculation shows that A is abelian and that the following holds: for every g in G, either $a^g = a$ for all a in A, or $a^g = a^{-1}$ for all a in A. Thus, $|G : C_G(A)| \leq 2$, so in many situations one can assume that A is central, and of course G/A is periodic. Thus, all the real problems reside in the domain of periodic groups, and the Tarski groups show that they are somewhat intractable in general. However, for locally finite groups we have:

Theorem 1 [1]. *Every locally finite CF-group is abelian-by-finite and BCF.*

The proof is rather long, uses fairly standard techniques from infinite group theory, and I shall say no more about it here.

Just how important is the local finiteness condition? Howard Smith and I had a look at the interaction of the *CF* property with local gradedness, which is a condition designed to exclude only the very worst types of groups (from the finiteness conditions point of view) such as those with finitely generated infinite simple subgroups and the like. First, a slight extension of Theorem 1:

Theorem 2 [7]. *Let G be a CF-group such that every periodic image of G is locally finite. Then G is abelian-by-finite.*

Perhaps the most interesting result in this area is the following:

Theorem 3 [7]. *Every locally graded BCF-group is abelian-by-finite.*

It would be nice to know whether locally graded *CF*-groups are abelian-by-finite. However, this does seem to be an exceptionally hard problem. For instance, in investigating the possibilities, one soon meets the following type of situation:

Problem 1. Does there exist a finitely generated, infinite, periodic, residually finite group G such that every subgroup of G is either finite or of finite index?

Clearly, every group with this property is locally graded, *CF*, and not abelian-by-finite. A positive answer to Problem 1 would yield a negative answer to the following well-known question (see [2]), which I first heard of from Avinoam Mann:

Problem 2. Does every infinite residually finite group contain an infinite abelian subgroup?

One has a feeling that the answer to Problem 1 should be "no": surely, residually finite groups cannot be that bad?

On the positive side, we have:

Theorem 4 [7]. *Every nilpotent CF-group is BCF and abelian-by-finite.*

There is another strand of problems that crop up quite naturally. Let G be a locally finite BCF-group, so that G is abelian-by-finite. Is there an abelian subgroup of bounded index? More precisely, we ask :

Problem 3. Does there exist a function $f : \mathbb{N} \to \mathbb{N}$ such that every locally finite core-n group has an abelian (normal) subgroup of index at most $f(n)$?

This is of interest for finite groups, where it reduces to a question about finite p-groups. Suppose that the answer is known to be yes for finite p-groups, and let G be a finite core-n group of order $p_1^{\alpha_1} \dots p_r^{\alpha_r}$, with standard notation, where $p_1 < \dots < p_r$. For each i, let P_i be a Sylow p_i-subgroup and X_i its core in G. Suppose that $p_i \leq n$ for $i \leq s$, and $p_{s+1} > n$. Then all subgroups of each P_i with $i > s$ are normal in G, and the P_i are abelian since they are Hamiltonian (there is no loss in assuming that all of these p_i are odd). By our assumption, for $i \leq s$ each X_i has an abelian subgroup Y_i of n-bounded index in X_i, and therefore in P_i since $|P_i : X_i| \leq n$. Finally, the subgroup generated by $Y_1, \dots, Y_s, P_{s+1}, \dots, P_r$ is abelian and of n-bounded index: this because s is n-bounded.

Lennox, Smith and I have considered the special case of core-p p-groups. This ought to be a very restricted class of groups, and in many ways it is.

Theorem 5 [4]. *For every odd prime p, every finite core-p p-group G is nilpotent of class at most* 3, *the derived group G' is of exponent at most p, and G has an abelian normal subgroup of index at most p^5.*

These bounds are probably too large, though the best example we know is the following.

Example 2. For an odd prime p, let G be the group

$$\langle a, b, c \mid [a, b] = c^p,\ [b, c] = a^p,\ [c, a] = b^p,\ [x, y, z] = 1 \text{ for } x, y, z \in \{a, b, c\}\rangle.$$

Then G is core-p but has no abelian subgroup of index p, though it does have an abelian subgroup of index p^2.

The proof of Theorem 5 uses the concept of breadth to a large extent, togther with standard techniques. We envisage greatly increased technical difficulties when we move on to considering core-p^n p-groups, even though we expect to find results similar to those embodied in Theorem 5, at least when p is large compared with n. The case $p = 2$ in Theorem 5 is unresolved, partly because there is no bound on the nilpotency classes of core-2 2-groups. However, they are, at least, metabelian.

We move on now to a brief mention of some generalizations of the CF-property.

Definition 3. (i) A group G has property S_1, A_1 or C_1 respectively, if every subgroup, every abelian subgroup or every cyclic subgroup respectively of G has finite index over its core in G.

(ii) A group G has property S_2, A_2 or C_2 respectively if the index $|H : H \cap H^x|$ is finite for every element x of G and every subgroup, every abelian subgroup or every cyclic subgroup H respectively.

Thus $S_1 = CF$, and we have the following result about the interconnections between the six properties:

Theorem 6 [3]. (i) $S_1 \Rightarrow A_1 \Rightarrow C_1$, $S_2 \Rightarrow A_2 \Rightarrow C_2$, $S_1 \Rightarrow S_2$, $A_1 \Rightarrow A_2$, $C_1 \Rightarrow C_2$.

(ii) *Apart from the implications given in* (i) *(and those that are formal logical consequences of them), there are no further implications among the six conditions, even with the additional property of nilpotency.*

However, with the extra hypothesis of finite generation, we have:

Theorem 7 [3]. *Every finitely generated C_2-group satisfies C_1, and every finitely generated A_2-group satisfies A_1.*

Properties S_2 and S_1 have not yielded to attack, and indeed we pose our final problem about them:

Problem 4. Is every finitely generated S_2-group an S_1-group?

It is conjectured in [3] that the answer should be in the positive. Certainly, the indications are strong. For example, if there is a finitely generated S_2-group outside S_1, there exists [3] a periodic group G of this type generated by a finitely generated subgroup H and a single element in such a way that $H/\operatorname{Core}_G(H)$ is infinite. This appears more than somewhat unlikely.

Finally, the dualization game should be attackable for the other finiteness conditions in discussed in [5], and we offer this as a suggestion to the reader.

References

[1] Buckley, J. T., Lennox, J. C., Neumann, B. H., Smith, H., Wiegold, J., Groups with all subgroups normal-by-finite. J. Austral. Math. Soc. Ser. A. To appear.

[2] The Kourovka Notebook (Unsolved Problems in Group Theory), Russian Academy of Sciences, Novosibirsk 1992, Problem 11.56.

[3] Lennox, J. C., Longobardi, P., Maj, M., Smith, H., Wiegold, J., Some finiteness conditions concerning intersections of conjugates of subgroups. Glasgow Math. J. To appear.

[4] Lennox, J. C., Smith, H., Wiegold, J., Finite p-groups in which subgroups have large cores. In: Infinite Groups 1994, Proc. Internat. Conference, Walter de Gruyter, Berlin–New York 1995, 163–169.

[5] Neumann, B. H., Groups with finite classes of conjugate subgroups, Math. Z. 63 (1955), 76–96.

[6] Ol'shanskii, A. Yu., Geometry of defining relations in groups, Nauka, Moscow 1989.

[7] Smith, H., Wiegold, J., Locally graded groups with all subgroups normal-by-finite. J. Austral. Math. Soc. Ser. A. To appear.

First-Order Group Theory

John Wilson

Abstract. We shall describe some results concerning first-order properties of simple groups. We begin with a brief and informal introduction to the first-order theory of groups, and then we show how ultraproducts can be used to illustrate various facts about simple groups. These lead us to a question about commutators in finite simple groups. Then we shall prove a theorem of Ulrich Felgner, which asserts that the finite simple non-abelian groups are just the finite groups satisying a certain first-order sentence. Finally, we show how information about finite simple groups can be used in the classification of certain infinite simple groups.

1991 Mathematics Subject Classification: 20E32, 20D05, 03C20

1. First-Order Properties of Groups

In the first-order theory of groups we are only allowed to make statements of certain types about groups. Each of the following statements asserts that the group G is nilpotent of class at most 2; however the second statement is not permissible in the language of first-order group theory:

$$(\forall x \forall y \forall z)[x, y, z] = 1;$$

$$(\forall x \in G')(\forall z)([x, z] = 1).$$

The following first-order statements (in which $\wedge$ means "and" and $\rightarrow$ means "implies") assert respectively that G has order at least 3 and that G has no elements of order 3:

$$(\forall x_1 \forall x_2)(\exists y)(y \neq x_1 \wedge y \neq x_2);$$

$$(\forall x)(x^3 = 1 \rightarrow x = 1).$$

The assertion that every element of G' is a product of two commutators may be expressed in the following first-order statement:

$$(\forall x_1 \forall x_2 \forall x_3 \forall x_4 \forall x_5 \forall x_6)(\exists y_1 \exists y_2 \exists y_3 \exists y_4)([x_1, x_2][x_3, x_4][x_5, x_6] = [y_1, y_2][y_2, y_4]).$$

The statement

$$(\forall x)(\exists n \in \mathbb{N})(x^n = 1),$$

Offprint from
Infinite Groups 94, Eds.: de Giovanni/Newell

which expresses the fact that G is a torsion group, is not a first-order statement.

All of the above statements concern equalities and inequalities of group elements, connected by logical symbols such as $\wedge$ and $\rightarrow$, and preceded by quantifiers. The feature which distinguishes the first-order statements is that the quantifiers range over elements of the whole group, and not over elements of a subgroup, or over a family of subsets or over the set of integers. Sometimes with a little ingenuity a statement which is not first-order can be replaced by an equivalent one which is; and some properties of groups, such as the property of being a torsion group, cannot be expressed at all in a first-order statement. Restricting attention to first-order properties of groups puts a different perspective on familiar topics, and leads to problems which seem interesting and sometimes difficult.

More formally, we consider properties of groups which can be formulated within the first-order predicate calculus with equality. Thus we are concerned with finite ***formulae***, involving variables representing group elements, the symbols $\cdot$, $^{-1}$ and 1 representing group multiplication, inverse and identity, the equality symbol $=$, the logical symbols $\rightarrow$ ("implies"), $\wedge$ ("and"), $\vee$ ("or"), $\neg$ ("not"), and quantifiers ranging over all elements of the group. We normally omit the symbol $\cdot$ representing multiplication, and to make formulae more comprehensible we introduce harmless shorthand such as $\neq$, x^2, $[x, y]$, and $\bigvee_{1 \leqslant i < j \leqslant 4}$, with the obvious meanings.

A first-order ***sentence*** is a formula in which each variable is preceded by a quantifier. (Though this definition is somewhat imprecise, the examples above and below should make its intended meaning clear. A fuller description of all matters relating to the first-order language of group theory can be found in the book [6].) A (first-order) ***theory T*** is just a collection of sentences, and the groups satisfying the sentences in T are called the ***models*** of T. A class of groups is called ***axiomatizable*** if it is the set of all models of some theory, and ***finitely axiomatizable*** if it is the set of models for a theory comprising just one sentence.

We write $G \models \varphi$ to mean that the sentence φ holds in the group G. Two groups G, H are called ***elementarily equivalent*** (and we write $G \equiv H$) if exactly the same first-order sentences hold in G, H.

To illustrate some of the above ideas, we give some examples of finitely axiomatizable classes. The class of groups of order at most n and the class of groups of order at least n are axiomatized by the sentences

$$(\forall x_1 \ldots \forall x_{n+1})(\bigvee_{1 \leqslant i < j \leqslant n+1} x_i = x_j)$$

and

$$(\forall x_1 \ldots \forall x_{n-1})(\exists y)(\bigwedge_{1 \leqslant i \leqslant n-1} y \neq x_i).$$

Every finitely based variety of groups is finitely axiomatizable: the variety determined by words $v_1, \ldots, v_r$ in the free group on $X_1, \ldots, X_m$ is axiomatized by the sentence

$$(\forall x_1 \ldots \forall x_m)\Big(\bigwedge_{1 \leqslant i \leqslant r} v_i(x_1, \ldots, x_m) = 1\Big).$$

The sentence

$$(\forall x_1 \ldots \forall x_{n+1})\Big(\Big(\bigwedge_{1 \leqslant i \leqslant n} (\forall y)\, \gamma(x_i, x_{i+1}, y)\Big) \to \Big(\bigvee_{1 \leqslant i \leqslant n} (\forall y)\, \gamma(x_{i+1}, x_i, y)\Big)\Big),$$

in which $\gamma(u, v, y)$ denotes the formula

$$[u, y] = 1 \to [v, y] = 1,$$

axiomatizes the class of groups in which chains of centralizers of elements have length at most n.

Let $H = \{h_1, \ldots, h_n\}$ be a finite group, and set $h_i h_j = h_{\mu(i,j)}$ for all $i, j \leqslant n$. Define formulae $\theta_H, \varphi_H, \psi_H$ as follows:

$$\theta_H(x_1, \ldots, x_n) : \ \Big(\bigwedge_{i \neq j} (x_i \neq x_j) \wedge \bigwedge_{i,j} (x_i x_j = x_{\mu(i,j)})\Big);$$

$$\varphi_H : \ (\exists x_1 \ldots \exists x_n)\, \theta_H(x_1, \ldots, x_n);$$

$$\psi_H : \ (\exists x_1 \ldots \exists x_n)(\forall y)\Big(\theta_H(x_1, \ldots, x_n) \wedge \Big(\bigvee_i y = x_i\Big)\Big).$$

Clearly $G \models \varphi_H$ if and only if G has a subgroup isomorphic to H, and $G \models \psi_H$ if and only if G is isomorphic to H. An obvious modification of this argument shows that if S is a finitely presented simple group then the class of groups in which S may be embedded is finitely axiomatizable, and so therefore is the class of groups having no subgroup isomorphic to S. On the other hand, as already mentioned, the class of groups having no subgroup isomorphic to $\mathbb{Z}$ (i.e., the class of torsion groups) is not axiomatizable.

Now we give some less obvious examples of finitely axiomatizable classes. It was shown by Mal'cev [8] that for $n \geqslant 3$ the class $\{\mathrm{PSL}_n(K) \mid K \text{ a field}\}$ is finitely axiomatizable. This result was extended by Simon Thomas in his dissertation [12]. Let L be an untwisted Lie type, so that

$$\mathrm{L} \in \{\mathrm{A}_n\ (n \geqslant 1), \mathrm{B}_n\ (n \geqslant 2), \mathrm{C}_n\ (n \geqslant 3), \mathrm{D}_n\ (n \geqslant 4), \mathrm{E}_6, \mathrm{E}_7, \mathrm{E}_8, \mathrm{F}_4, \mathrm{G}_2\}.$$

Thomas showed that the class $\{\mathrm{L}(K) \mid K \text{ a field}\}$ is finitely axiomatizable.

On the other hand, the class of finite groups and the class of abelian simple groups are not axiomatizable — this may be seen by taking ultraproducts as in Section 3 below. If $\mathcal{X}$ is a class of finite groups then there is a theory — for example the theory $\{\neg\psi_H \mid H \text{ finite}, H \notin \mathcal{X}\}$ where ψ_H is defined as above — of which $\mathcal{X}$ is the class of ***finite*** models. For special classes of groups there are more natural theories. The class of finite nilpotent groups and the class of finite soluble groups are the classes of finite models of the theories

$$\{(\forall x \forall y)(x^m = 1 \wedge y^n = 1 \rightarrow [x, y] = 1) \mid m, n \text{ coprime}\}$$

and

$$\{(\forall x \forall y)(x^m = 1 \wedge y^n = 1 \wedge (xy)^r = 1 \rightarrow x = 1) \mid m, n, r \text{ coprime}\}.$$

(The latter characterization is a paraphrase of one of the consequences listed by John Thompson in [13] of his determination of the minimal simple groups.) It seems hard to decide whether there exist two single sentences whose classes of finite models are the class of finite nilpotent groups and the class of finite soluble groups.

In some of our later results we shall be concerned with ***axiomatic characterizations of elements*** in certain groups. Trivial instances are characterizations involving formulae such as the following, for elements of order 4, products of two commutators, and elements of the c'th term of the upper central series of a group:

$$g^4 = 1 \wedge g^2 \neq 1;$$

$$(\exists x_1 \exists x_2 \exists x_3 \exists x_4)(g = [x_1, x_2][x_3, x_4]);$$

$$(\forall y_1 \ldots \forall y_c)([g, y_1, \ldots, y_c] = 1).$$

It is easy to prove the following result (see [15], Section 2).

Lemma 1.1. *Let* $G = \mathrm{Alt}(n)$ *where* $n \geqslant 9$. *Then* g *is a product of two disjoint transpositions if and only if* g *is an involution and all elementary abelian normal 2-subgroups of* $C_G(g)$ *have order at most* 4.

This leads to a first-order characterization of products of two disjoint transpositions in alternating groups of degree at least 9: g is such an element if and only if the formula

$$\iota(g) \wedge (\forall x_1 \ldots \forall x_4)((\bigwedge_{1 \leqslant i \leqslant 4} \iota(x_i) \wedge [x_i, g] = 1)$$

$$\wedge((\forall y)([y, g] = 1 \rightarrow (\bigwedge_{i,j \leqslant 4} [x_i, x_j^y] = 1))) \rightarrow \bigvee_{1 \leqslant i < j \leqslant 4} x_i = x_j)$$

holds; here $\iota(h)$ is a formula stating that h is an involution.

The result of Simon Thomas which was mentioned above depends on a first-order characterization of Steinberg generators of an untwisted Chevalley group $\mathrm{L}(K)$ independent of the field K.

Of course there are severe limits on the types of elements which can be characterized by first-order formulae. For a general group G, one cannot characterize the elements of G', the cyclic subgroup generated by an element g, or the normal closure of g. The elements of these subgroups are products of elements from obvious generating sets; however we do not know how many factors are needed, and quantification over $\mathbb{N}$ is not permitted.

2. Ultraproducts

Let I be an infinite index set. A ***filter*** on I is a set $\mathcal{F}$ of subsets of I such that

(i) $\emptyset \notin \mathcal{F}$;

(ii) if $S_1 \in \mathcal{F}$ and $S_1 \subseteq S_2$ then $S_2 \in \mathcal{F}$;

(iii) if $S_1, S_2 \in \mathcal{F}$ then $S_1 \cap S_2 \in \mathcal{F}$.

An ***ultrafilter*** is a filter which satisfies additionally

(iv) for each $S \subseteq I$ either $S \in \mathcal{F}$ or $I \setminus S \in \mathcal{F}$.

Clearly for $x \in I$ the family $\{S \subseteq I | x \in S\}$ is an ultrafilter; such an ultrafilter is called a ***principal*** ultrafilter. An ultrafilter which is not a principal ultrafilter must evidently contain the filter of all subsets of I with finite complements, and must be a maximal filter with respect to this property. Therefore the existence of ***non-principal ultrafilters*** follows from Zorn's lemma. We fix a non-principal ultrafilter $\mathcal{U}$.

Now let $(G_i \mid i \in I)$ be a family of groups indexed by I. Write C for the cartesian product $\prod_{i \in I} G_i$, consisting of all vectors (g_i) with $g_i \in G_i$ for each $i \in I$, and let K be the normal subgroup $\{(g_i) \in C \mid \{i \mid g_i = 1\} \in \mathcal{U}\}$ of C. The ***ultraproduct*** $\prod G_i/\mathcal{U}$ is defined to be the quotient group C/K. Thus two elements of C have the same image in $\prod G_i/\mathcal{U}$ if and only if their entries agree on a set which belongs to $\mathcal{U}$.

Much of the significance of ultraproducts comes from the following fundamental theorem.

Theorem 2.1. *If θ is a first-order sentence then $\prod G_i/\mathcal{U} \models \theta$ if and only if $\{i \mid G_i \models \theta\} \in \mathcal{U}$. In particular, if each G_i satisfies θ, then so does $\prod G_i/\mathcal{U}$.*

For a proof of this theorem and a discussion of the role of ultraproducts in algebra, we refer the reader to the article [3] by Eklov. The next result illustrates the importance of ultraproducts in the model theory of groups.

Theorem 2.2. *Let $\mathcal{Y}$ be a family of groups and let G be a group satisfying all sentences which hold in all groups in $\mathcal{Y}$. Then G is elementarily equivalent to an ultraproduct of groups in $\mathcal{Y}$.*

Proof. Each sentence is equivalent to one whose variables are taken from a fixed countable set X of variables. The set of sentences whose variables are taken from X and which hold in G is countable; we list the sentences in this set as $\theta_1, \theta_2, \dots$. For each $n \in \mathbb{N}$ define $\varphi_n = \bigwedge_{1 \leqslant i \leqslant n} \theta_i$. Thus each φ_n holds in G. For each n there is a group $G_n \in \mathcal{Y}$ satisfying φ_n, since otherwise $\neg\varphi_n$ would hold in all groups in $\mathcal{Y}$ and so would hold in G. For each $m \in \mathbb{N}$, the sentence θ_m holds in all groups G_n with $n \geqslant m$, and so holds in the ultraproduct $\prod_{\mathbb{N}} G_n/\mathcal{U}$, for any non-principal ultrafilter $\mathcal{U}$ on $\mathbb{N}$. Therefore $G \equiv \prod_{\mathbb{N}} G_n/\mathcal{U}$, as required.

Here is another fact about ultraproducts; we leave the proof as an easy exercise.

Exercise 2.3. *Let I be the disjoint union of finitely many subsets $I_1, \ldots, I_r$. Then $I_j \in \mathcal{U}$ for some j. Moreover $\mathcal{V} = \{S \cap I_j \mid S \in \mathcal{U}\}$ is an ultrafilter on I_j, and $\prod_I G_i/\mathcal{U} \cong \prod_{I_j} G_i/\mathcal{V}$.*

3. Simple Groups

All attempts to find a family of first-order sentences expressing the fact that a group is simple are bound to fail. An ultraproduct of the cyclic groups of prime order would satisfy these sentences and would be abelian; and it would be infinite since by Theorem 2.1 for each integer n it would satisfy a sentence expressing the fact that it contains at least n elements. If we restrict attention to non-abelian simple groups the situation does not improve.

Proposition 3.1. *The class of non-abelian simple groups is not axiomatizable; i.e. a group satisfying all sentences holding in all non-abelian simple groups is not necessarily simple.*

Proof. Let $I = \{n \in \mathbb{N} \mid n \geqslant 5\}$ and let $G_n = \mathrm{Alt}(n)$ for each n. Define

$$x_n = (12)(34), \quad y_n = \begin{cases} (1, 2, \ldots, n) & \text{for } n \text{ odd,} \\ (1, 2, \ldots, n-2)(n-1, n) & \text{for } n \text{ even.} \end{cases}$$

Let $\mathcal{U}$ be a non-principal ultrafilter on $\mathbb{N}$ and let x, y be the images in $\prod G_n/\mathcal{U}$ of (x_n), (y_n). Suppose that y is a product of d conjugates of x. Then we may equate components in some set $S \in \mathcal{U}$; we find that y_n is a product of d conjugates of x_n for all $n \in S$. However a product of d conjugates of x_n can move at most $4d$ points and so cannot equal y_n if $n > 4d$. Thus S is finite, and we have a contradiction.

With a little more effort one can prove the following result.

Proposition 3.2 (Felgner). *If $G \equiv \prod \mathrm{Alt}(n_i)/\mathcal{U}$ where $n_i \geqslant 5$ for all $i \in I$ and if G is infinite then G is not simple.*

The proof of this result is similar to the proof of Proposition 3.1, but since G is elementarily equivalent to, and not isomorphic to, the ultraproduct, we cannot construct elements as images of carefully chosen vectors. Instead we must use first-order characterizations of elements of $\mathrm{Alt}(n)$. We have already seen that if $n \geqslant 9$ then there is a first-order formula φ_s characterizing the elements x which are products of two disjoint transpositions, and it is not hard to find a formula φ_l which characterizes the involutions y in $\mathrm{Alt}(n)$ which leave at most 4 points fixed. (See [15], Section 2.) For such elements x, y it is clear that y is not a product of less than $\frac{1}{4}(n-4)$ conjugates of x. Thus, given an integer d, we see that if $n > 4d+4$ then $\mathrm{Alt}(n)$ satisfies the sentence

$$\xi_d = (\exists u \exists w)(\varphi_s(u) \wedge \varphi_l(w)) \wedge (\forall x \forall y)(\varphi_s(x) \wedge \varphi_l(y) \to \neg\eta_d(x, y)),$$

where $\eta_d(x, y)$ is a formula expressing the fact that y is a product of d conjugates of x. If $\{i \mid n_i \leqslant 4d+4\}$ belongs to $\mathcal{U}$ then the ultraproduct has order at most $|\mathrm{Alt}(4d+4)|$ by Theorem 2.1, and so therefore does G. This is a contradiction. Thus $\{i \mid n_i > 4d+4\}$ belongs to $\mathcal{U}$, and we conclude that the ultraproduct satisfies the sentence ξ_d, and so therefore does G. Since this applies for each d, we conclude that $(\exists u \exists w)(\varphi_s(u) \wedge \varphi_l(w))$ holds in G, and that, for such elements, w is not a product of conjugates of u, so that G cannot be simple.

Here is another result, whose proof is very similar to the proof of Proposition 3.1.

Proposition 3.3. *Let $\mathcal{Y}$ be a family of non-abelian simple groups. If for each $d \in \mathbb{N}$ there exist a group $S_d \in \mathcal{Y}$ and an element $s_d \in S_d$ which is not a product of d commutators, then there is a group which is an ultraproduct of groups in $\mathcal{Y}$ (and which therefore satisfies all sentences of all groups in $\mathcal{Y}$) and which is not perfect.*

The proof of Proposition 3.2 shows that the failure of the class of non-abelian simple groups to be axiomatizable is closely linked to the lack of a universal bound for the number of factors required in expressions for one non-trivial element as a product of conjugates of another. Proposition 3.3 shows that the existence of non-perfect groups satisfying the sentences holding in all non-abelian simple groups is related to the impossibility of expressing elements as products of a bounded number of commutators. In the next section, we shall see that this latter difficulty disappears if we restrict attention to finite non-abelian simple groups.

4. Finite Simple Groups: Classification and Commutator Lengths

In 1951, Ore [10] made the conjecture that every element of a finite non-abelian simple group S is a commutator in S. This conjecture seems hard to establish, even after the classification of the finite simple groups. The groups $\mathrm{Alt}(n)$ for $n \geqslant 5$ and $\mathrm{PSL}_n(K)$ for $n \geqslant 2$ were shown by N. Ito [7] and R. C. Thompson [14] to behave in accordance with the conjecture, and the conclusion of the conjecture has been verified for various other simple groups. We shall describe the proof of a somewhat weaker assertion.

Theorem 4.1. *There is an integer k such that every element of a finite non-abelian simple group S is a product of k commutators in S.*

The proof of this result depends on the classification of the finite simple groups, according to which every finite simple groups appears in the following list.
Alternating groups: $\mathrm{Alt}(n),\ n \geqslant 5$;

Chevalley groups:

	Untwisted	Twisted
Classical	$A_n(q), B_n(q), C_n(q), D_n(q)$	${}^2A_n(q), {}^2D_n(q)$
Exceptional	$E_6(q), E_7(q), E_8(q),$ $F_4(q), G_2(q)$	${}^2B_2(q), {}^3D_4(q), {}^2G_2(q),$ ${}^2E_6(q), {}^2F_4(q)$

for certain prime powers q and integers $n \geqslant 1$;

26 sporadic groups.

An elegant strategy for confirming Theorem 4.1 was proposed by Felgner in [4], but a complete proof was only given last year. The proof relies heavily on a recent result of Malle, Saxl and Weigel [9]:

Theorem 4.2 (Malle, Saxl and Weigel). *If S is a simple classical group not of type D_{2m} then S has a conjugacy class $R = R^{-1}$ such that all elements of S are products of 3 elements of R.*

Thus products of 2 elements of R are also products of 3 elements of R, and hence all elements of S are products of 4 elements of R. Since $R = R^{-1}$ it follows that all elements of S are products of 2 commutators. Given this result, there is an *ad hoc* proof (described in [15]) that elements of classical groups of type D_{2m} are products of 3 commutators.

Since there are only finitely many sporadic groups and since elements of simple alternating groups are commutators, it will now be sufficient to show that there is a bound on commutator lengths for each of the seven families of exceptional groups listed above. Suppose that this is not the case, for the family ${}^{\epsilon}X_n$, say; then by Proposition 3.1 there is an ultraproduct H of groups of type ${}^{\epsilon}X_n$ which is not perfect. However the following result was proved by F. Point [11]:

Theorem 4.3. *For each Lie type L, any ultraproduct of groups of type L is a group of type L.*

Moreover, it is known that if H is a Chevalley group of type L over a field K then H is simple unless K is a very small finite field. In particular, H is perfect, and Theorem 4.1 follows from this contradiction.

5. Felgner's Theorem

In view of the results of Section 3 above, the following theorem of Ulrich Felgner is perhaps surprising.

Theorem 5.1. *There is a first-order sentence τ such that a finite group G satisfies τ if and only if G is non-abelian and simple.*

I would like to give a proof of this result. First, define σ to be the sentence

$$(\forall x \forall y)(x \neq 1 \wedge C_G(x,y) \neq \{1\} \rightarrow \bigcap_{g \in G} (C_G(x,y)C_G(C_G(x,y)))^g = \{1\}).$$

It is easy to see that, at some cost in terms of comprehensibility, this can be written out without mention of centralizers.

Claim 1. If G is simple and non-abelian then $G \models \sigma$.

Choose $x, y \in G$ with $x \neq 1$, set $H = C_G(x,y)$ and suppose $H \neq \{1\}$. Clearly $\bigcap_{g \in G}(HC_G(H))^g \lhd G$. If $\bigcap_{g \in G}(HC_G(H))^g \neq \{1\}$ then $HC_G(H) = G$ and H is normal in G, so that $H = G$ and x is in the centre of G, a contradiction.

Claim 2. If G is a finite non-trivial group and $G \models \sigma$ then G has a simple normal subgroup S such that $C_G(S) = \{1\}$.

Write N for the product of the minimal normal subgroups of G and write $N = S \times W$, with S simple. Every finite simple group can be generated by two elements (see Aschbacher and Guralnick [1]). Let $S = \langle x, y \rangle$ where $x \neq 1$, and write $H = C_G(x,y)$. We have $W \leqslant H$ and $S \leqslant C_G(H)$, so that $N \leqslant HC_G(H)$ and $N \leqslant \bigcap_{g \in G}(HC_G(H))^g$. Since $G \models \sigma$ we must therefore have $H = \{1\}$, as required.

Claim 3. If G, S are as in Claim 2 then every element of G' has the form $sc_1c_2c_3$ where $s \in S$ and c_1, c_2, c_3 are commutators; moreover G is simple if and only if $G = G'$.

The action of G on S by conjugation provides an embedding of G in $\mathrm{Aut}(G)$ as a subgroup containing the group $\mathrm{Inn}(S)$ of inner automorphisms. It follows from the classification of the finite simple groups that there is a series

$$1 \lhd K_0 \lhd K_1 \lhd K_2 \lhd K_3 = \mathrm{Aut}(S)/\mathrm{Inn}(S)$$

with K_0 abelian and K_i/K_{i-1} cyclic for $i = 1, 2, 3$. (See Gorenstein [5], Theorem 4.237 and the discussion which precedes it.) So it suffices to show that if $W \lhd H$ and H/W is cyclic then each element of H' has the form $w[h_1, h_2]$ with $w \in W'$ and $h_1, h_2 \in H$, and this is easy.

Completion of the proof. At this point we have to invoke Theorem 4.1; let k be the integer given by this theorem and let δ be a sentence expressing the fact that every element of a group is a product of k commutators. From Claim 1 and Theorem 4.1, every finite non-abelian simple group satisfies $\sigma \wedge \delta$. On the other hand, if G is a finite group satisfying $\sigma \wedge \delta$, then G is evidently perfect, and so simple by Claim 3. Therefore the sentence $\tau = \sigma \wedge \delta$ has the required properties.

The reader will notice that Claim 3 contains more information than we have used. This extra information will be useful in the discussion of simple pseudofinite groups in Section 7.

6. Pseudofinite Groups

The ideas concerning finite simple groups in Sections 4 and 5 allow us to treat some infinite groups, since all that is necessary is that the groups satisfy the first-order sentences valid in finite groups.

The motivation for the next definition comes from field theory. A field K is called ***pseudofinite*** if it is perfect, has just one extension of degree n for all n, and if every absolutely irreducible variety defined over K has a K-valued point. Pseudofinite fields play an important role in field arithmetic. Examples of such fields are many infinite subfields of the algebraic closure of a finite field (those whose Galois group has finite p-Sylow subgroups for all p); and for each finite extension F of $\mathbb{Q}$ there is a pseudofinite field whose subfield of algebraic numbers is F.

It was shown by J. Ax [2] in 1968 that a field K is pseudofinite if and only if it is an infinite model for the theory of finite fields, i.e., if and only if it satisfies all sentences valid in all finite fields. This characterization inspired Ulrich Felgner to introduce the following

Definition. A ***pseudofinite group*** is an infinite model for the theory of finite groups.

Here are some sentences which must hold in pseudofinite groups:

(i) $(\forall x_1 \forall x_2)(x_1^n = x_2^n \to x_1 = x_2)) \leftrightarrow (\forall x \exists y)(x = y^n)$.

This is simply the assertion that the map $x \mapsto x^n$ is injective if and only if it is surjective.

(ii) $C_G(x) \leqslant C_G(x^y) \to C_G(x) = C_G(x^y)$.

This holds in finite and pseudofinite groups since conjugate subgroups of finite groups have the same order.

(iii) It was shown by Higman that the group with the presentation

$$\langle x, y, z, w \mid x^y = x^2, y^z = y^2, z^w = z^2, w^x = w^2 \rangle$$

is non-trivial but has no finite non-trivial images. Therefore the sentence

$$(\forall x \forall y \forall z \forall w)(x^y = x^2 \wedge y^z = y^2 \wedge z^w = z^2 \wedge w^x = w^2 \rightarrow x = 1)$$

holds in pseudofinite groups but not in all groups.

It follows from (i) above that $\mathbb{Z}$ is not a pseudofinite group; on the other hand, $\mathbb{Q}$ is a pseudofinite group. From Theorem 2.1, ultraproducts of finite groups are pseudofinite; and by Theorem 2.2 every pseudofinite group is elementarily equivalent to an ultraproduct of finite groups.

It is easy to obtain examples of pseudofinite groups from pseudofinite fields. Let K be a pseudofinite field, and $n \geqslant 2$. Both $\mathrm{SL}_n(K)$ and $\mathrm{PSL}_n(K)$ are pseudofinite; and $\mathrm{PSL}_n(K)$ is also simple. More generally, if K is a pseudofinite field and L a Lie type such that $\mathrm{L}(K)$ exists, and if $G \equiv \mathrm{L}(K)$, then G is a simple pseudofinite group.

7. Simple Pseudofinite Groups

Ulrich Felgner suggested in [4] that a classification of the simple pseudofinite groups might be possible and made some important steps towards the classification. He conjectured that the pseudofinite simple groups are precisely the groups (isomorphic to) $\mathrm{L}(K)$ for some Lie type L and field K. In [15], I have proved a somewhat weaker result than this:

Theorem 7.1. *If G is a pseudofinite simple group then $G \equiv \mathrm{L}(K)$ for some pseudofinite field K and Lie type* L.

If $G \equiv \mathrm{L}(K)$ *with* L *untwisted* then $G \cong \mathrm{L}(K_1)$ for some $K_1 \equiv K$, from the result of S. Thomas mentioned in Section 1. Thus all that is missing in a determination of the simple pseudofinite groups up to isomorphism is an analogue for twisted groups of this result of Thomas.

I shall now describe the proof of Theorem 7.1. It is divided into two steps, the first of which was carried out by Felgner.

First step in the proof. We show that *if G is as in the theorem then $G \equiv \prod G_i/\mathcal{U}$, for some family $(G_i \mid i \in I)$ of finite simple groups.*

Let G be a simple pseudofinite group. Thus we have $G \equiv \prod G_i/\mathcal{U}$, where each G_i is finite. Since G is simple, the ultraproduct satisfies the sentence σ defined in Section 5, and so by Theorem 2.1 we have

$$J_1 = \{i \in I \mid G_i \models \sigma\} \in \mathcal{U}.$$

Let

$$J_2 = \{i \in J_1 \mid G_i \text{ not simple}\}.$$

If $i \in J_1$ then G_i satisfies the condition given by Claim 2 of the proof of Theorem 5.1; thus if $i \in J_2$ then by Claim 3 we have $G_i \models \gamma$, where γ is a first-order sentence asserting that

the set of products of $k + 3$ commutators is a proper normal subgroup

and k is the integer given by Theorem 4.1. If $J_2 \in \mathcal{U}$ then it follows that $G \models \gamma$, and that G is not simple. We conclude that $I \setminus J_2 \in \mathcal{U}$ and that $J_1 \setminus J_2 = (I \setminus J_2) \cap J_1 \in \mathcal{U}$, so that $\prod G_i/\mathcal{U}$ is isomorphic to the ultraproduct of the G_i with $i \in J_1 \setminus J_2$. This completes the first step.

From the classification of the finite simple groups (together with Exercise 2.3) we can assume that one of the following holds:

(a) for each i, $G_i \cong \mathrm{Alt}(n_i)$, where $n_i \geqslant 5$;

(b) for each i, $G_i \cong {}^{\epsilon}\mathrm{X}_{n_i}(q_i)$, where $\epsilon \in \{1, 2, 3\}$ is fixed, $\mathrm{X} \in \{\mathrm{A}, \mathrm{B}, \ldots, \mathrm{G}\}$ is fixed, the integers n_i, q_i vary, but all q_i are odd;

(c) as in (b), but with all q_i even.

By Proposition 3.2, case (a) does not arise.

Second step in the proof. It remains to treat (b), (c). For clarity we just consider case (b). It suffices to show that if G is simple then the integers n_i are bounded, since then by Exercise 2.3 we can assume that they are equal and Theorem 4.3 shows that the ultraproduct is a Chevalley group. Therefore we are concerned with classical groups. We want to imitate the argument of Proposition 3.2, by finding first-order characterizations of suitable pairs i, j of elements for which we know that expressions of j as a product of conjugates of i must have many factors.

Let V be an n-dimensional vector space, over a field of characteristic not equal to 2. Suppose that $u, w : V \to V$ are linear maps and that there are subspaces W_1, W_2, U of V such that the following conditions are satisfied:

(A) $V = W_1 \oplus W_2$, $|\dim(W_1) - \dim(W_2)| \leqslant 8$ and $\dim(V/U) \leqslant 8$;

(B) $w|_{W_1} = \mathrm{id}$, $w|_{W_2} = -\,\mathrm{id}$, and either $u|_U = \mathrm{id}$ or $u|_U = -\,\mathrm{id}$.

We shall be interested in the case when $u, w \in \mathrm{SL}(V)$, and on expressions of the image of w modulo the centre of this group as a product of conjugates of the image of u. Suppose therefore that $\pm w$ is a product $u^{g_1} \ldots u^{g_d}$ of d conjugates of u.

Case 1. $u|_U = \mathrm{id}$, $w = u^{g_1} \ldots u^{g_d}$. Then $u^{g_i}|_{Ug_i} = \mathrm{id}$, so that w acts as the identity on $\bigcap_i Ug_i$ and hence $W_2 \cap (\bigcap_i Ug_i) = \{0\}$. Clearly $\bigcap_i Ug_i$ has codimension at most

$8d$, and so $\dim(W_2) \leqslant 8d$. Therefore from (i) we have

$$n = \dim(V) \leqslant 8 + 2\dim(W_2) \leqslant 16d + 8.$$

Other cases. Easy calculations yield that $n \leqslant 16d + 8$ in all of the other cases which arise.

Now let S be a finite simple classical group of odd characteristic. Then there exist a vector space V_S over a field of odd characteristic and a form f_S on V_S, such that $S \cong H'_S/Z$, where $H_S = \{h \in \mathrm{GL}(V_S) \mid h \text{ preserves } f_S\}$ and Z is the centre of H'. In [15] we proved the following result.

Theorem 7.2. *There exists a first-order formula $\varphi(x, y)$ such that every group S as above with* $\dim V_S \geqslant 37$ *has the following properties*:

(i) *there exist involutions i, j in S such that $\varphi(i, j)$ holds*;

(ii) *if i, j are involutions in S and $\varphi(i, j)$ holds, then there are preimages u, w of i, j in H'_S and subspaces W_1, W_2, U satisfying the conditions* (A), (B) *above.*

It follows that j is not a product of d conjugates of i for $\dim(V_S) \geqslant 16d + 8$.

It is clear that this result is precisely what is needed to complete an argument, similar to that in Proposition 3.2, to show that a simple ultraproduct of simple classical groups of odd characteristic is isomorphic to such an ultraproduct in which the classical groups have bounded degrees. There is another result in [15] which yields the same conclusion for classical groups of even characteristic. These two results form the main content of the proof of Theorem 7.1. Their proofs are of course purely group-theoretic; they are rather technical, and they depend on information about centralizers of involutions in classical groups.

It is worth emphasising that the characterizations of elements by first-order formulae given in Theorem 7.2 are valid for all classical groups S of odd characteristic; they are independent of the underlying (pseudo-)finite field and the degree and type of S.

Such characterizations of elements by first-order sentences are of course harder to obtain than some characterizations of more traditional types; but the attempt to find them seems to be both instructive and illuminating.

References

[1] Aschbacher, M., Guralnick, R., Some applications of the first cohomology group. J. Algebra 90 (1984), 446–460.

[2] Ax, J., The elementary theory of finite fields. Ann. of Math. (2) 88 (1968), 238–271.

[3] Eklov, P.C., Ultraproducts for algebraists. In: Handbook of Mathematical Logic, 105–138. North-Holland, Amsterdam–New York–London 1977.

[4] Felgner, U., Pseudo-endliche Gruppen. In: Proceedings of the 8th Easter Conference on Model Theory, 82–96. Humboldt–Universität, Berlin 1990.

[5] Gorenstein, D., Finite Simple Groups. An Introduction to Their Classification. Plenum Press, New York, 1982.

[6] Hodges, W., Model Theory. Cambridge University Press, Cambridge 1993.

[7] Ito, N., A theorem on the alternating group $\mathcal{A}_n$ $(n \geq 5)$. Math. Japonicae 2 (1951), 59–60.

[8] Mal'cev, A. I., Elementary properties of linear groups. In: The metamathematics of algebraic systems, collected papers, 1936–1967. North-Holland, Amsterdam 1971.

[9] Malle, G., Saxl, J., Weigel, T., Generation of classical groups. Geom. Dedicata 49 (1994), 85–116.

[10] Ore, O., Some remarks on commutators. Proc. Amer. Math. Soc. 2 (1951), 307–314.

[11] Point, F., Ultraproducts of Chevalley groups. Submitted for publication.

[12] Thomas, S., Classification theory of simple locally finite groups. Ph.D dissertation, University of London 1983.

[13] Thompson, J. G., Nonsolvable finite groups all of whose local subgroups are solvable (Part I). Bull. Amer. Math. Soc. 74 (1968), 383–437.

[14] Thompson, R. C., Commutators of matrices with prescribed determinant. Canad. J. Math. 20 (1968), 203–221.

[15] Wilson, J. S., On simple pseudofinite groups. J. London Math. Soc. (2). To appear.

From Finite to Infinite Groups: a Historical Survey

Guido Zappa

1991 Mathematics Subject Classification: 20-03

In the following we propose to show how in the historical development of the theory of groups there has been a gradual passage from an almost exclusive interest in finite groups to a much wider vision, in which infinite groups also play a relevant role. Thus our survey will outline the development of group theory from its origins to its status at the end of the Second World War, a time in which the theory of infinite groups was solidly founded and in a phase of rapid development.

The Origin of the Group Concept

A very deep investigation concerning the origin of the concept of group is given in the book "Die Genesis des abstrakten Gruppentheorie" by Hans Wussing, published in 1969. This author finds elements of "implizite Gruppentheorie" (implicit group theory) not only in the geometry of the first decades of the last century, but also in the works of Euler on the remainders of powers of a natural number with respect to a given prime number, and in the research of Gauss concerning congruences and composition of algebraic forms.

On the other hand, the notion of a permutation group can also be found in some sense in the works of Lagrange, Ruffini and Abel on the solution of algebraic equations. A precise definition of a permutation group did not appear until the works of Galois and Cauchy.

To understand the way in which Galois introduced the notion of a group, one must carefully distinguish between the words "permutation" and "substitution" in his writings.

Given n symbols $x_1, \ldots, x_n$, all the possible orders in which these symbols can be placed he called "permutations" on $x_1, \ldots, x_n$, and the operations on these symbols

Offprint from
Infinite Groups 94, Eds.: de Giovanni / Newell

which move one permutation to another he called "substitutions". A set H of permutations on these symbols such that every substitution taking *a given* permutation of H onto *a prescribed* permutation of H assigns to *each* permutation of H a permutation of H, is what Galois called a "permutation group" on $x_1, \ldots, x_n$. The substitutions obtained in this way (that is the set of all substitutions stabilizing H) form what Galois called "group of substitutions" associated with H.

One of the difficulties one encounters reading the works of Galois lies in his use of the word "group" in both of the above mentioned contexts. Later the concept of a "permutation group" in the sense of Galois was no longer looked at, and only his so-called "substitution groups" were considered. These latter are now called "permutation groups".

Before the posthumous publication of the last works of Galois, which appeared in 1847 as a result of Liouville's initiative, Cauchy had already written several papers on substitution groups. Instead of "permutation" in the sense of Galois, he used the term "arrangement", and instead of "substitution group" he used "system of conjugate substitutions". The study of substitution groups was continued by Betti, Serret and especially by Jordan, who in his "Traité des substitutions" of 1870 gave an organic exposition of the whole theory.

On the other hand, by 1854 Cayley had already given a more general definition of groups. In his memoir entitled "On the theory of groups as depending on the symbolic equation $\vartheta^n = 1$" he said:

"A set of symbols $1, \alpha, \beta, \ldots$ all of them different, and such that the product of any two of them (no matter in what order), or the product of any one of them into itself, belongs to the set, is said to be a *group*. It follows that if the entire group is multiplied by any one of the symbols, either as further or nearer factor, the effect is simply to reproduce the group ..."

This was the first time in which it was not assumed that the set considered is finite. However, the definition is not completely clear. It is not evident how the definition given by Cayley should imply the existence and uniqueness of the right and left inverse. The examples given by Cayley in his paper all refer to finite groups. Because of this ambiguous definition the term "group" has often been used in the last century with different meanings.

The Study of Relevant Classes of Infinite Groups

The first investigations in which infinite groups were considered explicitly are due to Jordan. In 1867, he published in the "Comptes Rendus" of Paris an article entitled "Sur les groupes des mouvements" and one year later a memoir with the same title in "Annali di Matematica Pura ed Applicata". He was inspired by the studies of the mineralogist August Bravais on crystallographic lattices and the movements taking a lattice onto itself. Jordan considered in particular all movements which can be obtained by combining certain given movements, that is the group generated by these. He proved that the groups considered by him can only be of 174 different types.

In 1870 two young mathematicians, the German Felix Klein and the Norwegian Sophus Lie, were in Paris and become acquainted with Jordan. This was the starting point of the development of the theory of continuous groups of transformations due to Lie, and of that of infinite discontinuous groups of transformations, to which Klein made fundamental contributions. With these theories we clearly enter the area of infinite groups, but as yet only as transformation groups.

The Origin of the Abstract Group Concept

The research of the German Walter Dyck contained in the memoir "Gruppentheoretische Studien", published in the "Mathematische Annalen" (1882/83) had a more abstract character. Dyck was the first to introduce a group by means of a set of generators and a set of relations between them. First he considered the group G generated by n operations $A_1, \dots, A_n$ without any relation (that is, in modern terminology, the free group on $A_1, \dots, A_n$). If $\bar{G}$ is any group generated by n elements $\bar{A}_1, \dots, \bar{A}_n$, Dyck constructed an epimorphism of G onto $\bar{G}$ which maps A_1 in $\bar{A}_1$, A_2 in $\bar{A}_2$, etc. Moreover, he proved that the operations of G to which the identity of $\bar{G}$ corresponds form a normal subgroup of G. He also gave an interesting geometric representation of a group with n generators.

The works of Dyck appeared in a period in which the concept of an abstract group was gaining in strength. A first sign of this notion can be found in lectures given by Dedekind in Göttingen during 1856/57; however these were not published and remained generally unknown. They were finally published in 1981, on the initiative of W. Scharlau.

In 1879, in an article published in the "Journal von Crelle", Frobenius and Stickelberger gave an axiomatic theory of finite abelian groups, and in 1887 Frobenius, inspired by a paper of Kronecker of 1870, gave an axiomatic definition of a finite group in an article devoted to a new proof of Sylow's theorem and published in the same journal.

In 1893 H. Weber analyzed, in a paper published in the "Mathematische Annalen", the axioms of group theory, making a distinction between finite and infinite groups. For finite groups the axioms of associativity and cancellation are sufficient, while for infinite groups it is necessary to assume also the existence of the right and left inverse. Weber's book of 1895 "Lehrbuch der Algebra" was the first treatise in which groups of elements were considered. On the other hand, the classical treatise of Burnside of 1897 still considered groups of operations.

The first book on the theory of abstract groups is published in 1905 by J. A. de Séguier. But it dealt only with finite groups. One had to wait until 1916 to find a treatise in which the theory of groups is developed in various chapters without distinction between the finite and the infinite case. This is the book "Abstraktnaja teorija grupp" written in Russian by O. Ju. Schmidt and published in Kiev. Schmidt is the master of the Russian school of group theory, which is well-known to have produced very important results on infinite groups.

A relevant work for the abstract theory of groups is the one by Burnside in 1902, in which he put the celebrated question named after him: let G be a group generated by finitely many elements such that the orders of all its elements divide a given natural number; must G be finite?

However, to find an author who gives many important contributions to the theory of infinite groups one has to wait until 1917. The Dane J. Nielsen, reconsidering the works of Dyck, developed the theory of free groups. Among other things he proved that, if G is a free group with finite basis, then every non-trivial subgroup of G is free, and exhibited a set of generators and defining relations for the group of all automorphisms of a free group with a finite basis.

The Development of the Theory of Infinite Groups

Between 1920 and 1930 there was a strong development in the study of infinite abstract groups. First of all, there were relevant results on infinite abelian groups. Following the works of F. Levi (1917) and Schmeidler (1920) on countable abelian groups, Hans Prüfer made fundamental contributions to the theory of torsion abelian groups between 1923 and 1925. Clearly, it was enough to consider primary groups. Prüfer showed that every abelian p-group (where p is a prime) such that the orders of its elements are bounded, is a direct sum of cyclic p-groups. Recall that an element of an abelian p-group is said to have ***infinite height*** if it is divisible by each power of p. Prüfer also proved that every countable abelian p-group without elements of infinite height is a direct sum of cyclic subgroups.

Beginning in 1924, O. Schreier gave very important results concerning both finite and infinite groups. The theory of group extensions is due to him. The aim of this theory, as is well-known, is to construct every group G containing a normal subgroup N isomorphic to a given group, and such that G/N is isomorphic to a given group F. In this connection, Schreier introduced the well-known notions of sets of automorphisms and of factor sets.

Schreier also improved the knowledge of free groups. He showed that the rank of a free group (that is the cardinality of a set of free generators) is an invariant of the group, and that every subgroup of a free group is free, generalizing the already quoted theorem of Nielsen. Moreover, given a free group of finite rank n and a subgroup of finite index j, he obtained the rank of the subgroup as a function of n and j.

Schreier also extended the theorems of Jordan–Hölder to the infinite case, and showed that two normal or principal series of a group always admit isomorphic refinements.

Already in 1911, R. Remak had proved that any two decompositions of a finite group as a direct product of irreducible subgroups are isomorphic. In 1928, Schmidt generalized this theorem to a large class of infinite groups.

More Developments between 1930 and 1940

During the decade 1930–1940 there was further important progress in the theory of infinite groups. Since there are so many results in this area, we will restrict ourselves to those which seem especially relevant to us.

Concerning primary abelian groups, Ulm studied between 1933 and 1935 the case in which elements of infinite height exist. Let G be a countable primary abelian group which is reduced (that is without divisible subgroups). The elements of infinite height form a subgroup G^1 of G; the elements of G^1 having infinite height in G^1 form a subgroup G^2 of G^1, etc. We can continue this construction by transfinite induction. The first ordinal number τ such that $G^\tau = 1$ is called the ***type*** of G. The factor groups $G^\alpha / G^{\alpha+1}$ (for all ordinals $\alpha < \tau$) are called the ***Ulm factors*** of G. Ulm showed that if two countable primary abelian reduced groups have the same type and isomorphic Ulm factors, then they are isomorphic. Zippin proved in 1935 the existence of countable primary abelian groups of type τ, for every ordinal number τ which is countable. In this period also R. Baer made important contributions to the theory of primary abelian groups. Among other facts, he showed, in a paper of 1935, that if the type of a primary abelian reduced group is > 1, the group cannot be decomposed as a direct sum of indecomposable groups.

In the decade 1930–1940 the theory of torsion-free abelian groups and that of mixed abelian groups seem to have been less developed than that of torsion abelian groups. We just cite here some relevant results. In 1931 Baer characterized the torsion-free abelian groups of rank 1 (that is such that any two distinct elements are linearly dependent). Later the knowledge of torsion-free abelian groups of finite rank n (that is such that $n + 1$ distinct elements are always linearly dependent) was improved by the works of Kurosh, Derry and Mal'cev. Regarding mixed abelian groups, we mention an article of Baer of 1936, concerning sufficient conditions for a mixed abelian group to be a direct sum of a torsion and a torsion-free group.

Concerning infinite non-abelian groups, we mention above all the works of Kurosh of 1934 on the free decomposition of a subgroup of a free product, and the work of Baer and Levi of 1936, where it was proved that any two free decompositions of a group always admit isomorphic refinements.

We also mention that in 1938, Dietzmann, Kurosh and Uzkov extended Sylow's theorem to infinite groups. To be precise, they showed that if G is a group and P is a maximal p-subgroup (p prime) of G with only finitely many conjugates, then all maximal p-subgroups of G are conjugate, and their number is of the form $1 + kp$, where k is a non-negative integer. Moreover, Dietzmann in 1939 gave the first example of an infinite p-group with trivial centre. In this period also S. N. Černikov obtained various theorems about infinite soluble and infinite nilpotent groups.

Books on Infinite Groups

Finally we mention the main books which appeared in this period concerning infinite groups. Subsequent to Schmidt's book of 1912, which we have already mentioned, we have to wait until 1937 to find another treatise on groups not restricted to the finite case, when the work "Lehrbuch der Gruppentheorie" by H. Zassenhaus appeared. In particular, this book contained the theorems of Schreier on the existence of a common refinement for normal and principal series of subgroups of a group, and the theory of extensions of groups also due to Schreier.

Similar remarks apply to the book "Gruppi astratti" of Gaetano Scorza, published in 1942, three years after the author's death.

In Moscow in 1944 Kurosh published his book "Teorija grupp" in Russian. Its first translation into another language was the German edition of 1953. It can be said that in this book all the main results known at that time concerning infinite groups were clearly exposed. Two chapters were devoted to infinite abelian groups, one to free groups and free products. We can say that, at this stage, the contemporary history of infinite groups began.

Memory of Two Eminent Group Theorists

Our survey comes to an end. We have outlined how the imposing edifice the theory of infinite groups has become was constructed step by step during a long period, by many researchers, with constant, humble and continuous work. I would like to pay homage here to the memory of two great algebraists who, as we have seen, have been pioneers in this theory: Reinhold Baer and Alexander Kurosh. The first, a German, taught in Germany, Great Britain, the United States, again in Germany and finally in Switzerland, he was virtually a world citizen; the other resided almost permanently in the country known at that time as Soviet Union. I knew both personally, and I could appreciate at first hand the greatness of their mathematical genius, as well as the depth of their human qualities.

List of Participants

Bernhard Amberg, Fachbereich 17 - Mathematik, Universität Mainz, Saarstraße 21, 55122 Mainz, Germany

Stylianos Andreadakis, Department of Mathematics, University of Athens, Panistemiopolis, 15781 Athens, Greece

Bernard Aupetit, Département de Mathématiques et de Statistique, Université Laval Ste-Foy, Quebec G1K 7P4, Canada

Christine Ayoub, Department of Mathematics, Pennsylvania State University, 120 Ridge Ave, State College, PA 16803-3522, USA

Raymond G. Ayoub, Department of Mathematics, Pennsylvania State University, McAllister Building, University Park, PA 16802-6401, USA

Marco Barlotti, Dipartimento di Matematica, Università di Firenze, Via Lombroso 6/17, 50134 Firenze, Italy

Russell D. Blyth, Department of Mathematics, St. Louis University, St. Louis, MO 63103-2007, USA

Brunella Bruno, Dipartimento di Matematica Pura ed Applicata, Via Belzoni 7, 35131 Padova, Italy

Robert Bryce, Department of Mathematics, Australian National University, Canberra 0200, Australia

Joseph Buckley, Department of Mathematics, Western Michigan University, Kalamazoo, MI 49008-5152, USA

Dietrich Burde, Mathematisches Institut der Universität Bonn, Wegelerstraße 10, 53115 Bonn, Germany

Colin M. Campbell, Mathematical Institute, University of St. Andrews, North Haugh, St Andrews KY16 9SS, Scotland, UK

Giuseppina Casadio Zappa, Via Quintino Sella 45, 50136 Firenze, Italy

Carlo Casolo, Dipartimento di Matematica e Informatica, Via Zanon 6, 33100 Udine, Italy

Charles Cassidy, Département de Mathématiques et de Statistique, Université Laval Ste-Foy, Quebec G1K 7P4, Canada

Francesco Catino, Dipartimento di Matematica, Università di Lecce, Via Provinciale Lecce-Arnesano c.p. 193, 73100 Lecce, Italy

Maria Rosaria Celentani, Dipartimento di Matematica e Applicazioni, Università di Napoli "Federico II", Via Cintia, 80126 Napoli, Italy

Yu Chen, Dipartimento di Matematica, Università di Torino, Via Carlo Alberto 10, 10123 Torino, Italy

Maria Cristina Cirino Groccia, Dipartimento di Matematica e Applicazioni, Università di Napoli "Federico II", Via Cintia, 80126 Napoli, Italy

Gabriella Corsi, Dipartimento di Matematica, Università di Firenze, Viale Morgagni 67/A, 50134 Firenze, Italy

Giovanni Cutolo, Dipartimento di Matematica e Applicazioni, Università di Napoli "Federico II", Via Cintia, 80126 Napoli, Italy

Francesca Dalla Volta, Dipartimento di Matematica, Università di Milano, Via Saldini 50, 20133 Milano, Italy

Ulderico Dardano, Dipartimento di Matematica e Applicazioni, Università di Napoli "Federico II", Via Cintia, 80126 Napoli, Italy

Rex S. Dark, Department of Mathematics, University College, Galway, Ireland

Oliver Dickenschied, Fachbereich 17 - Mathematik, Universität Mainz, Saarstraße 21, 55122 Mainz, Germany

Dikran Dikranjan, Dipartimento di Matematica e Informatica, Via Zanon 6, 33100 Udine, Italy

Graham Ellis, Department of Mathematics, University College, Galway, Ireland

Ahmad Erfanian, Department of Pure Mathematics, University College P. O. Box 78, Cardiff CF1 1XL, UK

Temple Fay, Department of Mathematics, University of Southern Mississippi, Hattiesburg, MS 39406, USA

Valeria Fedri, Dipartimento di Matematica, Università di Firenze, Viale Morgagni 67/A, 50134 Firenze, Italy

Maria Rosaria Formisano, Dipartimento di Matematica e Applicazioni, Università di Napoli "Federico II", Via Cintia, 80126 Napoli, Italy

Norberto Gavioli, Dipartimento Pura ed Applicata, Università de L'Aquila, Via Vetoio, 67010 Coppito (L'Aquila), Italy

Anna Luisa Gilotti, Dipartimento di Matematica, Università di Bologna, Piazza di Porta S. Donato 5, 40127 Bologna, Italy

Francesco de Giovanni, Dipartimento di Matematica e Applicazioni, Università di Napoli "Federico II", Via Cintia, 80126 Napoli, Italy

Rüdiger Göbel, Fachbereich 6 - Mathematik, Universität Essen, 45141 Essen, Germany

Karl W. Gruenberg, School of Mathematical Science, Queen Mary College, Mile End Road, London E1 4NS, UK

Fritz Grunewald, Mathematisches Institut der Universität Bonn, Wegelerstraße 10, 53115 Bonn, Germany

Chander Kanta Gupta, Department of Mathematics and Astronomy, University of Manitoba, Winnipeg, Manitoba, Canada R3T 2N2

Brian Hartley, Department of Mathematics, University of Manchester, Manchester M13 9PL, UK

Frieder Haug, Mathematisches Institut, Universität Tübingen, Auf der Morgenstelle 10, 72076 Tübingen, Germany

Hermann Heineken, Mathematisches Institut, Universität Würzburg, Am Hubland, 97074 Würzburg, Germany

Wolfgang Herfort, Technische Universität, Wiedner Hauptstrasse 8-10, 1151 Wien, Austria

Burkhard Höfling, Fachbereich 17 - Mathematik, Universität Mainz, Saarstraße 21, 55122 Mainz, Germany

Waldemar M. Hołubowski, Institute of Mathematics, Silesian Technical University, Ul. Kaszubska 23, 44-101 Gliwice, Poland

Mark Hopfinger, Department of Mathematics, St. Louis University, St. Louis, MO 63103-2007 , USA

Ted Hurley, Department of Mathematics, University College, Galway, Ireland

Enrico Jabara, Via Torino 153, Venezia, Italy

Louise-Charlotte Kappe, Department of Mathematical Sciences, SUNY at Binghamton, Binghamton, NY 13902-6000, USA

Wolfgang Kappe, Department of Mathematical Sciences, SUNY at Binghamton, Binghamton, NY 13902-6000, USA

Manfred Karbe, W. de Gruyter & Co., Genthiner Straße 13, 10785 Berlin , Germany

Lev S. Kazarin, Department of Mathematics, Yaroslavl' State University, ul. Sovetskaja 14, 150000 Yaroslavl', Russia

Otto H. Kegel, Mathematisches Institut, Universität Freiburg, Albertstraße 23 b, 79104 Freiburg i. Br., Germany

László G. Kovács, School of Mathematical Sciences, Australian National University, Canberra 0200, Australia

Leonid A. Kurdachenko, Department of Algebra, University of Dnepropetrovsk, Gagarin-Prospekt 72, Dnepropetrovsk, Ukraine

Tommaso Landolfi, Dipartimento di Matematica e Applicazioni, Università di Napoli "Federico II", Via Cintia, 80126 Napoli, Italy

Felix Leinen, Fachbereich 17 - Mathematik, Universität Mainz, Saarstraße 21, 55122 Mainz, Germany

John C. Lennox, Department of Pure Mathematics, University College P. O. Box 78, CF1 1XL Cardiff, UK

Antonella Leone, Dipartimento di Matematica e Applicazioni, Università di Napoli "Federico II", Via Cintia, 80126 Napoli, Italy

Alexander Lichtman, Department of Mathematics, University of Wisconsin-Parkside, Wood Road Box 2000, Kenosha, WI 53141-2000, USA

Andrea Lucchini, Dipartimento di Elettronica per l'Automazione, Università di Brescia, Via Branze, 25123 Brescia, Italy

Maria Silva Lucido, Dipartimento di Matematica Pura ed Applicata, Via Belzoni 7, 35131 Padova, Italy

Olga Macedońska, Institute of Mathematics, Silesian Technical University, Ul. Kaszubska 23, 44-101 Gliwice, Poland

Mario Mainardis, Dipartimento di Matematica e Informatica, Via Zanon 6, 33100 Udine, Italy

Mercede Maj, Dipartimento di Matematica e Applicazioni, Università di Napoli "Federico II", Via Cintia, 80126 Napoli, Italy

Avinoam Mann, Department of Mathematics, The Hebrew University, Givat Ram, Jerusalem 91904, Israel

Consuelo Martínez, Departamento de Matematicas, Universidad de Oviedo, 33007 Oviedo, Spain

John McDermott, Department of Mathematics, University College, Galway, Ireland

Federico Menegazzo, Dipartimento di Matematica Pura ed Applicata, Via Belzoni 7, 35131 Padova, Italy

Marta Morigi, Dipartimento di Matematica Pura ed Applicata, Via Belzoni 7, 35131 Padova, Italy

Franco Napolitani, Dipartimento di Matematica Pura ed Applicata, Via Belzoni 7, 35131 Padova, Italy

Martin L. Newell, Department of Mathematics, University College, Galway, Ireland

Werner Nickel, Lehrstuhl D für Mathematik, RWTH Aachen, Templergraben 64, 52056 Aachen, Germany

Chiara Nicotera, Dipartimento di Matematica e Applicazioni, Università di Napoli "Federico II", Via Cintia, 80126 Napoli, Italy

Veacheslav Obraztsov, Industrial Str. D8 Ko 110, 156016 Kostroma, Russia

Virgilio Pannone, Dipartimento di Matematica, Università di Firenze, Viale Morgagni 67/A, 50134 Firenze, Italy

Gemma Parmeggiani, Dipartimento di Matematica Pura ed Applicata, Via Belzoni 7, 35131 Padova, Italy

Richard E. Phillips, Department of Mathematics, Michigan State University, East Lansing, MI 48824-0001, USA

Rossana Piemontino, Dipartimento di Matematica e Applicazioni, Università di Napoli "Federico II", Via Cintia, 80126 Napoli, Italy

Peter Plaumann, Mathematisches Institut der Universität Erlangen-Nürnberg, Bismarckstraße, 91054 Erlangen, Germany

Boris Plotkin, Institute of Mathematics, Hebrew University, Givat Ram, Jerusalem 91904, Israel

Eugene Plotkin, Department of Mathematics and Computer Science, Bar-Ilan University, 52900 Ramat Gan , Israel

Yakov D. Polovickii, Boltshevistkaya St. 198 - 33, 614068 Perm, Russia

Orazio Puglisi, Dipartimento di Matematica, Università di Trento, 38050 Povo (Trento), Italy

Salvatore Rao, Dipartimento di Matematica e Applicazioni, Università di Napoli "Federico II", Via Cintia, 80126 Napoli, Italy

Evagelos Raptis, Department of Mathematics, University of Athens, Panistemiopolis, 15781 Athens, Greece

Luis Ribes, Department of Mathematics and Statistics; Carleton University, Ottawa, Ontario, Canada K15 5B6

James B. Riles, Department of Mathematics, St. Louis University, St. Louis, MO 63103-2007, USA

Silvana Rinauro, Dipartimento di Matematica, Università della Basilicata, Via N. Sauro, 85100 Potenza, Italy

Edmund F. Robertson, Mathematical Institute, University of St. Andrews, North Haugh, St Andrews KY16 9SS, Scotland, UK

Derek J.S. Robinson, Department of Mathematics, University of Illinois, 1409 W. Green Street, Urbana, IL 61801-2917, USA

Alessio Russo, Dipartimento di Matematica e Applicazioni, Università di Napoli "Federico II", Via Cintia, 80126 Napoli, Italy

Susan E. Schuur, Department of Mathematics, Michigan State University, East Lansing, MI 48824-0001, USA

Carlo M. Scoppola, Dipartimento di Matematica, Università di Roma "La Sapienza", Piazzale A. Moro 5, 00185 Roma, Italy

Dan Segal, All Souls College, Oxford OX1 4AL, UK

Vladimir I. Senashov, Computing Center SO RAN, 660036 Krasnoyarsk, Russia

Luigi Serena, Istituto di Matematica, Facoltà di Architettura, Università di Firenze, Via dell'Agnolo 14, 50122 Firenze, Italy

Vladimir Shpilrain, Fakultät für Mathematik, Ruhr-Universität, 44801 Bochum, Germany

Pavel Shumyatsky, Department of Mathematics, Technion-Israel Institute of Technology, Haifa 32000, Israel

Howard Smith, Department of Mathematics, Bucknell University, Lewisburg, PA 17837-2005, USA

Panagiotis Soules, Department of Mathematics, University of Athens, Panistemiopolis, 15781 Athens, Greece

Lucia Serena Spiezia, Dipartimento di Matematica e Applicazioni, Università di Napoli "Federico II", Via Cintia, 80126 Napoli, Italy

A. G. R. Stewart, Mathematics Department, University of Zimbabwe, Box MP 167 MT, Pleasant Harare, Zimbabwe

Andrzej Strojnowski, Wydriat Matematyzki, ul. Banacha 2, 02-097 Warsaw, Poland

Igor Ya. Subbotin, 1617 N. Fuller Ave, Los Angeles, CA 90046, USA

Vital J. Sushchanskii, Department of Mathematics and Mechanics, Kiev University, Vladimirska St. 64, Kiev, Ukraine

Yaroslav P. Sysak, Department of Mathematics and Mechanics, Kiev University, ul. Dobrokhotova 28, 252142 Kiev, Ukraine

Maria Chiara Tamburini, Facoltà di Scienze, Università Cattolica del Sacro Cuore, Via Trieste 17, 25121 Brescia, Italy

Umberto Tiberio, Dipartimento di Matematica, Università di Firenze, Viale Morgagni 67/A, 50134 Firenze, Italy

Cesarina Tibiletti, Dipartimento di Matematica, Università di Milano, Via Saldini 50, 20133 Milano, Italy

Sean Tobin, Department of Mathematics, University College, Galway, Ireland

M. J. Tomkinson, Department of Mathematics, University of Glasgow, University Gardens, Glasgow G12 8QV, UK

Jiří Tůma, Mathematics Institute, Academy of Science, Zitná 25, 11567 Praha 1, Czech Republic

Dimitrios Varsos, Department of Mathematics, University of Athens, Panistemiopolis, 15781 Athens, Greece

Nikolai Vavilov, Fakultät für Mathematik, Universität Bielefeld, Universitätstraße, 33615 Bielefeld, Germany

Giovanni Vincenzi, Dipartimento di Ingegneria Informatica e Matematica Applicata, Università di Salerno, Baronissi (Sa), Italy

Samuel M. Vovsi, Department of Mathematics, Trenton State College, Trenton, NJ 08650-4700, USA

Gary L. Walls, Department of Mathematics, University of Southern Mississippi, Hattiesburg, MS 39406, USA

James Ward, Department of Mathematics, University College, Galway, Ireland

James Wiegold, Department of Pure Mathematics, University College, P. O. Box 78, Cardiff CF1 1XL, UK

John S. Wilson, Department of Pure Mathematics and Mathematical Statistics, University of Cambridge, 16 Mill Lane, Cambridge CB2 1SB, UK

Giovanni Zacher, Dipartimento di Matematica Pura ed Applicata, Via Belzoni 7, 35131 Padova, Italy

Pavel A. Zalesskii, Institute of Engineering Cybernetics of the Academy of Sciences of Belarus, 6 Sunganov St., 220605 Minsk, Belarus

Guido Zappa, Via Quintino Sella 45, 50136 Firenze, Italy

Efim Zelmanov, Department of Mathematics, University of Wisconsin-Madison, 480 Lincoln Drive, Madison 53706, USA

Irene Zimmermann, Karlstrasse 69, 79104 Freiburg, Germany

List of Lectures

Invited Lectures

B. Amberg
Some New Results about Factorized Groups

R. Göbel
Automorphism Groups - Some Methods Borrowed from Model Theory and Group Theoretic Results

K. W. Gruenberg
The Genus of Nilpotent Groups

C. K. Gupta
Some Non-Finitely Based Varieties of Groups and Group Representations

B. Hartley
Periodic Locally Soluble Groups with Chernikov Sylow Subgroups

O. H. Kegel
Abelian Subgroups of Locally Finite Groups

J. C. Lennox
Centrality in Soluble Groups

F. Menegazzo
On Groups in Which Every Subgroup is Subnormal

D. J. S. Robinson
Normalizers of Subnormal Subgroups

D. Segal
Subgroup Growth and Related Problems

Ya. P. Sysak
Some Examples of Factorized Groups and Their Relation to Ring Theory

M. J. Tomkinson
Infinite Extraspecial Groups

J. Wiegold
Finiteness Conditions for Groups

J. S. Wilson
First Order Group Theory

G. Zappa
Des Groupes Finis aux Groupes Infinis: Revue Historique

Afternoon Talks

C. M. Campbell
Defining Relations for Hurwitz Groups

Y. Chen
On Classificational of Rational Subgroups of Reductive Algebraic Groups over Integral Domains

O. Dickenschied
On the Adjoint Group of a Radical Ring

D. Dikranjan
Isomorphisms between Homeomorphic Topological Groups

A. Erfanian
Some New Results on Growth Sequences

T. Fay
Categorical Compactness for a Class of R-Groups

F. de Giovanni
Locally Finite Products of Locally Nilpotent Groups

F. Haug
The Universality of the Automorphism Group of the Real Line

H. Heineken
Groups Admitting an Automorphism Fixing All Non-Normal Subgroups

W. Herfort
On Certain Fixed-Point-Free Automorphisms of a Free Pro-p-Group

B. Höfling
On Subgroups of Certain Products of Two Locally Nilpotent Groups

W. Holubowski
Symmetric Words in Nilpotent Varieties

L. C. Kappe
Metabelian Groups with All Cyclic Subgroups of Bounded Defect

L. S. Kazarin
On the Derived Length of a Solvable Product of Two Groups

L. G. Kovács
Torsion-free Varieties of Metabelian Groups

L. A. Kurdachenko
On Groups with Certain Restrictions on Conjugacy Classes

F. Leinen
Hypercentral Unipotent Finitary Linear Groups

O. Macedońska
On Fixed Points for Permutation of Generators in 2-Generator Groups

C. Martínez
Power Subgroups of Profinite Groups

W. Nickel
Computing in Nilpotent Quotients of Finitely Presented Groups

V. Obratzsov
Some Applications of a New Embedding Scheme

R. E. Phillips
Finitary Linear Groups

B. Plotkin
Varieties of Representations of Groups, Quasivarieties and Pseudovarieties of Groups

Ya. D. Polovickii
Infinite Groups with Chernikov Classes of Conjugate Subsets of Certain Types

E. Raptis
On Profinite Topology of the Automorphism Group of a Free Group

L. Ribes
Conjugacy Separability of Free Products of Groups with Amalgamation

S. Rinauro
Intersection and Join of Almost Normal subgroups

E. F. Robertson
Theorems of Reidemeister–Schreier Type

V. I. Senashov
Applications of Layer-Finite Group Theory

V. Shpilrain
Monomorphisms of Free Groups

P. Shumyatsky
On Centralizers in Locally Solvable Periodic Groups

P. Soules
Groups with Small Commutator Subgroups of Two-Generator Subgroups

L. S. Spiezia
A Combinatorial Property Related to Engel Conditions

A. Strojnowski
Virtually Abelian Torsion-Free Groups with Trivial Centers

I. Ya. Subbotin
Groups with Pronormal Subgroups of Normal Subgroups

J. Tůma
HNN-Extensions and Lattices of Subgroups

N. Vavilov
The Structure of Unitary Groups over Rings

D. Varsos
On the Automorphism Group of Certain Graph Products

S. M. Vovsi
Growth of Relatively Free Groups

G. L. Walls
On a Class of R-Groups

P. A. Zalesskii
Normal Subgroups of Free Constructions of Profinite Groups

E. Zelmanov
On Narrow Groups

List of Contributors

Bernhard Amberg, Fachbereich 17 - Mathematik, Universität Mainz, Saarstraße 21, 55122 Mainz, Germany

James C. Beidleman, Department of Mathematics, University of Kentucky, Lexington, KY 40506-0027 , USA

Dikran Dikranjan, Dipartimento di Matematica e Informatica, Via Zanon 6, 33100 Udine, Italy

Martyn R. Dixon, Mathematics Department, University of Alabama, Tuscaloosa, AL 35487-0350, USA

Manfred Dugas, Department of Mathematics, Baylor University, Waco, TX 76798-7328, USA

Silvana Franciosi, Dipartimento di Matematica e Applicazioni, Università di Napoli "Federico II", Via Cintia, 80126 Napoli, Italy

David J. Garrison, Loral Federal Systems-Owego, 1801 State Route 17C, Owego, NY 13827, USA

Francesco de Giovanni, Dipartimento di Matematica e Applicazioni, Università di Napoli "Federico II", Via Cintia, 80126 Napoli, Italy

Rüdiger Göbel, Fachbereich 6 - Mathematik, Universität Essen, 45141 Essen, Germany

Brian Hartley, Department of Mathematics, University of Manchester, Manchester M13 9PL, UK

N. V. Kalashnikova, Department of Algebra, University of Dnepropetrovsk, Gagarin-Prospekt 72, Dnepropetrovsk, Ukraine

Louise-Charlotte Kappe, Department of Mathematical Sciences, SUNY at Binghamton, Binghamton, NY 13902-6000, USA

Lev S. Kazarin, Department of Mathematics, Yaroslavl' State University, ul. Sovetskaja 14, 150000 Yaroslavl', Russia

László G. Kovács, School of Mathematical Sciences, Australian National University, 0200 Canberra, Australia

Jan Krempa, Wydriat Matematyzki, ul. Banacha 2, 02-097 Warsaw, Poland

Leonid A. Kurdachenko, Department of Algebra, University of Dnepropetrovsk, Gagarin-Prospekt 72, Dnepropetrovsk, Ukraine

John C. Lennox, Department of Pure Mathematics, University College, P. O. Box 78, Cardiff CF1 1XL, UK

Olga Macedońska, Institute of Mathematics, Silesian Technical University, Ul. Kaszubska 23, 44-101 Gliwice, Poland

Avinoam Mann, Department of Mathematics, The Hebrew University, Givat Ram, Jerusalem 91904, Israel

Consuelo Martínez, Departamento de Matematicas, Universidad de Oviedo, 33007 Oviedo, Spain

Martin L. Newell, Department of Mathematics, University College, Galway, Ireland

M. F. Newman, Department of Mathematics, Australian National University, Institute of Advanced Studies, P. O. Box 4, Canberra ACT 2601, Australia

Boris Plotkin, Institute of Mathematics, Hebrew University, Givat Ram, Jerusalem 91904, Israel

Derek J. S. Robinson, Department of Mathematics, University of Illinois, 1409 W. Green Street, Urbana, IL 61801-2917, USA

Dan Segal, All Souls College, Oxford OX1 4AL, UK

Vladimir I. Senashov, Computing Center SO RAN, 660036 Krasnoyarsk, Russia

Pavel Shumyatsky, Department of Mathematics, Technion-Israel Institute of Technology, Haifa 32000, Israel

Howard Smith, Department of Mathematics, Bucknell University, Lewisburg, PA 17837-2005, USA

Andrzej Strojnowski, Wydriat Matematyzki, ul. Banacha 2, 02-097 Warsaw, Poland

Igor Ya. Subbotin, 1617 N. Fuller Ave, Los Angeles, CA 90046, USA

Yaroslav P. Sysak, Department of Mathematics and Mechanics, Kiev University, Ul. Dobrokhotova 28, 252142 Kiev, Ukraine

Witold Tomaszewski, Institute of Mathematics, Silesian Technical University, Ul. Kaszubska 23, 44-101 Gliwice, Poland

M. J. Tomkinson, Department of Mathematics, University of Glasgow, University Gardens, Glasgow G12 8QV, UK

Jiří Tůma, Mathematics Institute, Academy of Science, Zitná 25, 11567 Praha 1, Czech Republic

Vladimir V. Uspenskij, Mathematisches Institut der Ludwig-Maximilians Universität, Theresienstraße 39, 80333 München, Germany

Arthur Weinberger, Department of Mathematical Sciences, SUNY at Binghamton, Binghamton, NY 13902-6000, USA

James Wiegold, Department of Pure Mathematics, University College, P. O. Box 78, Cardiff CF1 1XL, UK

John S. Wilson, Department of Pure Mathematics and Mathematical Statistics, University of Cambridge, 16 Mill Lane, Cambridge CB2 1SB, UK

Guido Zappa, Via Quintino Sella 45, 50136 Firenze, Italy